CEDIA™

FUNDAMENTALS OF RESIDENTIAL ELECTRONIC SYSTEMS

V 1.2

Published by the Custom Electronic Design and Installation Association (CEDIA)

CEDIA would like to acknowledge the following contract service provider for this project:

Book Design and Layout
Anthony Sclavi, NAME LLC

The information in this text is general in nature and intended for instructional purposes only. Actual performance of activities described in this manual requires compliance with all applicable codes and operating, service, maintenance, and safety procedures under the direction of qualified personnel. References in this manual to patented or proprietary devices do not constitute a recommendation of their use.

ISBN: 978-0-9817394-2-7

FIRST EDITION

To purchase bulk quantities of this book please contact:
CEDIA EMEA, Unit 2, Phoenix Park, St Neots, Cambridgeshire PE19 8EP, UK
+44 (0) 1480 213744

Printed in UK

PHOTO CREDITS

Cover 1	Future Home	17	BEKINS	203	LUTRON ELECTRONICS
Cover 2	Electronics Design Group, Inc.	83	STARR SYSTEMS	211	NTX
Cover 3	La Scala Integrated Media	97	AUDIO IMAGES	212	LUTRON ELECTRONICS
Cover 4	Integrated Control Concepts, Inc.	99	TRIPHASE TECHNOLOGIES	212	ELECTRO-KINETICS
		131	SERVICE TECH INC.	213	KING SYSTEMS
		155	GLOBAL WAVE INTEGRATION	213	LA SCALA
		173	TRIPHASE TECHNOLOGIES	263	DEFINITIVE CONCEPTS, INC.

CEDIA would like to thank the following subject matter experts for their contributions to this text:

Eric Andkjar	*Electro-Kinetics, Inc.*
Steve Briggs	*Madison Media Institute*
Ian Bryant	*Simplified Concepts*
Christine Bucks	*Lutron Electronics*
Steven Castle	*TecHome Builder / GreenTech Advocates*
Sam Cavitt	*Paradise Theater*
K.W. Chan	*Nanpeng-Tech*
Dave Chic	*CEDIA*
Buzz Delano	*Delano Associates*
Matt Dodd	*Clavia Group*
Ken Erdmann	*Erdmann Electric*
Grayson Evans	*Cinema at Home*
Lewis Franke	*Texadia Systems*
Sergio Gaitan	*Multisistemas BVC, SA de CV*
Jeff Gardner	*CEDIA*
Larry Garter	*TechPlex*
Bob Hadsell	*Draper, Inc.*
Ron Kipper	*Ideal Industries*
Jeff Kussard	*Kussard Consulting*
Max Li	*Zenetech*
Eric Lee	*Integrated Control Experts*
Gerry Lemay	*Home Acoustics Alliance*
Daryn Lewellyn	*Madison Media Institute*
Michael Masten	*ADI*
Geoff Meads	*Presto A/V Ltd.*
Dave Pedigo	*CEDIA*
Steven Rissi	*Sensory Technologies*
Stuart Robinson	*Sound Integration*
Leslie Shiner	*The Shiner Group*
Bill Skaer	*Eric Grundelman's Cool AV*
Richard Stoerger	*Audio Design Associates*
Aaron Stroude	*ESPA*
Steve Sumners	*Sound Insights*
Dave Tkachuk	*Symbol Logic*
Gordon van Zuiden	*cyberManor, Inc*

CEDIA™

FUNDAMENTALS OF RESIDENTIAL ELECTRONIC SYSTEMS

V 1.2

CEDIA™

About CEDIA

The Custom Electronic Design & Installation Association (CEDIA) is an international trade association of companies that specialize in planning and installing electronic systems for the home. CEDIA provides top-ranked educational conferences, industry professional training, and certification focused on the installation and integration of residential electronic systems that consumers use to enhance their lifestyles. CEDIA's purpose is to build recognition and acceptance for this specialized field, and to speak up for its interests in addressing the industry, government, and the marketplace.

Comments Requested

Our industry is constantly evolving. No printed reference can remain current for long. In order to keep up with these changes, we invite comments from members of our industry. Please address them to:

CEDIA
Attn: Certification Dept.
7150 Winton Drive, Suite 300
Indianapolis, IN 46268

When you send comments to CEDIA, you grant CEDIA a non-exclusive right to use or distribute your comments in any way it believes appropriate without incurring any obligation to you.

FOREWORD

In 2015, CEDIA celebrated the 25th anniversary of its first official event held to serve the residential electronic systems industry. During that quarter-century, an incredible amount of innovation and creativity have combined to take technology in the home to new heights. Today the career paths created in this sector offer virtually unlimited possibilities to anyone willing to work hard, and never stop learning. This book, and the certification it supports, is just a sample of the resources CEDIA brings to the industry every day. They are the gateway to a variety of different career destinations, in virtually any part of the world. You are invited to take advantage of these resources; to learn, grow, and achieve higher certifications. Welcome to the exciting world of home technology, and we wish you well as you pursue your personal goals!

Ken Erdmann

Chair, CEDIA Certification Working Group

Larry Pexton

CEDIA Chairman of the Board

Chapter 1
INTRODUCTION TO THE INDUSTRY

INDUSTRY HISTORY AND OVERVIEW
CEDIA HISTORY AND MISSION
COMPANY TYPES WITHIN THE INDUSTRY
OTHER IMPORTANT STAKEHOLDERS
CAREER OPPORTUNITIES

INDUSTRY HISTORY AND OVERVIEW

In the Beginning...

Before there were any custom installers or system integrators, there were Hi-Fi shops. These were the 50's and 60's - the days of LP records, turntables, open-reel tape machines, receivers (first tubes, then transistors) and loudspeakers of all shapes and sizes. Everyone loved their stereos; from college kids on a budget to audiophiles who wanted the best money could buy.

The Origin of "Custom"

As early as the 50's, people were finding ways to "build in" their turntables, tube receivers and amplifiers, even speakers. In the 70's, customers started asking if their stereo receivers could power up speakers in a second or third room. Of course there were challenges to this concept, and some knowledge about basic electronics was needed. A little later, IR remotes were coming into use, and as we know they are a "line of sight" technology, so controlling a device from another room seemed impossible. A few resourceful people with some electronics savvy began to come up with "custom" solutions, because there were really no products being made for these applications - and the ball was rolling.

Multi-room Audio and the First Home Theaters

As customers started enjoying the convenience of whole-house music another exciting option emerged; combining a VCR and a projector to create an actual "theater" in the home. The late 80's and early 90's saw a lot of advancements, not only in the kind of systems being installed, but in the actual products being created to meet the needs of this new market.

Integration Comes of Age

As technologies evolved, custom installation evolved into what is now known as systems integration. Manufacturers responded to the demand for more control over more devices and before long homeowners were using one control system to address a wide variety of different systems in the home.

1990's

In the 90's, advances were seen in many different disciplines. Home theater took off like a rocket when digital projection and the DVD joined forces late in the decade. Control systems evolved quickly and other subsystems became part of the mix.

The 21st Century

Since 2000, advances have come rapidly in all areas of home technology. High Definition displays and sources made home theater a huge success. Not only have control systems become more capable and versatile, many more devices have become controllable. Today's integrated system might include audio, video, security, HVAC, cameras, door access, motorized devices...just about anything. One of the main developments making this possible is the fact that the home network, and its connection to the world via the internet, has brought us to a new level of communication – between people as well as between devices. And this platform means that anything that is network enabled has the ability to be part of the control ecosystem. The possibilities are endless, as you will see as you continue your learning path in the ever-evolving residential systems industry.

Today's residential systems integrator provides a variety of products and services. They strive to be their client's "go to" expert on all things electronic in the home, from the network to entertainment and subsystem control. They have successfully evolved to understand and offer solutions which include streaming media, sophisticated control and automation, and the ubiquitous mobile devices that customers have become so accustomed to. It is this constant evolution that makes the business so exciting and dynamic.

CEDIA HISTORY AND MISSION

The Founders; Their Passion and Vision

CEDIA began with a shared passion for home entertainment innovation and design that was recognized by visionary product designers and creative business leaders who also shared a love of technology. These individuals believed that an exciting and revolutionary industry was about to begin. 25 years later, they have been proven right. It is estimated that there are nearly 9,000 custom integration companies in the US alone, doing over $8 Billion in annual sales.[1]

These pioneers of the industry were there in the 80's when innovative products and ideas began to come together - balancing home décor and in-home entertainment like never before. They came from all corners of the US, but had a lot in common. Most importantly, they recognized that there was a growing demographic of luxury customers who were interested in custom solutions and cutting edge technologies – and were willing to pay for high quality design and installation services. Conversations began and ideas were shared.

Creative ideas for system design and effective business management techniques quickly

1: (2013). CEDIA Size and Scope of the Residential Electronic Systems Market ©.

became cornerstones of early dealer conferences among founding companies. "How to source products", and "how to effectively market and run a custom installation business" were big topics of discussion and shared vision. A breakthrough in marketing eventually came along with the creation of the first magazine to communicate what it was that these "custom installation" people were doing that differed from the typical audio retailer of the day. Audio Video Interiors published as a coffee-table style magazine, exuding elegance and emotion, connecting beautiful home décor with sophisticated electronic system installations. Suddenly, there were no speaker cabinets in the room; they were in the walls. Components no longer required a cabinet that didn't match with anything in the room. They were now hidden behind customized panels or in a rack in a different room. Remote controls could now control hidden products, and TV screens appeared and disappeared behind "automatic" panels controlled by the push of a button at the customer's coffee table or nightstand.

By 1989 they had really come together with a common belief in promoting and mutually supporting the business of custom designed and installed home entertainment systems. They realized the only way they would have a shared voice with customers and manufacturers was to form their own association. The associations and trade shows active at the time simply did not meet the specialized needs of this specialized group. It was time for them to go their own way and form a new alliance focused on custom solutions in the home. Many long and passionate discussions resulted in a new association with a name that proved to be appropriate for decades to come; the Custom Electronic Design and Installation Association – CEDIA.

Chris Stevens was a pioneer of custom installation as early as 1976 with his company, Phoenix Systems. In 1989 he founded AudioAccess, a manufacturer which helped shape the industry and was a CEDIA Founding Member. He went on to become CEDIA's first Membership Chairman. "The idea that always motivated us was that it was the people doing the work that made the difference between an average experience and a great experience for the customer. We wanted to harbor the energy of members who were dedicated to being masters at what they did, and we were able to leverage the shared learning from each other - which was really the true reason we founded the association. To this day, CEDIA is still about the difference that the dedicated professional makes in bringing beautiful and enjoyable new experiences into people's homes; and CEDIA's shared learning is a critical part of what makes this possible," said Stevens.

CEDIA Membership was the first official step towards building a sense of community among like-minded business owners, who would become the Founding Members of CEDIA. Over 30 of these original companies are still active members today.

The 90s - Association Growth and CEDIA EXPO

In 1990, CEDIA held its first EXPO in Amelia Island, Florida. Around 500 attendees represented installers, product manufacturers and industry affiliates. The focus was on fundamentals like multi-room audio and IR routing, new products, and sales opportunities. Perhaps the greatest value was found in the sharing of ideas among CEDIA members. As the decade progressed, CEDIA EXPO grew larger each year, and the foundation of education and learning clearly became the focal point. In 1994, CEDIA's first regional education events were held, where members could gather to concentrate on skill mastery and to share ideas among peers and colleagues, a culture that continues to exist today. In 1995 the Australia office opened to serve the Pacific Rim, and in 1996 the UK office opened to serve all of Europe, Africa and the Middle East (A Shanghai office would be added years later). CEDIA had officially become a global association.

Figure 1-1 - Product Demo at Early EXPO

As the industry grew, so did the association. New technologies brought new opportunities. It was truly a "rising tide floats all boats" environment of competitive collaboration. As the millennium approached, lighting, R,F and serial control and automation gained rapid acceptance and the incredible success of DVD elevated the home theater experience to new levels.

21st Century - The Internet and More Change

As in any age of technology, advancements continued at an accelerating pace as multi-room audio, distributed video, customized programming and touch screens and structured wiring became staples of business for CEDIA members. It was during this time that CEDIA recognized the need for a more consistent and higher level of learning for its members, and set out to develop CEDIA University, the industry's first formalized, vendor-neutral training. The goal was lofty; five "colleges" focusing on five different areas of study; Technician, Designer, Project Management, Business, and Customer Relations. As with all CEDIA initiatives, the expertise was drawn from member volunteers, who were eager to share their experience for the betterment of the industry.

As we moved through the 2000s, the internet and networking in the home drastically changed the delivery of music and video content, as well as control systems. This era was full of disruptive technologies, including the emergence of smart mobile devices. As this dizzying pace rolled on, control and automation simultaneously became simpler and more robust, making integration of subsystems like security, access, cameras, lighting and climate control even easier and more widespread.

Figure 1-2 CEDIA Training

As this new focus on the network became fully realized, CEDIA once again met the needs of the industry, with the introduction of a full track of learning dedicated to the home network, along with a certification to prove one's mastery of this important body of knowledge. In 2013 there were over 2,000 registrations for these networking classes at CEDIA EXPO in Denver. A new 5 day Boot Camp was introduced in 2013, focusing on advanced networking and systems integration.

Figure 1-3 - One Small Corner of Today's EXPO

2014 saw the introduction of a certification for those new to the industry, and this study guide to support this "gateway" program. For the first time there is now a program specifically for those coming into home technology from other industries as well as students looking for a great career.

CEDIA continues to establish new professional standards and techniques to ensure that its members possess the knowledge to exceed their customers' expectations in every way. From system design and technical knowledge to ethical business practices and personal professionalism, CEDIA Members are respected as the best in the industry. CEDIA continues to be the "voice of authority" for an entire global industry which has defined the Electronic Lifestyle for homeowners everywhere.

CEDIA Mission Statement

CEDIA's mission is to be a core component of its members' prosperity.
Built upon a strong volunteer foundation, CEDIA provides access to industry-leading education, certification, research, and consumer awareness.
CEDIA serves as the go-to home technology source for consumers and professionals alike.

COMPANY TYPES WITHIN THE INDUSTRY

In order to understand how the industry works, it is important to know who all the major players are and what their role is in the process. In your career you will likely deal with all of these different company types, and learn much more about how they work.

Systems Integrator

These companies are the heart and soul of the CEDIA channel. They are often small companies with highly skilled people who do highly complex projects. The average CEDIA member company of this type has only about 10 employees, which means many of them wear multiple hats within the process of selling, designing, and installing systems. These companies differ from a mainstream retailer in a number of ways. First, their primary goal is to create a customized solution which is based specifically on the client's wants, needs, and lifestyle. Once this "client discovery" has taken place, the process of design, engineering and installation can begin. The end result is a dependable, easy to use system which meets or exceeds the client's expectations and adds convenience, enjoyment, security, and efficiency to their home and their lives. This ability to design and install technology based primarily on the client's needs is what makes the industry unique. An integrator sees some profitability from the sales of product, but usually provides the most added value in design, engineering, programming, installation and support after the sale.

Custom Retailer

Some companies are larger, carry more inventory, and sell a wider variety of products and services than a small custom integrator. Their customer base may be larger and many customers may not require customized solutions. But they have the expertise and personnel to do custom installations, integrate various products and subsystems together, and provide services which a typical retail store can't offer. This type of company seeks to be profitable from both product sales and installation services.

Distributor

A distributor is a company which purchase product from manufacturers, stocks it, and sells wholesale to retailers and integrators, who then sell to the end user. Most distributors have multiple locations, either regionally or nationally, and sell only at the wholesale level. A distributor adds value for their customers in a number of ways. By stocking commonly used products they make it convenient to purchase just about anything needed for many projects "just in time" rather than ordering well in advance. This is especially valuable to smaller companies with less cash flow. Distributors also provide their expertise and design assistance when an integrator is using a product they may be less familiar

with or expanding into an area that is new to them. Another way distributors serve the industry is by hosting manufacturer training on a regular basis. This training helps manufacturers raise awareness of their product and gives the installers the confidence to specify and use those products. Buying from a distributor gives a small company access to thousands of products from hundreds of manufacturers, all available at one place, either at a physical location or online. A distributor gets its net profit from a small markup on all products and benefits from buying in large quantities.

Manufacturer's Rep

More specialized or high end products are often sold to the integrator through independent manufacturer rep firms. These companies typically carry a "line card" of products from manufacturers and work directly with integrators to help them specify and sell those products. They conduct both technical and sales training, and provide direct support as needed. They also work directly with the manufacturers to stay abreast of new products and technologies, so they can represent those manufacturers professionally to the dealer who sells and installs the equipment. Most rep firms are regional, and may carry a slightly different lineup of products in different states. Independent reps of this kind are paid a sales commission by the manufacturer.

Manufacturer

None of the previous business models would be possible without products designed and produced by manufacturers. Many of these are large multi-national companies whose main business is consumer oriented electronics. Some are smaller, more specialized companies, who make products mainly for the custom market. These companies display their products at trade shows like CEDIA EXPO and are constantly involved in Research and Development (R&D) to create new products which deliver higher quality, more convenience, and new experiences to the consumer. Manufacturers are often publicly held corporations who show profits based on product sales through a variety of channels.

Specialized Services

There is also a category of industry professionals who don't manufacture, sell, or install product but provide very specialized services to those who do. They may specialize in such services as acoustical design, documentation, project management, engineering, or control system programming. Often their services are used when a particular project has a scope beyond what a smaller company usually does. For instance an integrator has an unusually sophisticated theater project where the client expects highly accurate acoustical performance and sound isolation. The integrator may bring in someone who specializes in this discipline. Or, a company who occasionally does projects with control requirements beyond what they normally do may engage the services of a highly trained independent programmer on an "as needed" basis. Specialists of this kind bill in a variety of ways, including set quotes and hourly rates. They usually sell no physical product, just their expertise.

OTHER IMPORTANT STAKEHOLDERS

A stakeholder is anyone involved in a project who will be interacting to help achieve the ultimate goal of a satisfied homeowner. Even if they are not directly involved in electronic systems design/installation, it is critical that we communicate with them on a professional level, and relay information in a timely manner. These professionals may belong to their own industry's trade associations. CEDIA works closely with many of these organizations to raise industry awareness, identify trends, and to provide continuing education courses that help them maintain their certifications.

Client/Homeowner

Of course, the homeowner is the most important stakeholder of all. They are paying for a complex set of products and services and they will be the ultimate judge of the project's success. Different clients require different communication styles, but in all cases communication is key to success. In many cases, a very special skillset is needed because we are dealing with very wealthy people who expect an extraordinary level of service. The scope of customers served by the home technology industry may be growing, but the "luxury customer" will always be important and so will be the special skillset needed to serve them.

Architect

In many cases, an architect is the one who conceptualized the structure, the theme, and the way the home flows. The architect may have specified a space for a theater, or audio distributed to certain rooms, or televisions in certain spaces. Working in harmony with other professionals of this type is a skill that is critical to your success, especially in sales and design. CEDIA has many great initiatives to enhance the working relationship between integrators and architects. Many integrators have established an outstanding rapport with area architects, which has in turn brought them more high quality projects over the years and valuable referrals. The main associations for this segment of the industry are the AIA (American Institute of Architects) and the American Institute of Building Design (AIBD).

Interior Designer

Just as with architects, a good working relationship with an interior designer can go a long way to a smooth and successful project. Interior designers often have a very specific vision and a great deal of passion for how a home should look, function, and feel. As experts in the field of electronic systems, we should strive to help them realize this vision by providing solutions that are aesthetically pleasing, blend well or are completely invisible. In the case of a home theater, it is imperative that we communicate what is necessary to meet our functional goals, so that performance is not compromised. CEDIA works closely with the ASID (American Society of Interior Designers), the Interior Designers of Canada (IDC), and the International Interior Design Association (IIDA).

Builder/General Contractor

Home builders, remodelers, and general contractors are similar to custom integrators in many ways, they all work to enhance their clients' lifestyle, and to provide them with comfort, convenience, and safety. They are often creating (or re-creating) a one-of-a-kind home, to meet the customer's specific needs. They coordinate the work of many different trades and disciplines to accomplish this goal. If an integrator can become a major contributor to the success of this process, they will likely be called upon again by that builder or remodeler. Some very successful CEDIA member companies have built their business models based on their ongoing loyal relationships with these building and remodeling professionals. CEDIA has enjoyed a mutually beneficial and prosperous collaborative relationship with The NAHB (National Association of Home Builders) and NARI (National Association of the Remodeling Industry) for many years.

Other Trades on the Jobsite

The builder is generally the contractor who hires and supervises all of the other trades that work on the project, be it new construction or a remodel. These trades include electrical, plumbing, roofing, HVAC, kitchen cabinets, countertops, flooring and much more. As a technician, project manager, sales person, or designer you are likely to be on the jobsite with these subcontractors and it is important that you represent your company professionally and establish a good relationship with everyone working onsite.

CAREER OPPORTUNITIES

If you are reading this book it is probably because you are interested in a career somewhere in the residential systems industry, or just want to know more about the technologies involved. Since many companies doing this work are small, many people find themselves doing the work normally associated with different job descriptions. In larger companies, the division of labor is more defined and each person has more specific areas they focus on. Although these different career paths require different skill sets, they have one thing in common; the fundamentals presented in this book. So no matter what your career leads you to, this body of knowledge will form the foundation for your journey.

Electronic Systems Technician (EST)

An EST is the person who installs, configures, and maintains electronic systems in the field. This includes installing the infrastructure (cabling, etc), the equipment, the control systems, and user interfaces. Many ESTs do their work based strictly on a work order, which is created by someone else. But in a smaller company the same person may specify the equipment and design the system, and then install it all themselves. A good technician has an understanding of how signals and electricity work, allowing them to solve problems and troubleshoot issues in the field when necessary. Some people who love working with their hands, and the satisfaction of seeing their work through to completion, work their entire career in the field as a technician. They continue to learn and expand their skillset over the years and become more and more valuable as they become adept at system calibration, programming, troubleshooting, etc. They may become a Lead Technician or crew leader, supervising the work of other ESTs and mentoring them as they learn new skills.

Electronic Systems Designer

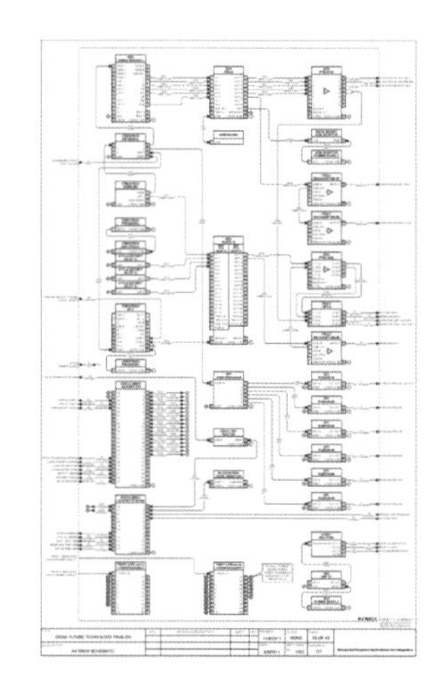

The role of a designer ranges from directly interviewing the client to evaluate their needs and lifestyle, to specifying the functionality and performance of a system, and designing the wiring scheme and user interface concepts. This varies somewhat from company to company. A designer must have full understanding of the equipment, how it works, and how it is controlled. A designer must stay up to date on emerging technologies and unique products which can be used to solve unique challenges. Since the most important part of our industry is now integration and control, a designer needs to have a solid knowledge of each and every subsystem that will be part of the larger integrated system.

Systems Engineer

In some companies, there are technical people who take the overall design and do the detailed engineering necessary for technicians to execute the plan accurately. This includes cable and connector details, platework, drawings, researching component functionality, etc. and communicating all of this clearly to the technician in the field. As with a designer this person needs to have extensive knowledge of every piece of equipment and every subsystem.

Project Manager

Custom design and installation is a complex process, with many moving parts and stakeholders. All of this needs to be organized based on timelines, human resources, and cost in order to ensure profitability and client satisfaction. Project management is a well-defined process and body of knowledge which is not unique to our industry, but taught in a variety of venues and applies to a variety of industries. Someone with industry knowledge can learn project management and become a very valuable asset to their company. Someone with project management experience can use those skills in our industry once they have mastered the fundamentals which are unique to what we do. Either way, a good project manager can mean the difference between a profit or loss on a project, especially a large or complex project. Project managers are in high demand and someone who enjoys managing multiple projects and people may find this to be a very satisfying career path.

Sales

Some people love to sell, and they are good at it. Sales requires a combination of technical knowledge, personality, great communication and resourcefulness. Some of the most successful sales people in our industry did not start out as a technician, but rather had sales experience and learned the fundamentals in order to apply their skills to work well with homeowners and builders to close deals and deliver projects to their company. As with all of the careers, the fundamentals needed make up the body of knowledge found in this book.

Business Owner

If owning and operating your own business is your dream, then your most important learning will take place in the realm of general business practice, marketing, cash flow, etc. – the same knowledge needed regardless of what product or service you intend to offer. CEDIA recommends business training for all industry professionals, even if they are not involved in the business on a day to day basis. In CEDIA classes we often see individuals who want to start a new business in our industry. CEDIA offers a wide variety of business training both in person and online. Remember to be successful you will need both technical and business knowledge. In the business section of this book there is an overview of "how to start" for those of you who want to be your own boss. It isn't for everyone but for some it is the perfect career path.

SUMMARY

No matter what your goals, or what kind of company you work for (or will work for), the fundamentals found in this book will provide the foundation for an exciting and rewarding career in residential electronic systems.

Questions

1. The first home theaters were made possible because of the introduction of:
   a. The DVD player
   b. The VCR
   c. The Laserdisc
   d. The DLP projector

2. The first CEDIA EXPO took place in _____.
   a. 1980
   b. 1990
   c. 2000
   d. 2010

3. A distributor:
   a. Stocks product
   b. Provides manufacturer training
   c. Sell wholesale to dealers and retalers
   d. All of the above

4. The person who installs, upgrades, and services electronic systems in the field is a (an)___:
   a. Electronic Systems Technician (EST)
   b. Electronic Systems Designer
   c. Systems engineer
   d. Control system programmer

5. The person who ensures success by tracking cost, time, and resources is the:
   a. Designer
   b. Programmer
   c. Administrator
   d. Project Manager

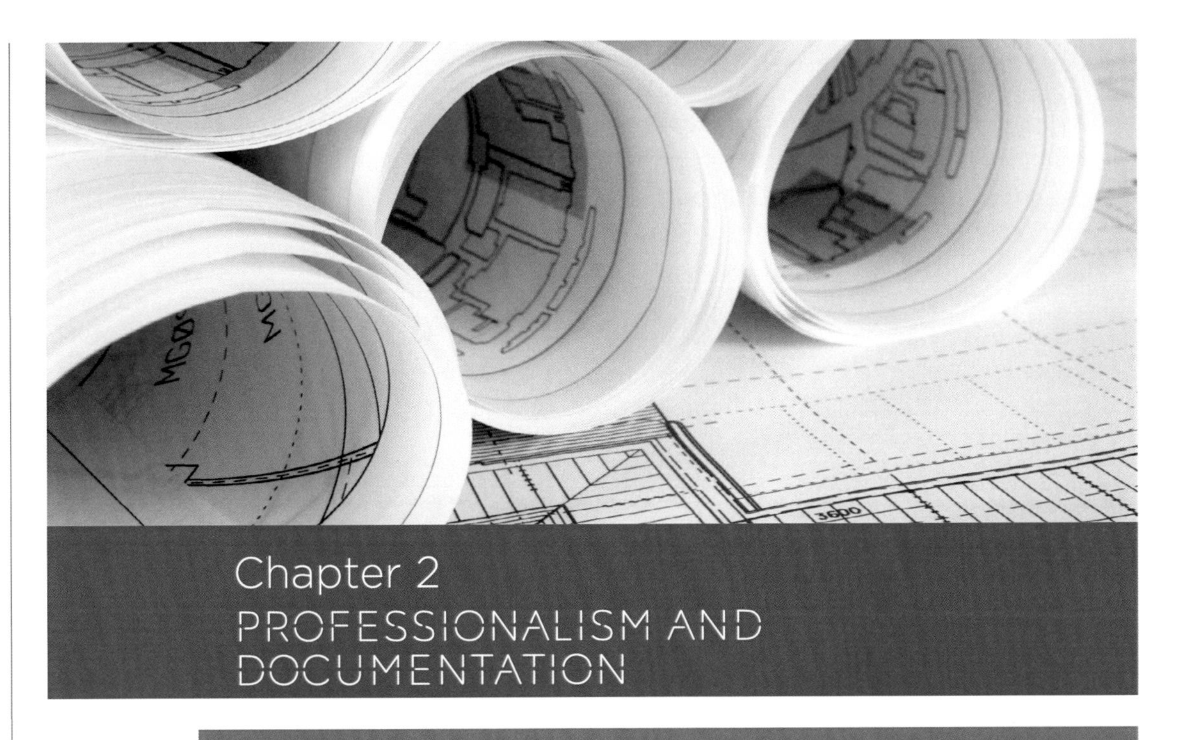

Chapter 2
PROFESSIONALISM AND DOCUMENTATION

WHY PROFESSIONALISM MATTERS

The True Value of Professional Standards

Since early days of the home technology industry, one of the most unique aspects of the business has been related to the fact that we are working in the homes of real people, crafting entertainment and technology solutions which are customized specifically to their needs. We add value to the products we sell by listening closely to our clients, and using technology to and enhance their lifestyle.

Successful integrators realize that there is a special responsibility associated with being invited into someone's home to work. There is a special level of trust associated with being part of planning and building a home. Being professional is not just "the right thing to do". It is also a key component in the overall success and profitability of a business.

Accuracy and Efficiency

Establishing best practices, policies and procedures, and the right company culture all contribute to success, by encouraging good communication and a common goal of great customer service. As we will see later, documentation is an important part of this pursuit of excellence. Accuracy and good planning lead to efficiency, which saves time and money.

Customer Satisfaction

The ultimate goal, of course, is customer satisfaction. Regardless of a project's budget, if the client is ultimately satisfied then the integration team has done their job. Exceeding client expectations makes them want to tell everyone what a great experience they had.

Referrals

This highly satisfied customer then becomes the best form of marketing there is. They show off their systems, and tell their friends about the company that designed and installed them.

Long Term Success

In the grand scheme of things, our real goal is long term viability; the ability to provide a good living for everyone in the company over a number of years. If there is one attribute shared by all companies who have remained successful for many years, it is their professionalism, both in business and on the jobsite. We will explore some general guidelines here which help maintain a high level of professionalism.

BUSINESS PROFESSIONALISM

Best Practices

The range of best practices related to professionalism is beyond the scope of this text but a few key points are worth mentioning.

Adherence to Job Scope

A well-defined scope of work is one absolutely necessary to ensure a successful project and a satisfied customer. The key to this is clear communication from the beginning to the end of the project. Requests for changes must be handled according to well established protocol. This way the client knows exactly what is happening, what impact the change has on the rest of the project, and how much it will cost (and why). "Scope Creep" causes misunderstandings and negatively impacts both the quality and profitability of the project.

Conflict Resolution

There will inevitably be a time when there is a difference of opinion between two stakeholders. How this is handled will say a lot about how you do business. Keeping a cool head and referring to documentation and facts is the best approach. When push comes to shove, it is best to be part of the solution, if at all possible. This reflects positively on your business and will likely result in good will and more business.

Integrity in All Transactions

CEDIA member companies and their employees are obligated to maintain high ethical standards in all dealings with customers, vendors, and other stakeholders. Not only is the reputation of your company dependent on this, so is the association and all other members. To ensure that all who deal with CEDIA members receive goods and services as agreed, and that all members adhere to a high standard in all dealings, an Ethics Committee made up of volunteers evaluate any complaint or report that is submitted and has the authority to terminate membership if the CEDIA Code of Ethics is violated.

CEDIA Code of Ethics

The CEDIA Code of Ethics has been in place since 1999, and applies to all member companies. The complete Code of Ethics can be found in the Appendix of this book.

General Guidelines

These general guidelines for employees, especially those on the jobsite and dealing with clients, will go a long way to demonstrating the overall professionalism of the company.

Accuracy

Believe it or not, displays have been hung on the wrong wall, holes have been cut in walls where there are no speaker cables, and YES – entire prewires have been completed in the wrong house! Verify the accuracy of the work order; make sure you are in the right place, in the right room, looking at the right wall with the right equipment. When in doubt, CALL!

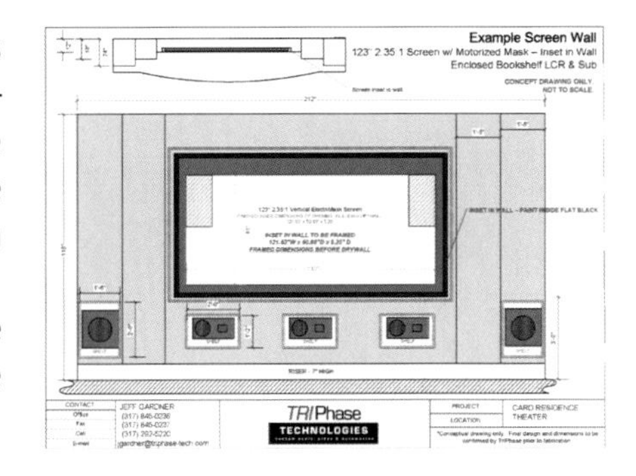

Designers and sales people are human. They make mistakes. Double check everything. Mistakes on paper are easily fixed. Not so once something is cut or installed.

Efficiency

Your company expects all tasks to be completed correctly and in a timely manner. Every extra hour spent on the job translates into lost profit. Get there on time, plan your course of action, lay out your tools and supplies in an organized fashion, and use your head to make the job go smoothly. Working smart makes everyone's job more enjoyable.

Attention to Detail

There is a lot of competition out there – plenty of companies and individuals who will do the job for less. Your company should define itself in the marketplace by its attention to detail. "Almost right" is not good enough. Make sure your installations are level, plum, in line with other details in the room, clean, precise, and picture perfect. As a matter of fact, employees should be encouraged to take pictures and put them up at the shop raising the level of accountability and allowing technicians to take pride in their work.

This applies not only to final installations, but also to prewiring. Builders and homeowners will see the quality and neatness of a prewire as an indication of your company's overall level of quality. A clean, crisp, neat prewire is like a perfectly made wiring harness in a quality automobile. It is a window to how the rest of the car is built. There should be an obvious difference between the way a low voltage prewire looks and the way the work of the other trades looks before drywall goes up. It may be hidden in a week, but in the meantime all of the important stakeholders WILL see that prewire, and form an opinion about the company that did it - especially the local building inspectors!

Protection of Assets

Your work van – a huge investment. Drive carefully, keep it clean. Treat it right and it will treat you right.

Most of what you touch in the course of a work day is pretty expensive. High quality tools and test equipment should be stored, carried, and used carefully. Displays, components – all products – must be handled with care. A damaged product may result in more than a financial loss. What if something is damaged and another one takes 2 weeks to be delivered? This could have a huge impact on the project.

Also worth talking about; the little items that add up. Wire ties, connectors, fasteners, hardware, drill bits, interconnects, power supplies, platework. These items may not be counted every week but they really do add up.

Profitability of any job is based on costs. But there are two types of costs, tangible and intangible. Tangible costs are the equipment and material that are used on the job. By treating equipment with care and using materials efficiently, the job costs will stay in line with the budget. Intangible costs are the lost opportunities when a job takes longer than expected. Time is money, and poor use of time on a job costs money and reduces the profitability of not only the job, but the company as a whole.

Respect

A true professional is respectful to all co-workers, from the CEO to the newest entry-level employee. This attitude will create an atmosphere conducive to good teamwork. If you need to talk about negative issues, take them to your supervisor in private. A positive attitude drives positive results.

General Behavior on the Jobsite

As we have discussed, anyone working in someone's home has a responsibility to behave in a professional manner and maintain a pleasant demeanor. For instance, you should focus on the job at hand and not be distracted by the client's possessions, memorabilia, etc. Always behave as if your every move is being recorded. A simple glance at mail or other personal items may be construed as inappropriate, and perception is everything. Don't be tempted to become too much like a guest, rather than a professional doing a job. It is always good to be conversational, but there is a thin line of too much familiarity that should not be crossed.

Company Policies

Your company should have established policies, not only related to human resources, but in all areas of installation techniques, labeling, and documentation. It is everyone's responsibility to know these policies and adhere to them.

When you are not sure.....ask! And know who to ask.

The culture of a company trickles down from the top. As you move up, there will be new employees that look to you for an example. Accept this role and always set the bar high for those who are learning.

Client Expectations

Everyone likes to do business with high-end customers. They appreciate quality and are willing to pay for it. But they are also accustomed to a very high level of service. They enjoy things like the complimentary car wash at the dealership, the personal greeting at the country club, and concierge service when they travel. They expect and appreciate the same kind of service from their home technology professional.

It is important to understand what all clients, especially high-end clients, expect from anyone who is working in their home. These points apply to everyone, but focus on the technicians who are frequently working in the home while the family is living there (final installation, service calls, etc.).

Timeliness

When you have an appointment in a customer's home you must be on time, every time. If you get stuck in traffic or are otherwise delayed, make a call to let them know you are running late and make sure this is not an inconvenience for them. They may have an appointment or conference call scheduled. High end customers by definition are high achievers, and their time literally is money to them. Don't waste any of it! Always keep this in mind, and they will appreciate it.

Cleanliness

It is imperative that you leave the home just as you found it. Shoe tracks, even fingerprints, will leave a negative impression. Shoe covers (booties) are a must when working in a finished home. Large dropcloths keep sawdust, small parts, wire clippings, etc. off of expensive flooring. A thick packing blanket is even better; it provides protection in case a tool or part is dropped. Clean hands, and even white gloves, make a big difference. And when it is time to pack up and leave, every single scrap and part needs to go with you. The client should be able to look around and see no sign that a technician was in their home.

Your company should have an established dress code and recommended attire. This is part of a clean, neat appearance. Remember you are the face of the company when you are with a client.

Respect for property and people

Every member of the team must show absolute respect for the customer's home and possessions, as well as everyone in the home. Not just family members but other vendors, housekeepers, etc.

You are a guest in the home and every action you take reflects on your company and on you. Once you have gotten in the habit of being polite and respectful, it will come naturally and become part of our everyday work ethic.

Minimal Intrusion

When working in a customer's home, make sure you park somewhere that will not pose an inconvenience, or drip fluids on their driveway. Stay focused on your work. You must be pleasant but you are not there to play with the kids or make small talk. Stay busy and focused on the work at hand.

Personal phone calls and texting can wait for a break or your lunch time. If you must make a business-related call, take it outside or at least keep it quiet and discrete. Always make sure you are not disrupting the activities or decorum of the home. When in doubt, ask.

Pleasant Attitude and Good Communication

This is really just common sense, but these traits are key to making a good impression, and representing your company properly. A smile and good attitude goes a long way. The best technician in the world can still make a negative impression if they have a negative attitude. If the client asks you to hold off on noise, stay out of a certain room, or make sure to be done by a certain time – DO IT!

Good communication

Upon arrival (and before you leave) there are always things that need to be communicated to the right person. Make sure you know who that person is. It might be a house manager or general contractor and not the customer. Tell them what you are doing, if you are finished, if you need to come back, and whether or not they can use the system. If the work is done and ready to use, show them exactly how everything works, especially if something has changed.

No Smoking

No exceptions.

The Documentation Process

Documentation is an important part of any electronic design project, and an important component in the general professionalism of a company. Accurate and complete documentation provides a high level of communication between stakeholders, a guide to installation, and the foundation for accurate billing and job costing. High quality documentation is an integral part of high quality work and customer service. It begins with the sales call and continues through planning, installation, customer training and service.

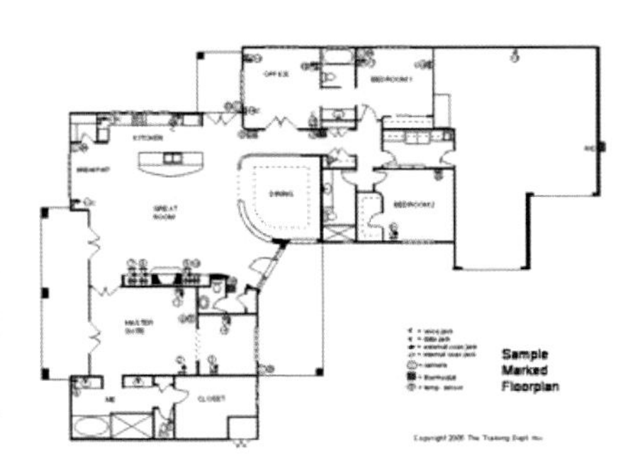

Documentation Types

There are three general categories of project documents:

- **Architectural Drawings** - Drawings or blueprints usually created by an architect for conveying a design concept and creating a building or space.
- **System Design and Engineering Documents** - Documents usually created by the A/V designer and/or engineer for creating an electronics system.
- **Project Management Documents** - Documents usually initiated by the project manager which utilize the elements of time, cost and scope to successfully complete a project.

Documentation Flow Chart

Companies should establish a documentation work flow. In the example flow chart shown, the box headings represent project phases while the smaller boxes indicate the documents needed for each of the phases.

Project Needs Assessment

During the initial client sales meetings the sales person will need to consider the following worksheets to help document all of the client's project information.

- **Interview Worksheet** - A document which shows all of the general client and project information.
- **Design Worksheet** - A document which shows all of the client's requested room-by-room device information.
- **Functionality Worksheet** - A document which shows the client's requested user-interface functionality information.

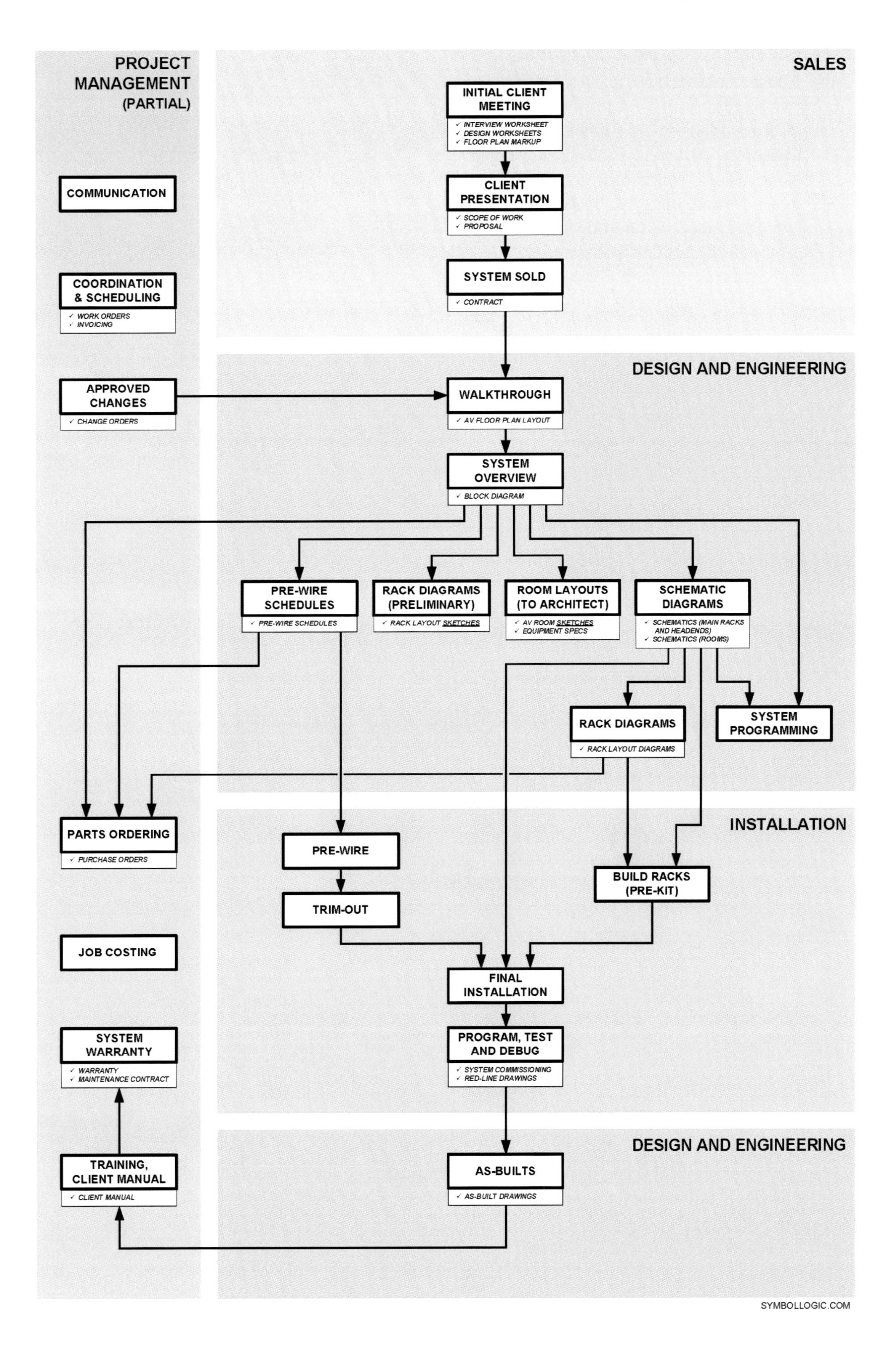
PROJECT MANAGEMENT (PARTIAL)
SALES
INITIAL CLIENT MEETING
✓ INTERVIEW WORKSHEET
✓ DESIGN WORKSHEETS
✓ FLOOR PLAN MARKUP
CLIENT PRESENTATION
✓ SCOPE OF WORK
✓ PROPOSAL
SYSTEM SOLD
✓ CONTRACT
COMMUNICATION
COORDINATION & SCHEDULING
✓ WORK ORDERS
✓ INVOICING
APPROVED CHANGES
✓ CHANGE ORDERS
DESIGN AND ENGINEERING
WALKTHROUGH
✓ AV FLOOR PLAN LAYOUT
SYSTEM OVERVIEW
✓ BLOCK DIAGRAM
PRE-WIRE SCHEDULES
✓ PRE-WIRE SCHEDULES
RACK DIAGRAMS (PRELIMINARY)
✓ RACK LAYOUT SKETCHES
ROOM LAYOUTS (TO ARCHITECT)
✓ AV ROOM SKETCHES
✓ EQUIPMENT SPECS
SCHEMATIC DIAGRAMS
✓ SCHEMATICS (MAIN RACKS AND HEADENDS)
✓ SCHEMATICS (ROOMS)
RACK DIAGRAMS
✓ RACK LAYOUT DIAGRAMS
SYSTEM PROGRAMMING
PARTS ORDERING
✓ PURCHASE ORDERS
INSTALLATION
PRE-WIRE
TRIM-OUT
BUILD RACKS (PRE-KIT)
JOB COSTING
FINAL INSTALLATION
PROGRAM, TEST AND DEBUG
✓ SYSTEM COMMISSIONING
✓ RED-LINE DRAWINGS
SYSTEM WARRANTY
✓ WARRANTY
✓ MAINTENANCE CONTRACT
DESIGN AND ENGINEERING
TRAINING, CLIENT MANUAL
✓ CLIENT MANUAL
AS-BUILTS
✓ AS-BUILT DRAWINGS
SYMBOLLOGIC.COM

Architectural Drawings

These drawings are usually created by the architect early-on in the building process. We use these drawings in our process to document our designs. They are created according to established practices and utilize symbols and icons which can be identified by looking at the legend, which always appears in this type of drawing. You will also see revision notes documenting any changes that have taken place and when they happened. In most cases these drawings will be in an electronic format as well as printed, and will be in landscape orientation. In this section we will discuss:

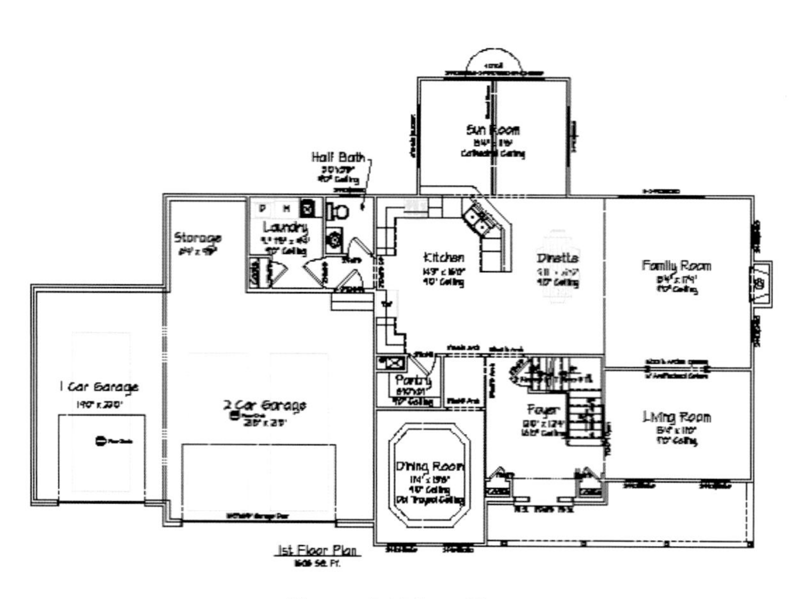

Figure 2-1 Floor Plan

Site Plan

This drawing typically shows the property's lot boundaries, easements, set-backs, out-buildings, and driveways.

Floor Plan

This drawing shows the view from directly above the floor and is the basis for many other architectural plans including the AV floor plan.

Reflected Ceiling Plan (RCP)

This drawing shows the view from above as if the floors had mirrors reflecting the ceiling view. The RCP is used to determine what is planned on being installed in the ceiling, such as lighting fixtures and heating and cooling vents.

Exterior Elevations

These drawings show the straight-on view from the outside of the building, typically from the north, south, east and west directions. These are helpful in determining floor heights and potential locations for outdoor loudspeakers and video cameras.

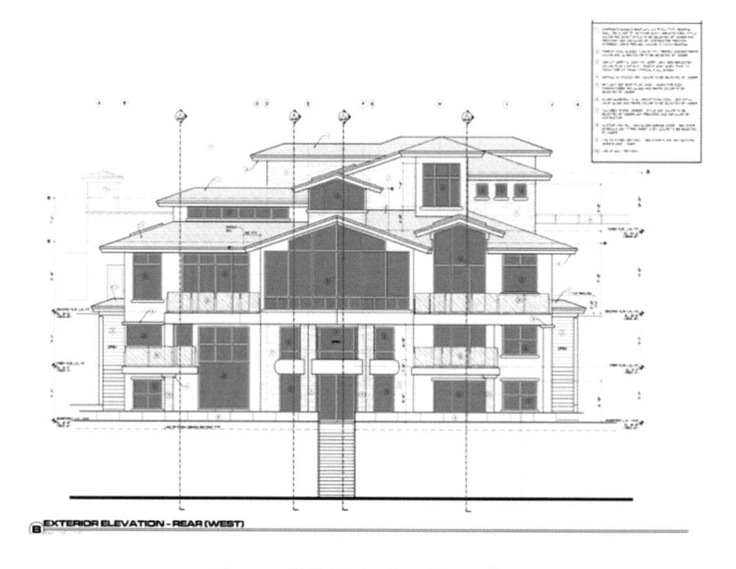

Figure 2-2 Exterior Elevation

Interior Elevations

These drawings show the straight-on view from the inside of a room, showing anything that might be on a wall. These are helpful for media room and theater installations which show important wall-mounted devices such as flat panel displays, video screens, loudspeakers, cabinetry and AV equipment racks.

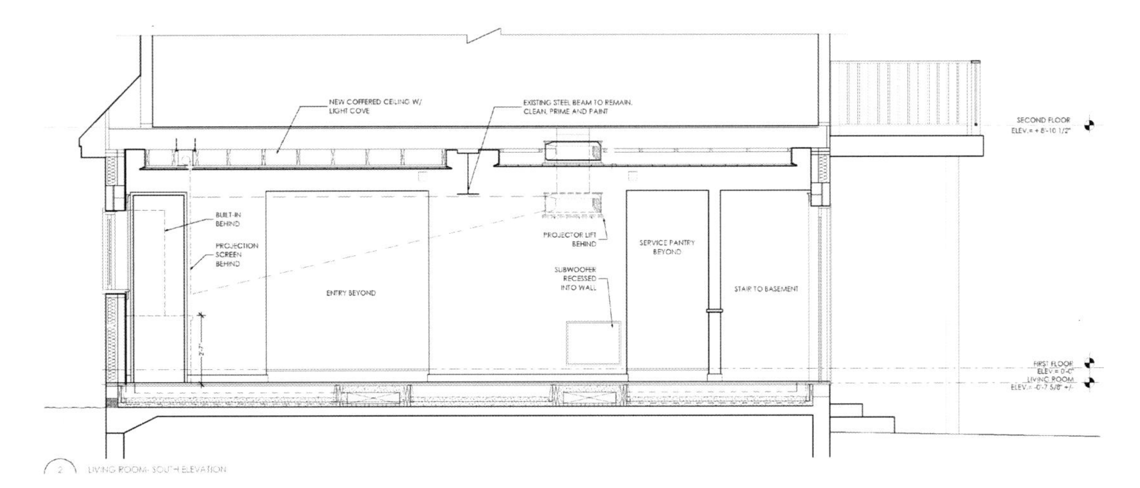

Figure 2-3 Interior Elevation

Section Drawings

These drawings show the building or room as if it were cut with a knife revealing the infrastructure of the floors, ceilings and walls. These are helpful for planning cable routes and determining device mounting locations.

Detail Drawings

These drawings show magnifications of small items and how they must be installed.

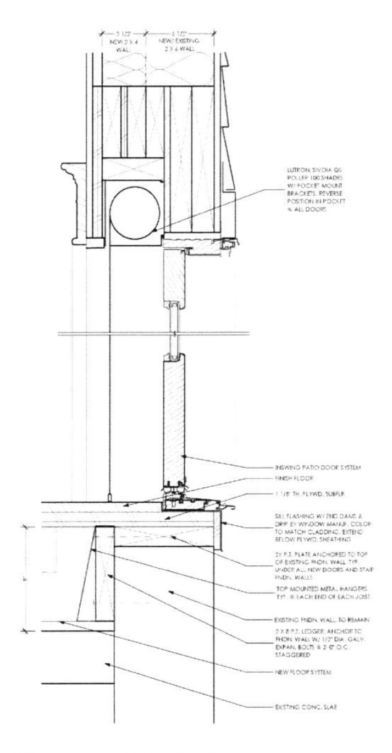

Figure 2-4 Detail Drawing

Electrical Plan

This plan is similar to the floor plan but includes details regarding the location and wiring of electrical devices such as outlets, switches, lighting fixtures and ceiling fans.

Mechanical Plan

This plan is similar to the floor plan but includes details regarding the location and pathways of mechanical and plumbing devices such as heating and cooling vents and ducts, and plumbing fixtures.

Design and Engineering Documents

These documents are created in the order shown below by the AV designer and/or engineer for the purposes of installing an electronics system.

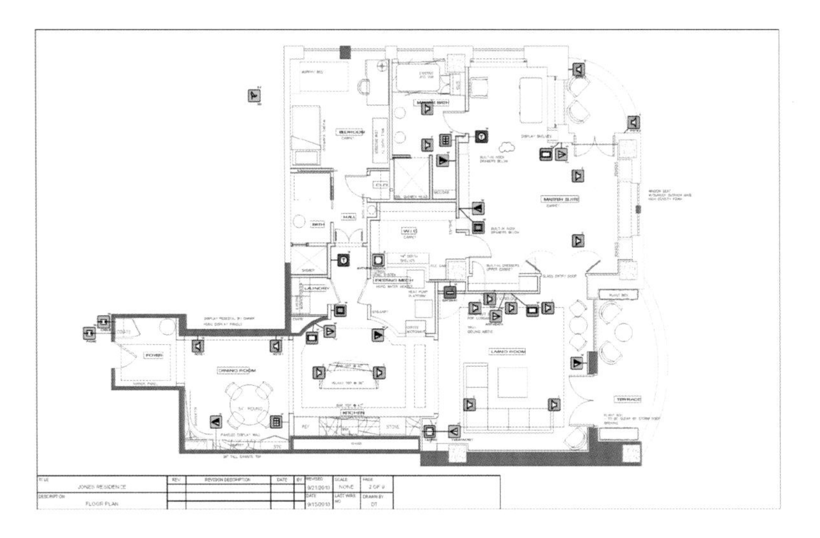

Figure 2-5 A/V Floor Plan

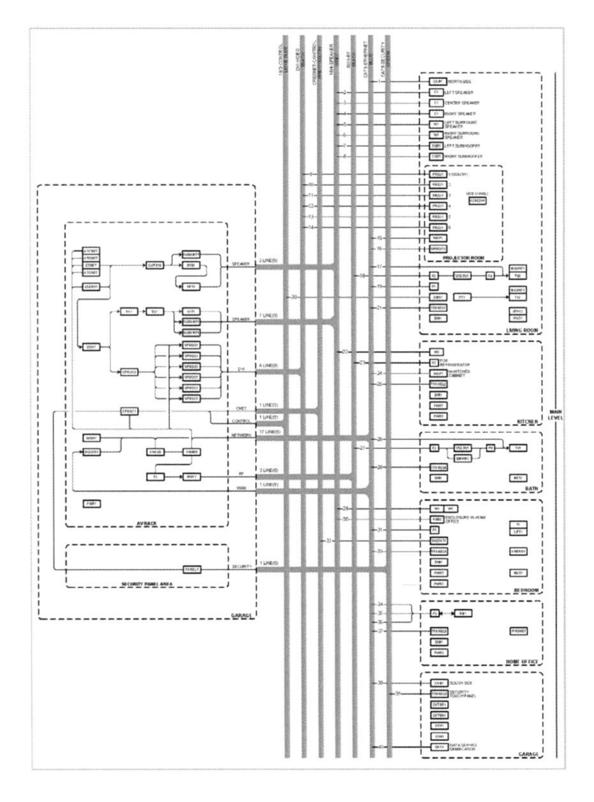

Figure 2-6 System Block Diagram

A/V Floor Plan

This design drawing shows the physical location of all key A/V devices and confirms to the client and other project stakeholders what is being installed. The drawing is created by using the architectural floor plan as the background with A/V floor plan symbols laid on top. This drawing is also very helpful in creating the system block diagram.

System Block Diagram

This design and engineering diagram reveals more information than can be shown on the A/V floor plan. It shows the big picture view of the system's entire device layout and wiring infrastructure. This diagram is essential in creating accurate prewire schedules and the detailed schematic wiring diagrams.

Prewire Schedule

The prewire schedules are often created from the system block diagram and can be shown with wires sorted by wire type, location or wire label. In each schedule, each row represents a unique wire run in the system. Schedule headings typically include: From, To, Wire Type, Color and Label.

Detailed Schematic Diagrams

These engineering drawings are point-to-point wiring diagrams for A/V racks, headends and room electronics. They are often created from the system block diagram and reveal more information than can be shown on the block diagram. Each equipment rack and set of room electronics will usually be shown on its own schematic page. A left to right signal flow helps prevent confusing wire crossovers and makes the diagram easy to follow.

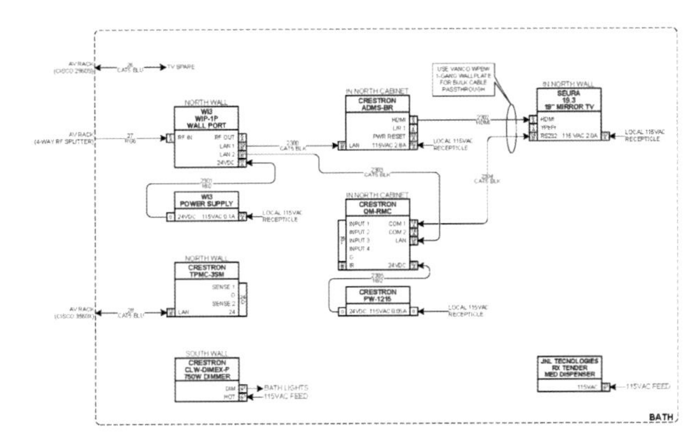

Figure 2-7 Room Schematic

Equipment Rack Drawings

These elevation drawings show the correct place ment of all rack-mounted equipment, overall size of rack, power and ventilation requirements and rack accessories.

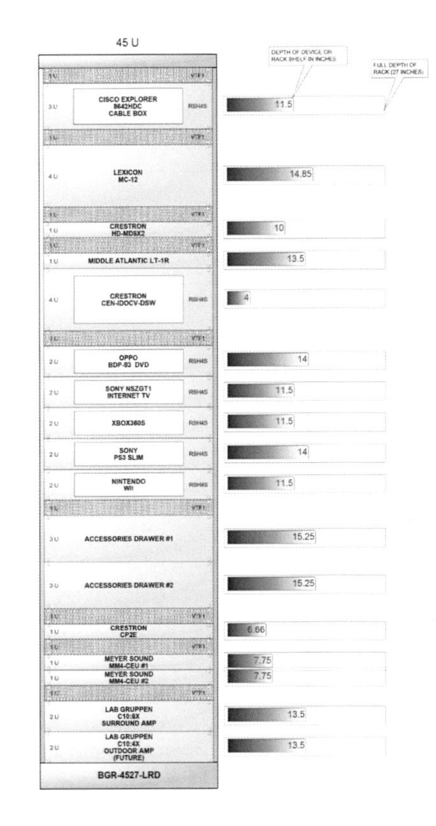

Figure 2-8 Equipment Rack Elevation

A/V Room Layout Drawings

These drawings show the location and size of room devices such as loudspeakers, video projector, screen and seating. Typically these drawings need only be sketches to be provided to the architect and cabinet designers in order for them to complete their final drawings. Included in the sketches should be all equipment locations and dimensions, power and ventilation requirements and verification that proper sound and visual sight lines are maintained.

Support Documents

These Documents are usually initiated by the project manager which utilize the elements of time, cost and scope to successfully complete a project. In this section we will discuss:

As-Built Drawings

These drawings are usually created from documents marked-up (red-lined) in the field by the installation technicians and show exactly how the system was built.

Scope of Work

This document is sometimes referred to as project scope and can be part of the proposal. The document refers to all specifications needed to complete the project and is helpful in reducing scope creep. Included in the scope of work are the client's needs, designer's vision, interior designer's desires, reality limitations, project possibilities, project cost, subcontractor and owner responsibilities and client set-asides.

Proposal

The proposal is a document which explains what the system will do and details the specifications, cost, payment terms, warranty and any service contracts. It generally includes a cover letter, company description and mission statement.

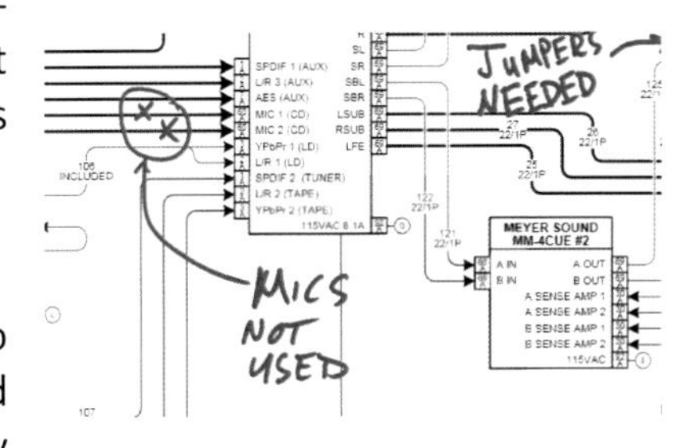

Figure 2-9 Red-line Revision

Contract

The contract is a document which states the installation company's overall services and responsibilities for the proposed system, including specific installation services, materials, design services, warranties, performance guarantees, client responsibilities, payment terms and conditions, and a signature and date section.

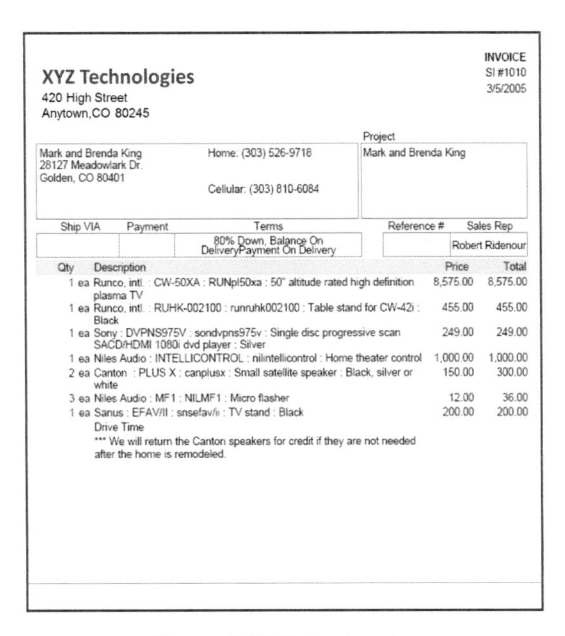

XYZ Technologies
420 High Street
Anytown,CO 80245

INVOICE
SI #1010
3/5/2005

Mark and Brenda King
28127 Meadowlark Dr.
Golden, CO 80401

Home: (303) 526-9718
Cellular: (303) 810-6084

Project
Mark and Brenda King

Ship VIA	Payment	Terms	Reference #	Sales Rep
		80% Down, Balance On DeliveryPayment On Delivery		Robert Ridenour

Qty	Description	Price	Total
1 ea	Runco, intl. : CW-50XA : RUNpl50xa : 50" altitude rated high definition plasma TV	8,575.00	8,575.00
1 ea	Runco, intl. : RUHK-002100 : runruhk002100 : Table stand for CW-42i : Black	455.00	455.00
1 ea	Sony : DVPNS975V : sondvpns975v : Single disc progressive scan SACD/HDMI 1080i dvd player : Silver	249.00	249.00
1 ea	Niles Audio : INTELLICONTROL : nilintellicontrol : Home theater control	1,000.00	1,000.00
2 ea	Canton : PLUS X : canplusx : Small satellite speaker : Black, silver or white	150.00	300.00
3 ea	Niles Audio : MF1 : NILMF1 : Micro flasher	12.00	36.00
1 ea	Sanus : EFAV/II : snsefav/ii : TV stand : Black	200.00	200.00
	Drive Time		
	*** We will return the Canton speakers for credit if they are not needed after the home is remodeled.		

Figure 2-10 Sales Invoice

Sales Invoice

Usually triggered by a signed contract, the sales order is a document sent to the client demonstrating the amount of money owed under the terms of the contract for the products and services sold.

Work Order

A work order is a document, usually triggered by a sale, indicating the location, date and time the work or service is to be carried out and the nature of the work to be done. The work order will include the client contact information, site address and phone number, job start and completion times, materials and products needed, as well as the expected completion time of any unfinished work.

Change Order

A change order is document indicating an amendment to a work order or contract which may change the project scope, timeline and contract price. It is often regarded as a mini contract where client and contractor approval for the changes must be signed–off and dated before the any work is completed. Additionally a change order will often affect many other project management and design documents.

Purchase Order

A purchase order (PO) is a document issued by a buyer and sent to a vendor for the purchase of products or services. The PO describes the exact product types, quantities and agreed-upon prices. The purchase order also contains the date of the order, PO number and ship-to-address and method of shipment, if applicable.

Subcontractor Agreement

The subcontractor agreement is a contract between a main contractor and another contractor (subcontractor) who will provide services for the main contractor. The agreement indicates the type of work the subcontractor will do and how it will be done, as well as a timeline and fee paid for the work.

System Commissioning

The System commissioning document confirms to all project stakeholders that substantial completion of the project has been achieved and that all design and performance goals have been met.

SUMMARY

No matter how skilled the sales people, designers, and technicians in a company are, without accurate documents (and the communication they enable) problems, mistakes, and scope creep are bound to make projects less successful and less profitable. Truly professional companies all utilize quality documentation. And conversely, those who have implemented great documents as part of their standard operating procedure always find it easier to deliver a quality product in a professional manner.

CEDIA offers advanced training on this topic, and for those situations which require a level of documentation beyond the in-house capabilities there are providers who specialize in creating all of these documents according to established best practices, especially for very large projects.

Project management is a sophisticated process which allows everything an integrator does to be carefully monitored and tracked, making the flow more predictable and the outcome more successful. The most successful companies are usually the ones with excellent project management in place. The Project Management Institute (PMI) has developed a widely recognized document known as the Project Management Body of Knowledge. These concepts can be applied to virtually any kind of project, and CEDIA's Project Management training applies them work done in our industry.

Questions

1. Professionalism results in:
   a. More efficiency
   b. Referrals
   c. Customer satisfaction
   d. All of the above

2. Documents which utilize the elements of time, cost, and scope to ensure success are:
   a. Architectural drawings
   b. Project management documents
   c. System design documents
   d. Floor plans

3. An amendment to a work order, which conveys modifications to cost, scope, and some times design, is known as a:
   a. Change order
   b. Work order
   c. As-built drawing
   d. Scope statement

4. A Cable Schedule should include the following information:
   a. Location ID
   b. Cable ID
   c. Check or initials to show each run has been tested
   d. All of the above

5. The type of drawing that shows the screen wall and acoustical panels as if the viewer is looking directly at the walls is a:
   a. Floor plan
   b. Exterior elevation
   c. Interior elevation
   d. Reflected Ceiling Plan (RCP)

Chapter 3
BUSINESS FUNDAMENTALS

INTRODUCTION

Most CEDIA member integration companies are small businesses. Most owners have a great deal of technical savvy but may not have a lot of formal business education. In this chapter we will provide an overview of some basic concepts of small business. Anyone involved in running their business is encouraged to seek out more advanced training in business, customer relations, and project management.

STARTING A BUSINESS

Starting your own business can be an exciting and rewarding experience. Many of us have started our business so that we can support our lifestyle by doing something we are passionate about. It can offer many other advantages such as setting your own schedule, being your own boss, and the potential for unlimited wages. But, becoming a successful entrepreneur requires thorough planning; creativity and hard work. This section will cover a few guidelines for your entrance into the entrepreneurial world.

There are some basic questions you need to ask yourself prior to starting your small business:

1. **What is your aversion to risk?** If you have never liked to take chances or to try something new, then owning a small business may not be for you.
2. **Do you have good negotiation skills?** Negotiating leases, contracts, insurance, and loan terms can save you money and maintain a positive cash flow. Selling also involves negotiations with your clients in order for them to feel comfortable with the final price.
3. **Are you an independent person?** Running a small business will require you to make decisions without the help of others and to trust that your decisions are the right ones.
4. **Do you have the support of others?** You should always have a strong support system before starting a small business. Getting a business mentor can be one of the best support methods available. A mentor has already gone through the pains and downfalls of starting a small business and can give you guidance on how to avoid the same issues.
5. **Are you creative?** Can you think outside the box? Can you bring something to the table that will bring clients to your business? Creative people can be the most successful as they are constantly innovating and finding better ways of developing their products and services.

The Business Plan

Every business should have a business plan. Developing a business plan can provide you with the roadmap to your success, but keep in mind that any business plan is just a prediction. You'll still need to implement it and adapt it to changing conditions. Business plans can vary, but most contain the following components:

1. Your mission statement – This defines what your business is, why it exists, and its reason for being. At a minimum, your mission statement should define who your primary customers are, identify the products and services you sell, and describe the geographical location in which you operate.
2. An explanation of how your product or service is different and sets you apart from your competition.
3. A market analysis of the local market, your competitors, and the potential market share you can realistically achieve.
4. SWOT analysis – This will be discussed further in the next section.
5. Financials – Including cash flow and revenue projections.

As you can see, business plans can (and should) be very detailed. There are many online resources that will provide you with free templates and outlines that can help you create the perfect business plan based on your requirements. You can also have a professional create the business plan for you but in either case, a well constructed Business Plan is a valuable tool that will provide you, potential investors and your banker with a better understanding of your roadmap for success.

SWOT Analysis

SWOT is an acronym for Strengths, Weaknesses, Opportunities and Threats. This type of analysis can provide vital insights on how to plan for the future, enter the market, take advantage of your strengths and eliminate your weaknesses. The SWOT uses a 2X2 matrix that is broken into Internal (Strengths and Weaknesses) and External (Opportunities and Threats) factors. The following diagram shows you a simple SWOT matrix:

Internal Factors	
Strengths - Advantages of proposition? Capabilities? Competitive advantages? USP's (unique selling points)? Resources, assets, people? Experience, knowledge, data? Financial reserves, likely returns? Marketing - reach, distribution, awareness? Innovative aspects? Location and geographical? Price, value, quality? Accreditations, qualifications, certifications? Processes, systems, IT, communications? Cultural, attitudinal, behavioral?	Weaknesses - Disadvantages of proposition? Gaps in capabilities? Lack of competitive strength? Reputation, presence and reach? Financials? Own known vulnerabilities? Timescales, deadlines and pressures? Cash flow, start-up cash-drain? Continuity, supply chain robustness? Effects on core activities, distraction? Reliability of data, plan predictability? Morale, commitment, leadership? Accreditations, etc.? Processes and systems, etc.?
Opportunities - Market developments? Competitors' vulnerabilities? Industry or lifestyle trends? Technology development and innovation? Global influences? New markets, vertical, horizontal? Niche target markets? Geographical, export, import? Market need for new USP's? Market response to tactics, e.g., surprise? Major contracts, tenders? Business and product development? Information and research? Partnerships, agencies, distribution? Market volume demand trends?	Threats - Political effects? Legislative effects? Environmental effects? IT developments? Competitor intentions - various? Market demand? New technologies, services, ideas? Vital contracts and partners? Obstacles faced? Insurmountable weaknesses? Employment market? Financial and credit pressures? Economy - home, abroad? Seasonality, weather effects?
External Factors	

Because of the personal nature of the SWOT analysis it is often better to bring in an outside facilitator to cut through the possible hurt feelings and peer pressure that this analysis can generate. When done correctly, it will provide you with a clear picture of your business now and in the future.

Licensing and Insurance

In most locations, you will need a retail merchant license, and perhaps a low voltage license. For most businesses, you will need to obtain liability insurance. Do not skimp on insurance to save money, as a good insurance policy can help minimize losses if jobs encounter obstacles that can cause significant financial loss.

Distributors and Manufacturers

You will be reselling products as well as providing services. Once you have established your business as a retail merchant, you will able to purchase products at wholesale to resell. This also means you can take advantage of a very important resource offered by distributors and manufacturers; manufacturer training. This training will give you product-specific knowledge about specifying, installing, and configuring products you will be using in your business. As a customer of a distributor you will be kept informed about upcoming training and you should take advantage of as much as possible.

SMALL BUSINESS FUNDAMENTALS

Financial Accounting

All businesses need to produce financial statements. Financial statements are not only required to prepare tax returns, but are used to help analyze the result of business decisions, such as pricing, staffing and production.

Whether you hire a bookkeeper or you do the books yourself, you still need to understand the basics of accounting to effectively run your business.

- Business owners need to make a commitment to set up an accurate bookkeeping system.
- Maintaining daily records helps you keep track of the financial condition of your business.
- A good accounting system will help you make business decisions, and analyze future opportunities.
- Learning about your bookkeeping system will help you avoid becoming a victim of theft and fraud.
- Using software will make it easier to track income and expenses, prepare documents, summarize your company's finances, and keep a set of backup records.

Your accounting system is made up of different accounts, used to code transactions to create a financial statement. The Chart of Accounts lists all of your accounts by name (and optionally by number).

Account types include:	Account Names include:
Balance Sheet Accounts	
Current Assets	Cash, Accounts Receivables, Inventory
Other Assets	Inventory, Prepaid Insurance
Fixed Assets	Vehicles, Office Furniture and Equipment
Current Liabilities	Accounts Payable, Credit Line Payable, Customer Deposits
Long Term Liabilities	Notes Payable
Equity Accounts	Stock, Retained Earnings, Distributions

Income Statement Accounts	
Income (Revenue)	Job Revenue, Service Revenue, Recurring Revenue
Costs of Goods Sold (Direct Expense)	Equipment Material, Field Labor Expenses
Overhead or SG&A (Selling , General & Administrative)	Rent Utilities, Office Labor Expenses

Transactions entered into the accounting system are used to create reports. There are two typical financial reports, the Balance Sheet and the Profit and Loss Statement (also referred to as the Income Statement or P&L)

A Balance Sheet represents the financial health of the company. It can be seen as a snapshot of a company at a given time, including money that the company owns and is owed compared to the money that the company owes. It has three main sections:

- **Assets**
  - » Typically broken out by current and fixed
- **Liabilities**
  - » Typically broken out by short term and long term
- **Equity**
  - » Including two sections: 1) Stock or Contributed Capital and 2) Retained Earnings

The formula for a Balance Sheet is:
A = L+E (Assets = Liabilities + Equity)

Retained Earnings represents the earning that the company has earned since inception, less any profits drawn out of the company through draws, distributions or dividends.

The Profit and Loss Statement (P&L) represents the financial activities of the company related to sales and the costs of those sales over a specific period of time. The P&L includes the following information:

- Income (or Revenue)
- Less: Cost of Goods Sold (COGS)
- Equals: Gross Profit
- Less: Expenses (Overhead)
- Equals: Net Income

The key to an accurate P&L is to follow an accounting principle called The Matching Principle. Simply stated, it means that the statement matches the expenses for a project with the income for that same project for a given period. This allows you to measure the true profitability of the jobs and the company as a whole.

The key metrics that businesses track is the gross profit and gross margin. The formula for the gross profit is: **Income less COGS = Gross Profit**

Basic Balance Sheet	
Assets	
Current Assets	$XXX
Fixed Assets	$XXX
Total Assets	$XXX
Liabilities	
Current Liabilities	$XXX
Long Term Liabilities	$XXX
Total Liabilities	$XXX
Equity	
Stock/Contributed Capital	$XXX
Retained Earnings	$XXX
Total Equity	$XXX
Total Liabilities and Equity	$XXX

Basic Income Statement	
Multi-Step Financial Statement	
Income	$XXX
(COGS – All Job Costs)	$(XX)
Gross Profit	$XXX
(Expenses – Overhead, non-job)	$(XX)
Net Income	$XXX

The gross profit of any project represents the money that is left over after all the direct project costs have been paid, including equipment installed and labor to install that equipment. The gross profit is used to cover the operating costs of the company, including the overhead costs, often referred to as SG&A or Selling, General and Administrative costs. These are the costs the company incurs that are not specific to any project, but part of the ongoing costs to operate the business. The gross profit not only needs to cover the overhead but leave additional money for profit.

As the gross profit is measured and tracked, it is used to determine the gross margin:

Gross Profit divided by Income = Gross Margin

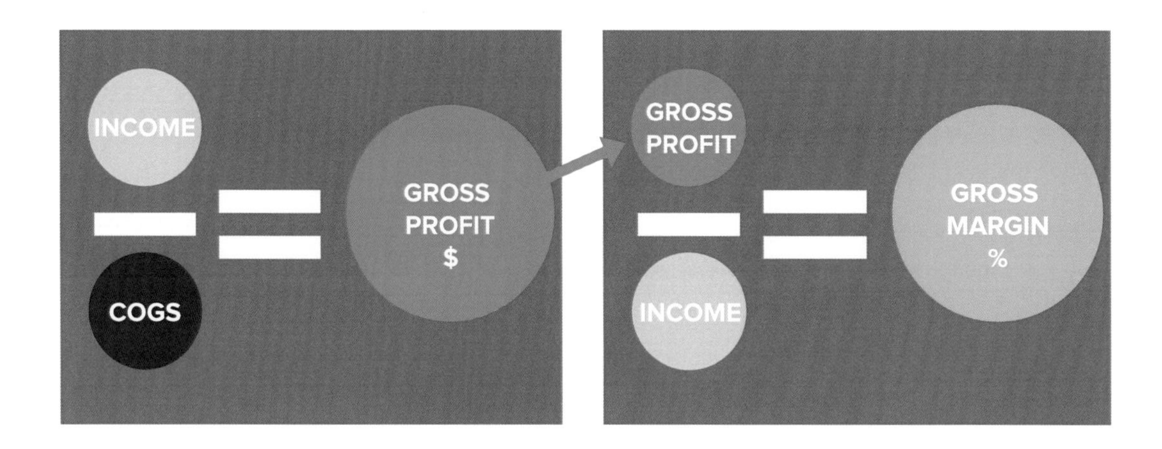

By using Gross Margin as a percentage, it allows business owners to compare profitability of projects of different sizes. For example if your company earns $20,000 in gross income, a $10,000 net profit equals 50% and that is great. If your company has $2,000,000 in gross income and only $10,000 net profit, that represents only 0.05%, which is not as good. Acceptable profits vary between business owners, but each business owner should have a goal in mind.

Marketing and Sales

There are several ways to produce revenue. Today's residential systems integrator provides a variety of products and services. They can sell and install packages of equipment from small projects to whole house audio video systems or home theatres.

These projects can be marketed and sold in different ways including Fixed Bid or Time and Materials (T&M). In a fixed bid project, the company reviews the scope of the project, estimates the cost of the equipment and labor and then adds a markup to cover overhead and produce a profit. In fixed bid projects, the company takes on the financial risk, assuming that the project can be completed within the budget (estimated costs.) Projects can lose money when a project costs more than anticipated. This could be due to installation errors, faulty equipment or scope creep. Scope creep occurs when more work is performed than originally outlined in the scope of work document. Scope creep can be avoided by making sure that all stakeholders have a clear understanding of the project goals and deliverables. If changes are made to the scope, a Change Order is created to increase the contract price. Additional work should not be performed without a signed change order.

With a T&M project, the customer assumes the risk and pays for the equipment and labor (at a specified labor billing rate,) plus a markup for overhead and profit. If scope creep occurs, the customer will be required to pay for the additional labor and equipment used in the project. This reduced risk makes T&M projects desirable, but if the stakeholders do not have a clear understanding of the project, as the price increases, so can customer dissatisfaction with the project and the company.

Whether the job is fixed bid or T&M, it is important to create a budget – an outline of the expected costs, broken down in a few categories, such as Prewire, Rough Equipment, Finish Equipment, Labor and Programming. Then, as the job progresses, it is crucial to track these job costs against the budget. Good project management keeps the job on budget and on time. A good project manager will make sure to purchase the right equipment for the project and provide the field technicians with an estimate for the hours allowed on the project. This not only helps prevent scope creep but will highlight any changes that can become a Change Order and increase the contract price.

Other sources of revenue are security monitoring contracts and network management agreements. These are examples of Recurring Monthly Revenue (RMR) models. They improve cash flow by having customers pay on a regular basis. Companies can create service agreements that include regular maintenance of the equipment. The client appreciates this service because it ensures that the installed equipment works as expected.

Websites and Social Networks

Clients can be found in many places. Creating an attractive website is an important way to connect with potential clients. The website should include detailed information about products and services offered. When showcasing products, it is important to not only show the features of a specific product but the benefits as wells. Also included in the website should be a portfolio of past work – a picture is worth a thousand words. A section of testimonials is another important aspect of any website.

Social media can also be used to market a company's products and services. As projects are completed, businesses can tweet and post pictures of implementation of successful projects.

The Sales Process

As potential clients are identified, the salesperson needs to help the client identify their needs and desires. Clients may need to see a demonstration of the equipment to help them decide what to purchase. It is important that any demo is short and focused and gives the client an idea of what is they can expect in their own home. Demos should not be confused with trainings; too many salespeople lose potential clients in demos that are too technical, too complicated, and too long. Be sure to practice any demo prior to meeting with the client to make sure that all equipment works as expected.

Once the potential client shows interest in the project, the salesperson (or business owner) creates a proposal, outlining the scope of the project and the sales price (either fixed bid or T&M estimate.) A good proposal should focus on design and client lifestyles. Presenting features of proposed equipment should be included in the Scope of Work document, but benefits should also be presented as part of the sales tool. The proposal should be simple, geared toward what will most interest the client.

By definition, CEDIA members are providing custom design and installation. This vastly different from simply selling products. It requires expertise in design, engineering and installation. Since this expertise is what a custom integrator is really selling, they should charge for these services. As margins on products dwindle and the internet commoditizes them even further, it is important to take this seriously. Some businesses charge a design retainer. This helps eliminate potential clients who are just price shopping or are not serious buyers. As an incentive, you can credit back a portion of the design retainer once the client signs a contract for the full project. Of course this will not work for all companies in all markets, but should be considered.

A good project with a good client not only provides revenue today but can be a good source of revenue in the future. It is crucial to follow up with clients after the sale and installation, to make sure they are happy with the project. Including an annual checkup in the sales proposal is a good way to get back in front of the client on a regular basis and assess their changing needs. As the technology changes, good clients can be a source of additional revenue – as they represent opportunities to go back and introduce newer products and technologies as they become available.

Customer service starts at the initial sales discussion and continues throughout the life of the client relationship. Developing strong and positive client relationships should be considered the lifeblood of your business. Set expectations at the beginning of every project to let your customers know how important their satisfaction is to your company. This is also the time to further clarify expectations by clearly communicating what is considered warranty (work with no charge) vs. maintenance (fee based service.)

At the completion of a job, be sure to ask for referrals; good clients can spread the word and help market your services by connecting you with additional potential clients.

Training and Certification

The expertise needed to ensure successful projects is something that must be gained through experience and ongoing training. Manufacturers offer a wealth of product-specific training and associations like CEDIA, ISF, HAA, CompTIA, and ESA all provide training that covers everything from the fundamentals to advanced calibration and system design. Failing to keep your team trained and current on technology will eventually lead to losing sales to those companies who work hard at staying well-trained and well-informed on emerging technologies.

Professional certification proves to peers and clients that your team members take training seriously and have proven their mastery of the tools and techniques of the industry. Many companies incentivize employees with incremental raises when higher certifications are earned. Certification documents are often displayed in plain view where customers can see them when visiting the office or showroom. Studies have shown that consumers recognize the significance of industry certifications and the value they add.

Filing and Paying Taxes

As your business grows, be sure to pay attention to taxes and other regulatory requirements. Know the laws related to your business and stay on top of any reporting deadlines. It is a good idea to put aside a portion of any profits to be able to pay taxes when due. Tax penalties can have a significant impact on any business, turning a once profitable business into one that is out of business.

CONCLUSION

Some very basic suggestions for starting a new systems integration company:

1. Create a Business Plan, including Mission Statement and SWOT Analysis
2. Get the proper licensure and retail business status (low-voltage license, retail merchant license, etc.)
3. Get set up with suppliers and distributors
4. Execute your plan, grow slowly, subcontract as needed
5. Join CEDIA, attend EXPO, get trained and certified
6. Stay current with technology
7. Adhere to proven business practices and professional ethics

No matter how technically savvy someone is, they will have a difficult time keeping a business viable without some formal business training. This training can be acquired through traditional universities or community colleges, or through specialize offerings by private companies or trade associations like CEDIA. It is critical that anyone responsible for the administration of the business, build on the very basic concepts shown here and become well versed in all of the practices needed for any successful business to survive and thrive.

Questions

1. A SWOT analysis outlines Strengths, Weaknesses, Opportunities, and ______.
   a. Technologies
   b. Taxes
   c. Threats
   d. Tendencies

2. Current assets are:
   a. Cash
   b. Accounts receivable (AR)
   c. Inventory
   d. All of the above

3. Gross profit divided by income =
   a. Gross margin
   b. Markup
   c. Net profit
   d. Real profit

4. Once your business is established and licensed as a retail merchant, you will be able to:
   a. Advertise
   b. Buy wholesale
   c. Get liability insurance
   d. Hire employees

5. The _________ represents the financial activities of the company related to sales and cost of sales over a specific period of time.
   a. Financial statement
   b. Mission statement
   c. P&L
   d. Retained earnings

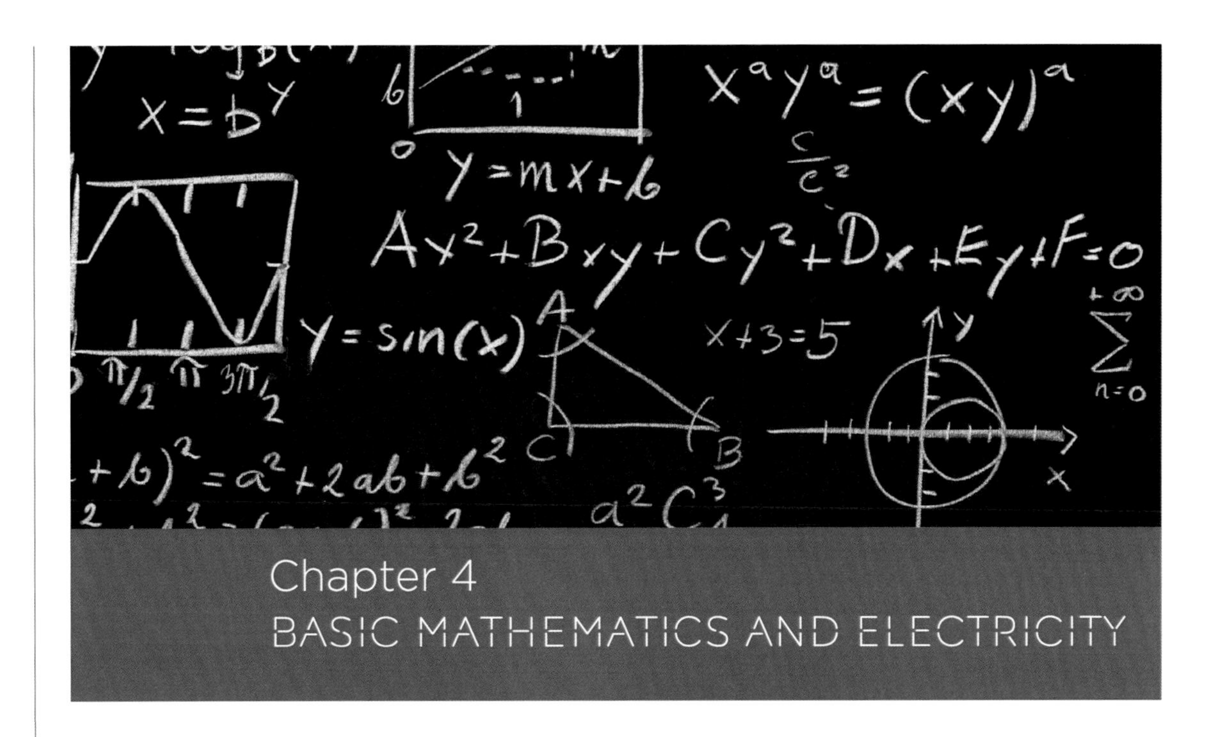

Chapter 4
BASIC MATHEMATICS AND ELECTRICITY

INDUSTRY RELATED MATHEMATICS
FUNDAMENTALS OF ELECTRICITY

INDUSTRY RELATED MATHEMATICS

Common Mathematic Functions

Basic calculations are something everyone learns in primary school and are applied daily on the jobsite. You should recognize all of the symbols and be able to do these easily, in various applications, in fractions and decimal notation. The following are some of the most common math functions that an entry-level EST will need to be familiar with for a career in residential AV systems.

- Addition (+) Example: 3 + 2 = 5
- Subtraction (-) Example: 3 - 2 = 1
- Multiplication (x) or (*) Example: 3 x 2 = 6
- Division (÷) or (/) Example: 3 / 2 = 1.5
- Equals (=) Example: 1.5 + 2.25 = 3.75, or 12 in. = 1 ft.
- Greater than (>) Example: 3 > 2
- Less than (<) Example: 23 < 45
- Exponent $x(^{y})$, where x is multiplied by itself y times. Examples: 102 = 100, or 53 = 125
- Root ($^{y}\sqrt{}$)x, where the root of x is a number, when multiplied by itself y times, equals x. Examples: 2√64 = 8, 3√9 = 3
- Pi (π), not a function, but a value equal to 3.14159265359...or 3.14

Geometric Shapes and Equations

The following are some of the most common geometric expressions, shapes and equations which are fundamental to any trade, and frequently used and referred to regarding solutions in residential A/V systems.

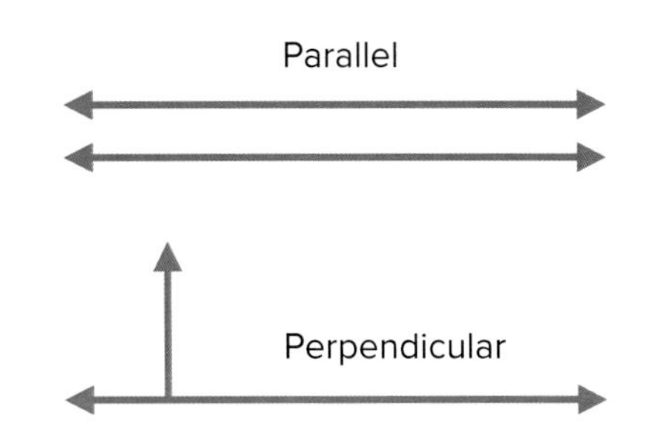

Parallel Lines; two lines which are always the same distance apart and never cross
Perpendicular Lines; two lines which cross at a right angle, or 90°.
Angle; when two lines intersect, the relationship is expressed in degrees.

- Two lines joined to make a straight line form a 180° angle
- Two lines that intersect at 90° form a right angle (perpendicular)
- Two lines forming an angle of less than 90° form an acute angle
- Two lines forming angle of more than 90° form an obtuse angle

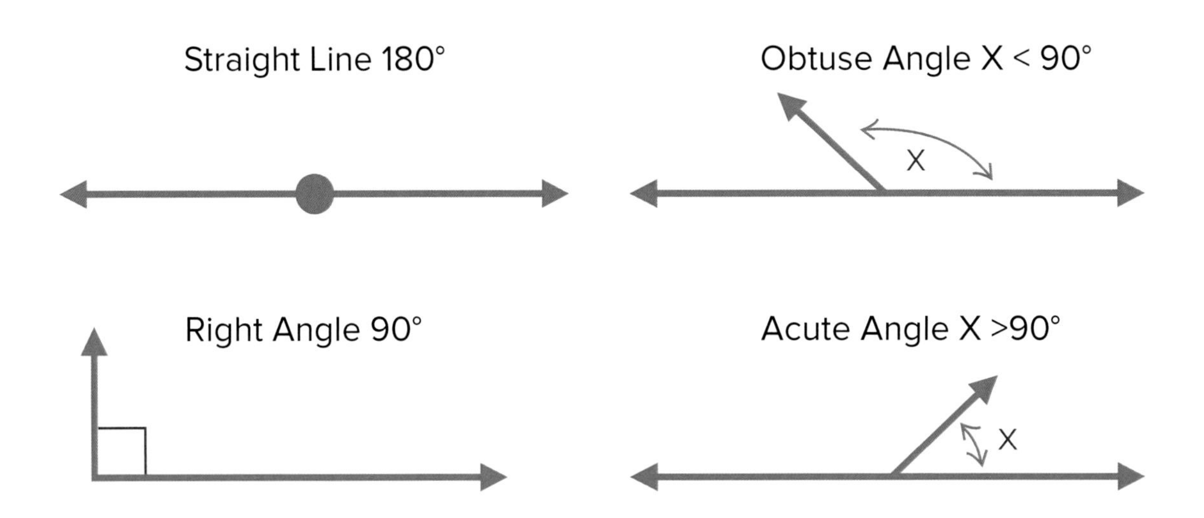

Triangle; a triangle is a plane figure with three sides and three corners, and the total of the angles in the corners is 180°

- Defined by side length (congruent sides have the same length)
  - » Scalene – no congruent sides
  - » Isosceles – two congruent sides
  - » Equilateral – three congruent sides (equilateral triangle)
- Defined by angles
  - » Acute - all angles > 90°
  - » Right - one angle = 90°
  - » Obtuse - one angle < 90°
- Area of a triangle
  - » (Base × Height)/2

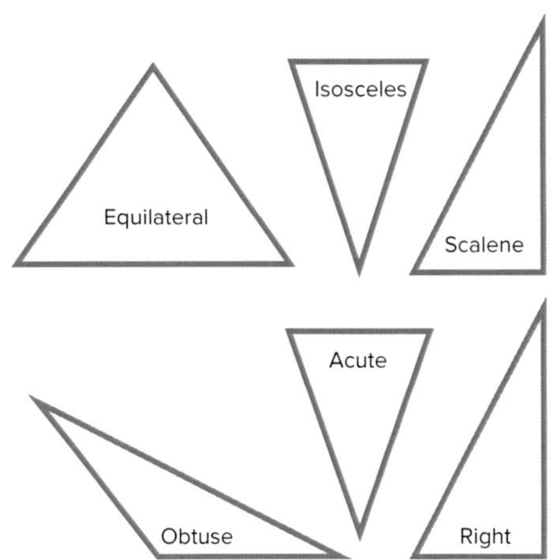

The Right Triangle and Pythagorean Theorem; This concept is credited to Pythagoras, a Greek philosopher of the 6th century B.C. It is an important formula to understand. The longest side of a right triangle is called the hypotenuse. It always opposes the right angled corner. We can find the length of that side if we know the lengths of the other two. In these formulas, C is the hypotenuse.

$$a^2 + b^2 = c^2$$

For instance, we can find the diagonal of any rectangular video display or screen if we know the height and width.

Square the length of side a, square the length of side b, and add them to together. The resulting number is the square of c, the hypotenuse (diagonal).

A simplified version of this can also be used to plot a right angle out from any surface. This is known as the "3, 4, 5 Rule". Mark out one of the short sides on the wall with 4 of any unit of length (ft, in, m). Then measure out 3 units from one end of that line, and 5 units out from the other and make those two intersect. When they do, the shorter line out from the original wall will be at exactly 90°. This can also be used to check an inside corner of a room to see if it is square, because;

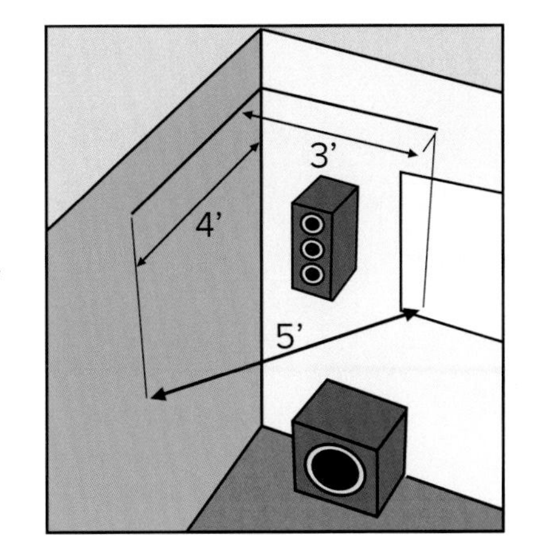

$$3^2 + 4^2 = 5^2$$

or

$$9 + 16 = 25$$

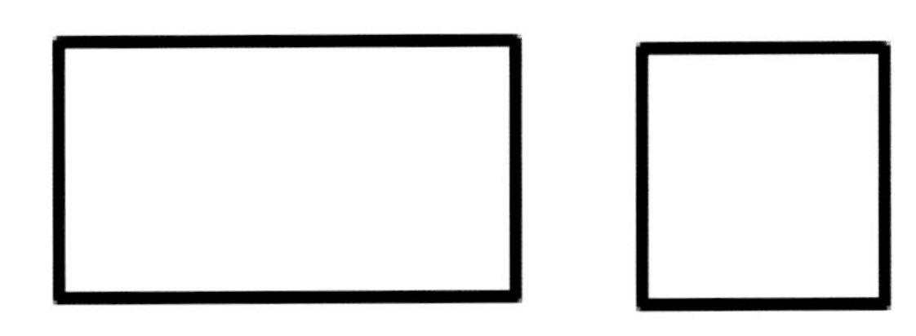

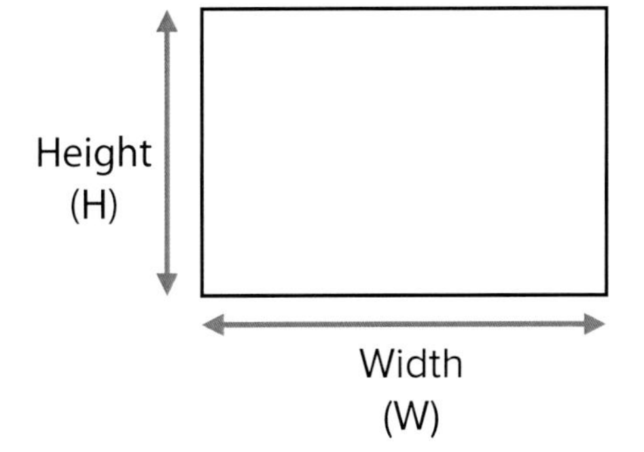

Rectangle; a four sided shape with two sets of parallel sides. All four corners are right angles.

Square; a square is a rectangle with four sides of the same length

Rectangle Equations; there will be situations where you need to know the area or total circumference of a rectangle or square.

- Area (A): A = h x w – expressed in square units (m^2, Ft^2)
- Circumference: sum of all four side lengths
- Rectangular Volume (V) in three dimension= Length x Width x Height - expressed in cubic units (m^3, Ft^3)

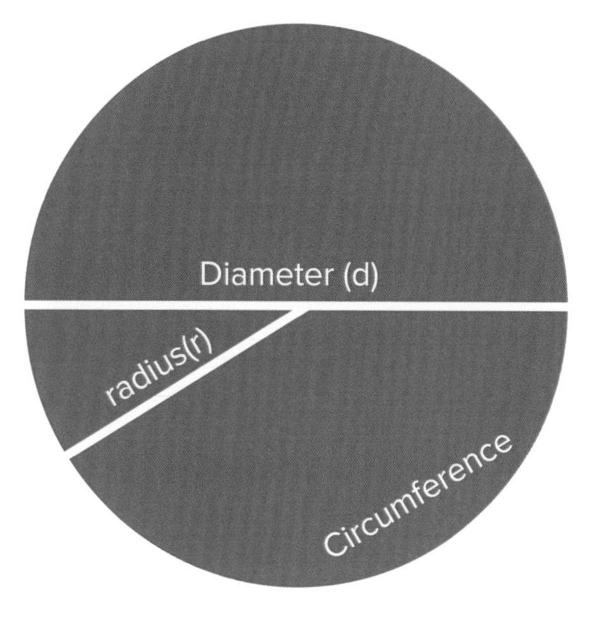

Circle; one continuous line, where every point on the line is equidistant from a common center point. This concept comes into play frequently in the industry; cables, loudspeakers, speaker configurations.

Circle terms;
• The **Radius** is the distance from the center to the outer circle
• The **Diameter** is the distance from one side, through the center , and to the other side. Twice the diameter.
• The **Circumference** is the distance all the way around the circle.

Circle Equations;
• Pi (π) is the ratio of circumference to diameter. So the circumference is approximately 3.14 x the diameter.
• Area (A): $A = \pi \times r^2$
• Circumference (C): $C = \pi \times d$ or $C = 2 \times \pi \times r$

Measurements and Conversions

The following are common measurements and conversions that are frequently used in the design and installation of electronic systems.

Metric and English Conversions:

The metric system makes conversion from one range of measurement to another very simple:
1 Kilometer (km) = 1,000 meters (m)
1 meter (m) = 100 centimeters (cm)
1 meter (m) = 1,000 millimeters (mm)

But you may find you need to convert to and from English (Imperial) measurements:
1 meter (m) = 3.28 feet (ft) = 39.4 inches (in)
1 foot (ft.) = 0.305 meters (m)

Feet-Inches to Decimal Feet Example:

In the US and some other regions, where inches and feet are the units of length, it is also sometimes necessary to divide feet into 10ths of a foot rather than inches (12 per foot). To convert inches into 10ths of an inch, follow these steps. Note: (') = feet (ft), (") = inches (in)
6'4" = 6' + 4"
4" = 4" x 1 ft/12 in
4/12' = 0.333'
So: 6'4" = 6.333'

Decimal Feet to Feet-Inches Example:

To reverse the process:
29.65' = 29' + 0.65'
0.65' = 0.65' x 12 in/1 ft
0.65' = 7.8"
0.8" = 8/10" (or 4/5")
So: 29.65' = 29' 7 4/5"

Example Calculations

Here are a couple of examples of using mathematics in the industry.

Percentage

When driving a lighting load with a low voltage magnetic transformer, it is common to supply 120 percent of the full load power due to transformer loss.

Calculate the power needed to drive a 55 Watt lighting load to full power when driven by a transformer.

Load power = 55 Watts max.
Supply power = 55 Watts x 120%
Supply power = 55 Watts x 1.20
Supply power = 66 Watts

Projection Screen Aspect Ratio

Calculate the aspect ratio of a projection screen when the screens height is 40 inches and the width is 94 inches.

The aspect ratio of a screen is the viewing surface width divided by the height, so:
Aspect Ratio = screen width/ screen height
Aspect Ratio = 96 inches/ 40 inches
96 divided by 40 = 2.4
Aspect Ratio = 2.4:1

Projection Screen Dimensions

Many times we need to know the height of a screen surface to calculate optimum viewing distance. If we know the screen width, we can simply divide it by the aspect ratio to get the height.

b / 1.78 = a
or, 100' divided by 1.78 = 56.2

If we know either the height or the width, and the aspect ratio, we find the other one.
Then using those two dimensions, we can calculate the diagonal using the Pythagorean Theorem.

$$a^2 + b^2 = c^2$$

Square the height.
Square the width.
Add them together and that will be the square of the diagonal.
So find the square root of that number and you have the diagonal.

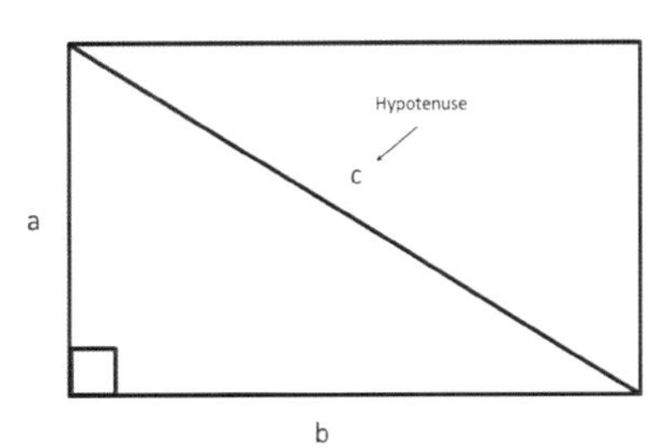

FUNDAMENTALS OF ELECTRICITY

What is Electricity?

Electricity is a broad term which applies to many disciplines. For our purposes we will discuss electricity as the presence and flow of electrical charge – in essence electrons moving from one atom to another. This movement carries with it energy, which can be used to many different things, from creating light and heat, to carrying data/audio/video and other signals.

Electrical Units

There are four main units of measurement everyone should understand. Once you know what they measure and how they are related, most concepts of electricity and basic electronics will be easy to grasp.

Volt

The Volt is a measurement of electrical potential, or the "pressure" behind the flow of electricity. It is named after Italian physicist Alessandro Volta, who invented the voltaic pile, possibly the first battery.

Ampere

The Ampere, or Amp, is the measurement of electrical current, the actual flow of electrons – the passing of electricity past a specific point. 6.28×10^{18} electrons flowing past this point in one second is one Ampere. It is named in honor of French mathematician and physicist Andre'-Marie Ampere, widely considered the father of electrodynamics.

Watt

The Watt is a measure of power. The combination of pressure (Volts) and current flow (Amps) create power. Power is what is need to do any kind of work; heat, light, motion, etc. We will see that the way Volts and Amps combine to create power is fundamental to our industry. This term is named after the Scottish engineer James Watt.

Ohm

The Ohm is a unit of electrical resistance. This is what slows or pushes back on the flow of electrons. Materials that are good conductors of electricity like copper have low resistance. The air and many non-metallic materials have high resistance and it is difficult for electricity to flow through them. This unit is named after German physicist Georg Simon Ohm. The unit is also often expressed by the Greek letter Omega (Ω)

Putting Them Together

Let's look at how these terms apply to electricity in the real world. It is sometimes helpful to think of an electrical current in terms of something we are more familiar with, the flow of water. In this analogy, the pressure behind the water faucet is what is there waiting to move the water as soon as a path is provided. This is similar to the Voltage. When the faucet is opened, that pressure moves water through the pipe or hose. How fast the water moves is dependent on the size of the "conductor". With the same pressure it will move faster through a large pipe than a small one. This is because the smaller pipe presents more resistance to the flow of the water. A second faucet, partially closed, would also present resistance, as would pumping the water uphill. The water pressure resembles Voltage, the amount of water that flows in a period of time is like Amperage, and the slowing of flow is like Ohms.

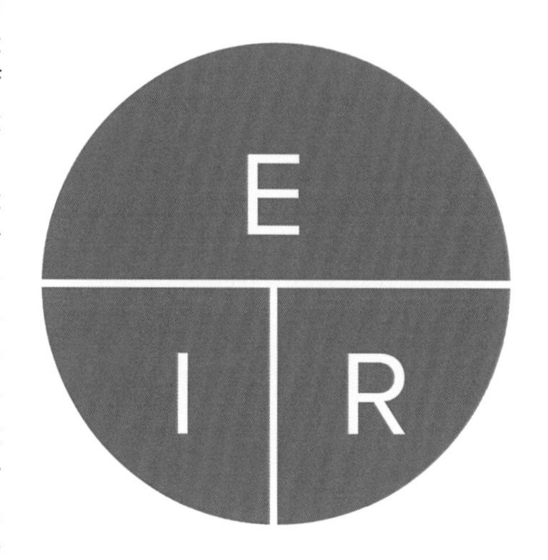

This relationship is defined in Ohm's Law, where:

I = Current, measured in Amps
E = Voltage, measured in Volts
R = Resistance, measured in Ohms

$$I = E \div R$$
$$E = I \times R$$
$$R = E \div I$$

Once we know the Amperage and Voltage in a circuit, we can calculate how much power is being used:

W = E x I
Or Watts = Volts x Amps

Electrical Circuits

An electrical circuit is a path that electricity follows from its source through conductors and at least load (resistance) and back to the source. Electricity which flows only in one direction is called Direct Current (DC). This is the type of current found in flashlights and other devices being powered by dry cells or batteries.

DC Circuit Example

In this example, a battery is the power source, a switch is in the circuit, and the load (resistance) is something like a light bulb. When the switch is open, the current cannot flow. This is called an open circuit. The battery still has power but it is not being used.

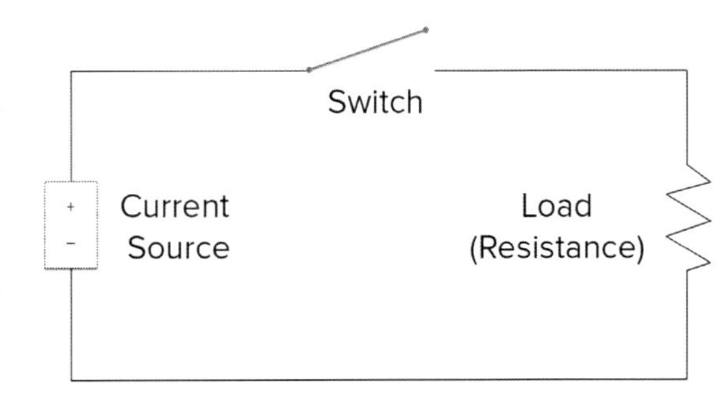

Figure 2-1 Open Circuit

Now we will look at how Ohm's Law applies to this circuit.

- For simplicity we will call the source (battery) 10 Volts DC.
- The load is a light bulb with a resistance of 250Ω
- So we can calculate the current in Amps;
  - » I = E / R (Ohm's Law)
  - » A = 10v / 250 Ω = .04 Amps (or 40 mA)
- We can now calculate how much power is being used;
  - » P = E x I (the Power Formula)
  - » W = 10 V x .04 A = 0.4 Watts (400mW)
  - » 10 volts x .04 amperes = 0.4 watts (or 400 milliwatts)

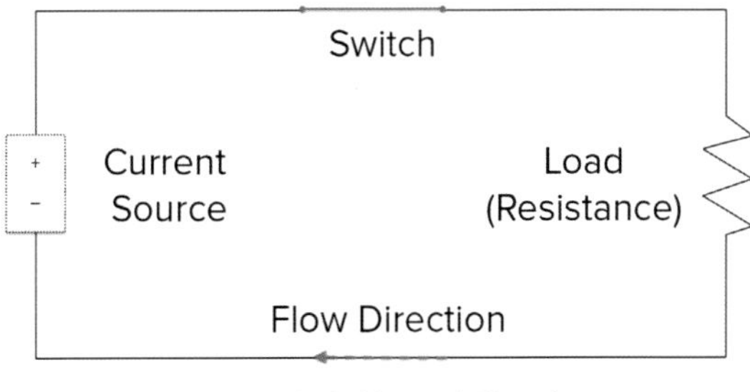

Figure 2-2 Closed Circuit

Note; We have shown an open circuit (no current flow) and a closed circuit (current flow through the load). What would happen if we place a conductor so that it directly connected the top and bottom line of the circuit, near the battery? This would be a "short circuit", or "short". In this case there would be no load, so the resistance would essentially be zero. When resistance goes down, current goes up. In this case the battery would be depleted of its power rapidly, and the light bulb would not receive any current.

Alternating Current (AC)

Electrical current that reverses direction in repeating cycles on the circuit is known as Alternating Current (AC). This is the type of electricity what is widely used for homes and businesses and lighting around the world. This type of electricity can be distributed at high voltages and "stepped down" as it gets nearer to its final destination. It is extremely versatile, and capable of carrying a great deal of power. An illustration of AC is shown here:

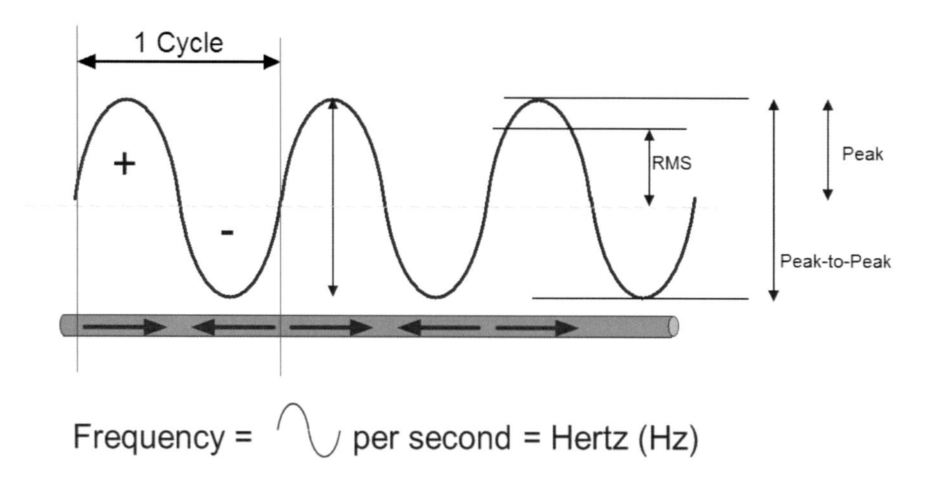

Each time the current flows one direction, reverses to the other direction, and reverses again to the start of the wave equals one cycle. The number of these cycles per second is its Frequency, or Cycles Per Second (CPS), often express in Hertz (Hz), named after another great physicist (this one German), Heinrich Hertz. He was also the first to prove the existence of electromagnetic waves. Throughout our industry, in electricity, audio, sound, and video you will see the unit Hertz (or cycles per second) used. In this diagram we see there are different ways to measure the amplitude of the wave (Peak, Peak-to-Peak, RMS). We will later see how this applies to amplifier power. Regardless of the line voltage (110, 117, 120, 220, 230 etc.), or the frequency (50Hz, 60Hz), Alternating Current is the kind of electricity found in homes everywhere. For more information on electricity around the world, see Chapter 18.

Also, AC is the type of electricity we deal with when we use analog audio or video signals. Sound in the air is composed of alternating regions of high and low pressure and an analog audio signal is analogous to that sound wave. Analog video signals are much more structured but still involve positive and negative components of a wave carried over the air or over a cable. Off air broadcasts, both radio and TV, have a carrier signal which is a simple wave on which additional information is attached. So the concept of alternating current is something we work with daily.

Resistance vs. Impedance

We have seen how a complete DC circuit has a source, conductor, and load. The load has resistance, measured in Ohms. In an AC circuit we have the same concept, but not quite as simple. If AC current is seeing a simple load like a light bulb, the resistive load is fixed. But when facing a more complex load, such as a loudspeaker, the current flow is affected by Reactance, which is made up of two components, Inductance and Capacitance. Without getting too technical, these two affect AC current flow in different ways based on frequency. So Reactance to an AC audio signal, which may be a combination of many different frequencies, will vary greatly. The voice coil of a loudspeaker has a fixed DC resistance but the overall load is variable because of the Reactance. So when we speak of Ohms in this case, we are speaking of average Impedance, or "nominal" Impedance. An 8Ω speaker, depending on many factors including frequency, may have an actual Impedance which varies from 2Ω to 20Ω. While Resistance is abbreviated "R", Impedance is represented by the letter "Z".

These illustrations show the difference between loads with and without Reactance in an AC circuit:

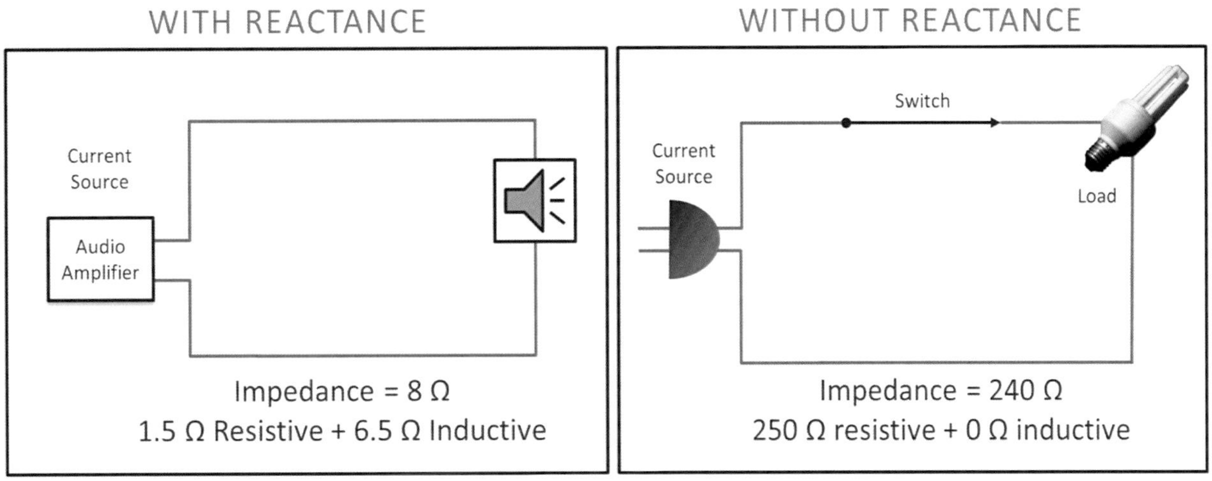

WITH REACTANCE

Current Source

Audio Amplifier

Impedance = 8 Ω
1.5 Ω Resistive + 6.5 Ω Inductive

WITHOUT REACTANCE

Switch

Current Source

Load

Impedance = 240 Ω
250 Ω resistive + 0 Ω inductive

The principles of electricity also apply to the world of electronics, where small devices and components manipulate small signals to enable functionality in the areas of radio, TV, telecommunication, audio, etc. The signals moving around on a printed circuit board or inside a microprocessor obey the same rules as line voltage.

In the field, ESTs and other electronic systems professionals depend on electricity to power their systems but in most jurisdictions the actual installation of electrical wiring and outlets must be done by a licensed electrician. In general, electricians deal with circuits above 90v, while low-voltage installers deal with non-lethal circuits below that threshold. This arbitrary point is based on the traditional ring voltage on telephone systems, at 90v. This is a generalization, so know your local codes and regulations.

Questions

1. What is the square footage of the room in the following diagram?
   a. 81
   b. 72
   c. 90
   d. 100

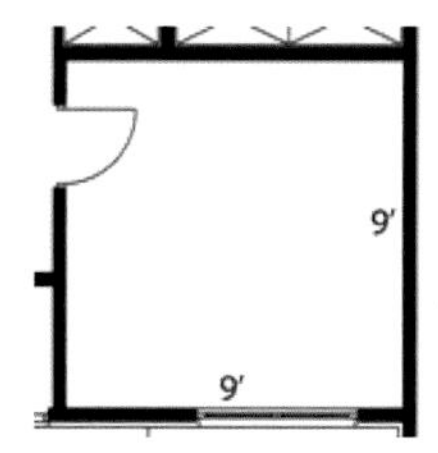

2. Convert 6 ½ inches to decimal equivalents in feet:
   a. 0.541 ft.
   b. 0.5 ft.
   c. 0.641 ft.
   d. 0.6 ft.

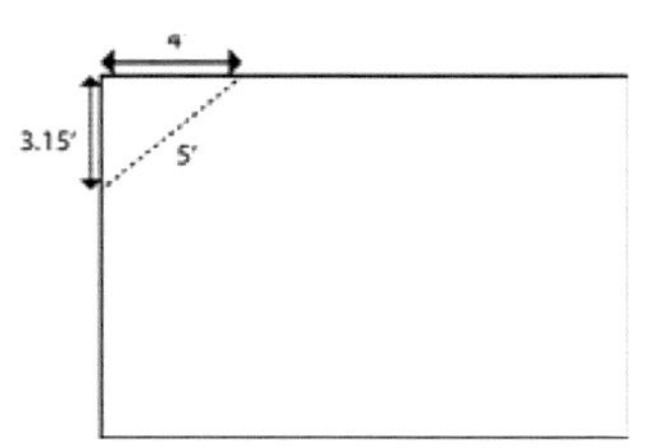

3. In the following diagram, are the walls square?
   a. Yes
   b. No

4. Assuming the speed of sound to be 1,130 ft/sec, what is the frequency of a sound wave with a wavelength of 2 ft.?
   a. F=565 Hz
   b. 56.5 Hz
   c. 2260 Hz
   d. 226 Hz

5. An amplifier channel is rated at 100 watts into an 8 ohm load. You decide to use two 8 ohm speakers, in parallel. What is the total load the amplifier channel will receive in this configuration?
   a. 8 ohms
   b. 4 ohms
   c. 16 ohms
   d. 2 ohms

Chapter 5
SAFETY, CODES, AND STANDARDS

CODES AND STANDARDS AROUND THE WORLD
JOBSITE SAFETY
FIRST AID AND MEDICAL EMERGENCIES
PROPER TOOL USE

CODES AND STANDARDS AROUND THE WORLD

People have been establishing building codes since ancient times. Nowadays there are hundreds of codes, standards, guidelines, regulations, etc. that affect every jobsite. It is important to understand each type, and how it works.

Code vs. Standard

The industry uses specific terms to express the authority and consequences of regulations. However, you may hear people referring to the same set of rules as either codes or standards, almost using the terms interchangeably. There are also recommended practices and other best practice guidelines published to encourage efficiency and quality.

We prefer to define the following terms in this way:

- **Code:** A formal body of rules specifying the requirements for the design and installation of allowable components for low-voltage electronics; commonly "shall" and "shall not" rules; often carrying criminal and/or civil penalties for infractions. Codes mainly deal with safety issues. Example: OSHA or H&H code which is enforced by authorities.

- **Standard:** A formal or informal set of industry or company-accepted guidelines that often serve to promote performance and efficiency. Example: TIA standard for communication cabling in the home.
- **Recommended Practice:** A less formal set of guidelines intended to aid in the design and installation process by establishing specific performance goals and techniques. Example: CEA/CEDIA CEB-22 Home Theater Recommended Practice - Audio Design.

Building Codes

You must remember that building codes vary from country to country. Therefore, you must educate yourself about local building codes that vary by province and municipality. For instance, areas like San Francisco or San Diego that could have earthquakes will most likely have specific codes than someplace like London or Sydney would not have.

Staying in Sync with Code Updates

All codes and standards are updated as new technology and installation situations change. It is your responsibility to remain knowledgeable about the changes that impact our industry. Stay abreast of upcoming changes to codes when designing and installing components. If applicable regulations are created or revised before a project is completed, you must ensure the design and installation adheres to the new codes and standards. Planning for this can save costly rework.

Authority Having Jurisdiction (AHJ)

Different projects may require you to work in various geographic locations, and it is likely you will fall under a different Authority Having Jurisdiction (AHJ). Per the National Electrical Code (NEC) which is similar to most codes in the world, the AHJ "has the responsibility for making interpretations of the rules, for deciding on the approval of equipment and materials, and for granting the special permission contemplated in a number of the rules." This means they have the last word regarding code enforcement.

It is critical that you identify the AHJ for each project since the AHJ sets the official standards to which you must perform. This may be a local building inspector, the police or fire department, a representative of a state or federal agency or someone designated by the building owner.

When the project is completed, the AHJ determines whether you have designed and installed a safe, compliant system or whether costly fines and rework will result. In many regions, the AHJ may be empowered to disconnect the electrical service to an installation determined to be unsafe.

JOBSITE SAFETY

Jobsite and electrical safety regulations are also strictly enforced in most countries. Improved jobsite safety enforcement has reduced on-the-job accidents and deaths greatly over the past few decades. Since each region has different specific regulations, we will focus on general guidelines. Consult your local codes and regulations.

It is also important to note that following safety regulations is good business. Of course, ensuring the safety of workers is in everyone's best interest, but violating safety regulations can also have a financial impact. Fines can be huge, and having a jobsite shut down for any reason affects the entire timeline and therefore profitability of a project.

Example of Health and Safety Threats

Many factors on site can pose a serious health or safety risk:

- Asbestos
- Improper electrical wiring
- Unsafe power tools
- Lead based paint
- Excessive sound pressure levels
- Scaffolding and ladders (falls)
- Heat, cold, noxious gases, lack of Oxygen
- Working at Height

It is your responsibility to ensure you and your team are aware of the local codes and standards before starting work on any building site.

The Perils of Getting it Wrong!

When you go over to your friend's house to help him put some shelves up in his garage, you tend to not have to wear hard hats, sign in to a book just inside the house and provide evidence that you are competent to be in the garage assisting with the task of putting up some shelves. If, however, you go to a large scale building operation where there are many tradesmen carrying out an array of different activities it is well known that personal protective equipment must be worn and if requested, proof of task competency must be shown to the authorities.

So what happens when arrive at a home which is undergoing renovation, and your job is to supervise the installation of an audio / video distribution system. There may be 3 or 4 other trades there, and a builder who is supervising the project. Although there are the necessary signs in place everyone is pretty laid back, some are wearing PPE (personal protective equipment) and others are not. Do the same Health and Safety Codes applicable on a large scale building site apply to a small site like this?

The likely answer to this is yes! In many countries, as far as the authorities and the insurance companies are concerned, if you have an accident or cause one you are still required to prove you are competent to be there. If you cannot prove this you are not only putting yourself at risk, you are also risking your employers business and your colleagues' jobs.

"Duty of Care" Liability

This is a term used in some regions to describe the concept that, under the common law organizations and individuals carry a responsibility for each other and others who may be affected by their activities. Where something goes wrong, individuals may, in some cases, sue for damages using the civil law if they are injured as a result of another person's negligence. Of course this varies by region, but in general there are usually implications within civil law that could put you and your company at risk of very expensive legal action.

Also, across the world the authorities are now focusing more on the smaller scale "basement renovation" type projects, as they become ever more popular. Because authorities have the power to shut down a site and indeed a company, it is imperative that we educate ourselves properly in all regulations which exist in our region.

Jobsite Safety Fundamentals

There are some basic terms and concepts that are common to all regions.

- Head and eye protection – also known as Personal Protection Equipment (PPE) – required in some areas, always a good idea
  - » Eye protection should always be used when using a hammer or any power tool
- Power tool safety
  - » Carry by the handle, not the cord
  - » Don't unplug by pulling the cord
  - » Disconnect from power before changing bits/blades
  - » Must be grounded or be "double insulated"
- Electrical safety
  - » Do not use damaged extension cords
  - » Do not hang cords over metal objects
- Extension ladder safety
  - » Height to landing point should be 4X the distance the base is out from that point (4 to 1 Rule)
  - » Extend top 3' or 1m beyond landing point
  - » Do not overextend ladder
- Step ladder safety
  - » Do not stand on top or last step
  - » Make sure legs are fully extended
  - » Always face the ladder

Double Insulated

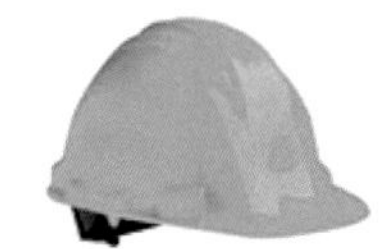

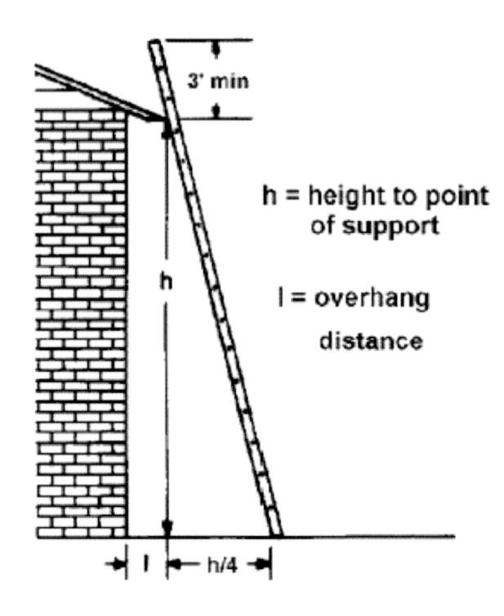

Figure 5-1 Extension Ladder Rules

FIRST AID AND MEDICAL EMERGENCIES

Your Responsibilities

It is your responsibility to practice common sense which may help to prevent the need for first aid, for instance:

- Stay alert, be well rested
- Stay straight and sober
- Never work alone in attics or confined spaces (the "buddy system")
- Know where the first aid kit is
- Know where the nearest medical facility is (this should be noted in the job file)
- Stay hydrated when working in heat
- Get trained in CPR. This short training could save a life

Priorities in the Event of an Emergency

Your priorities are to:

- Assess the situation – do not put yourself in danger
- Make the area safe
- Assess victims and attend first to any who are unconscious
- Send for help – do not delay.

What You Can Do First

- Check for a response - Gently shake the victim's shoulders and ask loudly, "Are you all right?". If there is no response, your priorities are to:
  - » Shout for help
  - » Open the airway
  - » Check for normal breathing
- If not breathing normally:
  - » Place your hand on the person's forehead and gently tilt the head back
  - » Lift the chin with two fingertips
  - » Look, listen and feel for normal breathing for no more than 10 seconds
  - » Look for chest movement
  - » Listen at the casualty's mouth for breath sounds
  - » Feel for air on your cheek

If the victim is not breathing normally, then you must get help and start chest compressions. This requires CPR from a trained individual.

Other Health and Safety Situations

There are many different situations which may arise on the jobsite; too many to cover here. The best course of action is to pursue formal training in first aid, CPR, etc. and be well versed in handling emergency situations on the job.

Heat Exhaustion and Heat Stroke

If a worker is overcome by heat, cool the subject with a wet towel and hydrate them with water. If they do not respond they may be experiencing heat stroke. In this instance emergency personnel should be called immediately.

Hearing Loss

Beware of extremely loud sounds and environments where elevate sound levels persist for long periods of time. Examples of situations that should be considered potentially dangerous to hearing:

- The noise is intrusive for most of the working day
- You have to raise your voice to have a normal conversation when about 2 m apart, for at least part of the day
- You use noisy powered tools or machinery for over half an hour a day
- The type of work is known to have noisy tasks, e.g. construction or demolition
- There are noises because of impacts or regular use of pneumatic impact tools

Another sign that something should be done about the noise is having muffled hearing at the end of the day, even if it is better by the next morning. If you have any ear or hearing trouble, let your employer know. Hearing protection is designed to reduce your risk to injury through excessive noise levels. Approved hearing protection should be used any time there is a risk.

Severe Bleeding

When you work with metal studs, or are pulling cable in and around metal studs, wear gloves to protect your hands from the sharp edges that may be exposed. Use caution in pulling cables to avoid ripping your jacket sleeves that may be scraped on the sharp edges. Use a punch rather than a saw if new routes need to be created. Always use grommets to protect cable from sharp edges. If an injury occurs resulting in severe bleeding:

- Apply direct pressure to the wound;
- Raise and support the injured part (unless broken)
- Apply a dressing and bandage firmly in place
- Get medical assistance

Broken Bones / Spinal Injuries

If a broken bone or spinal injury is suspected, obtain expert help. Do not move the casualties unless they are in immediate danger.

Burns

Burns can be serious, so if in doubt, seek medical help. Cool the affected part of the body with cold water until pain is relieved. Thorough cooling may take 10 minutes or more, but this must not delay taking the victim to hospital.

Certain chemicals may seriously irritate or damage the skin. Avoid contaminating yourself with the chemical. Treat as instructed on the materials package. Remove any contaminated clothing which is not stuck to the skin.

Eye Injuries

All eye injuries are potentially serious. If there is something in the eye, wash out the eye with clean water or sterile fluid from a sealed container, to remove loose material. Do not attempt to remove anything that is embedded in the eye.

If chemicals are involved, flush the eye with water or sterile fluid for at least 10 minutes, while gently holding the eyelids open. Ask the casualty to hold a pad over the injured eye and send them to hospital.

Electric Shock

Electrocution can be fatal, and should be taken very seriously.

- Don't touch the person if they're still in the source of electric current or you'll get a shock, too.
- Switch off the current at the source.
- Check their breathing. If they're unconscious but breathing normally put them in the recovery position.
- Call for emergency medical assistance.

Record Keeping

It is good practice to document any incidents involving injuries or illness. Include the following information in your entry:

- The date, time and place of the incident
- The name and job of the injured or ill person
- Details of the injury/illness and any first aid given
- What happened to the casualty immediately afterwards (eg went back to work, went home, went to hospital)
- The name and signature of the person dealing with the incident.

This information can help identify accident trends and possible areas for improvement in the control of health and safety risks.

PROPER TOOL USE

The low-voltage installer must correctly identify and properly use proper tools to be successful in their work. The following are some of the most-used tools for ESTs. The table briefly describes some appropriate and inappropriate uses for these tools.

Tool	Appropriate Use	Inappropriate Use
Screwdriver	Fastening and unfastening Phillips- and flat-head screws	Prying objects open; chiseling out holes in wood
Chisel	Chipping out openings in wood	Cutting wire or cable
Lineman Pliers	Twisting multiple conductors together (usually done by an electrician)	Crimping fasteners, driving in nails, tightening bolts
Hammer	Pounding nails, inserting shims	Hammering electrician's staples over coaxial or UTP cable
Black Electrical Tape	Insulating wire, fastening pull line to wire	Marking locations on walls or carpet; adhesive will instantly stain walls and carpet
Utility Knife	Cutting cable jacket	Prying objects open, loosening/tightening screws
Sheetrock Saw	Cutting holes in drywall	Cutting wire or lathe and plaster walls
Soldering Iron	Joining wire to connector	Burnishing scratches in cabinets or furniture
Crimping Tool	Terminating UTP cable	Hammering or prying

Questions

1. A formal set of industry or company-accepted guidelines intended to promote better performance and efficiency is called a___________.
   a. Standard
   b. Code
   c. Recommended practice
   d. Regulation

2. Which of these could be the Authority Having Jurisdiction on a project?
   a. State Board of Licensure
   b. City Building Inspector
   c. Building Superintendent
   d. Any of the above

3. ___________ should be used any time a power tool or hammer is in use.
   a. Steel toe shoes
   b. Eye protection
   c. Respirator
   d. Gloves

4. Which of these should be done when using a step ladder?
   a. Face the ladder
   b. Make sure the sides are fully extended
   c. Avoid standing on the top platform or highest step
   d. All of the above

5. Violating a ___________ can result in fines and/or stoppage of work.
   a. Standard
   b. Recommended Practice
   c. White paper
   d. Code

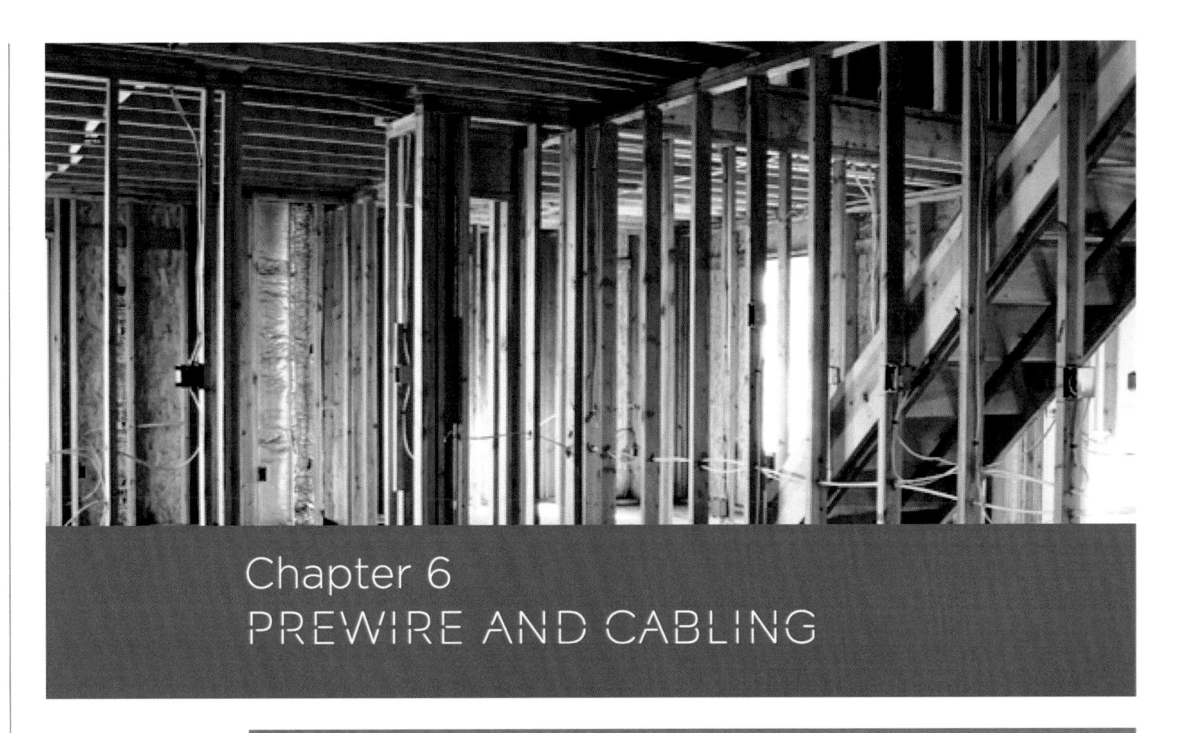

Chapter 6
PREWIRE AND CABLING

INTRODUCTION AND OVERVIEW

Electronic systems in the home depend on the infrastructure. While wireless technologies play an increasing role, there is no substitute for a well designed and installed cabling system. New construction provides an excellent opportunity to get the cabling in place which will allow for communication, control, and the easy interconnection of subsystems. Of course, not all projects are in new construction, but most of the basics covered here also apply to retrofit installation. Techniques vary by region. Wood frame construction is popular in North America and some other parts of the world. Metal studs and concrete construction dominate much of the residential industry globally. You are encouraged to be especially well trained in the methods used in your part of the world.

If you are pursuing a career as an EST (Electronic Systems Technician) it is especially critical that you know the best practices, codes, standards, and safety regulations in your country and locality. In the US, these areas as well as the installation of infrastructure are covered in detail in the ESPA (Electronic Systems Professional Alliance) program and certification. In other regions, CEDIA also offers training which is more specific to the region. In this chapter and the next we will cover the installation and testing of the low voltage wiring of the home in a general manner;

the base of knowledge needed regardless of what career path you may be following within the industry. Your company may follow slightly different practices but what is shown should be considered the recommended methods for getting the job done.

Jobsite Walk-Through and Prep

The Pre-Wire Phase of a project, also known as the Rough-in, is usually the first time the installation technicians visit the jobsite. This is when they get familiar with the jobsite and the project. In most cases a designer, project manager, or sales person has previously walked the site, finalized locations, and marked or tagged those locations.

Cable Installation Preparation

Part of the preparation, prior to the installation of cabling, is the need to place and mount any and all backboxes at which the cables terminate. In addition, paths must be created for the cable; and anchors and raceway must be installed to support and manage the cables.
All the cable must be gathered up and verified with documentation as to quantity, application and construction. In addition, all required fasteners, labels and tools should be transported to the location where the cable will be pulled. This step will vary greatly, depending on the type of construction.

Cable Installation

The last step in the Pre-Wire Phase is to install the cable. Cable installation can vary according to the type and amount of cable being installed. But, in each case cables should be labeled appropriately, care must be taken to protect cable from damage and adequate slack and trim length must be maintain for proper termination at a later time.

Cable must be secured and protected from jobsite damage, lengths logged and checked off against the documentation, and the jobsite must be cleaned up and cleared of any waste.

Retrofit Wiring

Throughout this chapter and the next there will be references to how the work described applies to retrofit installation. The cable, connectors, and platework are the same whether installed during new construction or in an existing home. However the techniques used to get cable from one point to another are vastly different. It is important to understand and sharpen your skills for both scenarios.

WALKTHROUGH AND PREP

Review Documentation

You must be familiar with the basic types of job documentation, since this is what you will be using to perform each phase of the job. Good documentation makes the management of the project more accurate, and more efficient. Project management helps to control costs, deliver quality, and keep the job profitable.

Make sure you discuss the scope of the job and review the job documentation with your project manager, designer, sales person, PRIOR to leaving for the job.

Go over any job documentation to make sure you understand it before you get to the job!

The most common documentation you will use includes:

- A project worksheet
- A marked set of architectural plans for the home
- An equipment schedule and cable schedule, or a combination of the two

- Change orders
- A pre-wire checklist (used during the pre-wire stage)

The Project Worksheet defines the job and will contain most if not all of the information shown below:

Client Information

- Address, phone, other contact information
- Job contact person
- Overview of work to be performed
- Dates work is scheduled to begin, estimated date of completion

The worksheet may also contain:

- Name of salesperson, designer, project manager
- Breakdown of work to be performed and equipment to be installed
- Customer provided equipment
- Subcontractors

You should have a copy of the worksheet (work order) as part of the job documentation.

A set of plans should be available. These will be a set of architectural plans that have been marked by hand, or a plan made on a computer and marked on the computer. The plan which will be referred to most often is the Floor Plan.

Different symbols are used to indicate what equipment is installed at each location. Each location is identified by a unique identifier (ID), typically a number. The ID is used on other documentation to identify the plan location.

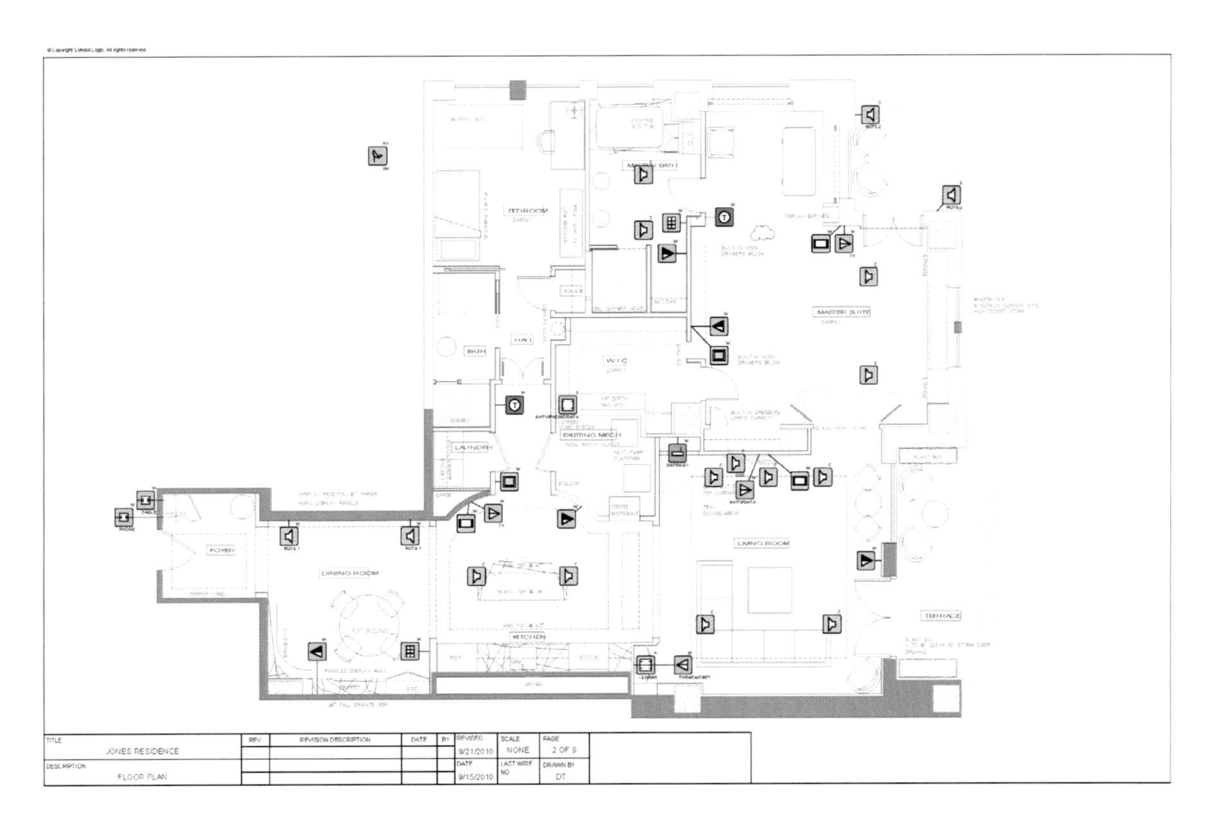

Figure 6-1 A/V Floor Plan

Some form of schedule(s) defining each location in the home, and what cable go to those locations, is needed to serve as a guide for the installation of boxes, brackets, and all cabling.

Sometimes there will be a separate Device Schedule (Equipment List) showing the location ID and what kind of rough-hardware to install.

There will also be a Cable Schedule, showing the type and ID of all cable that is to be run, where it originates, and where its destination is.

Many times these two documents are combined into one comprehensive Cable Schedule showing all of this information, and more. Regardless of how the information is delivered it should include the following:

- Rough-in hardware
- Where and how to mount the equipment
- Make a model number of the equipment
- Any installation notes needed for the equipment
- Cable type
- Where the cable originates
- Where the cable lands at the other end
- How the cable is terminated (connected) at each end
- Estimate of the length of cable required (always overestimate cable requirements by 10%).
- Platework and inserts needed

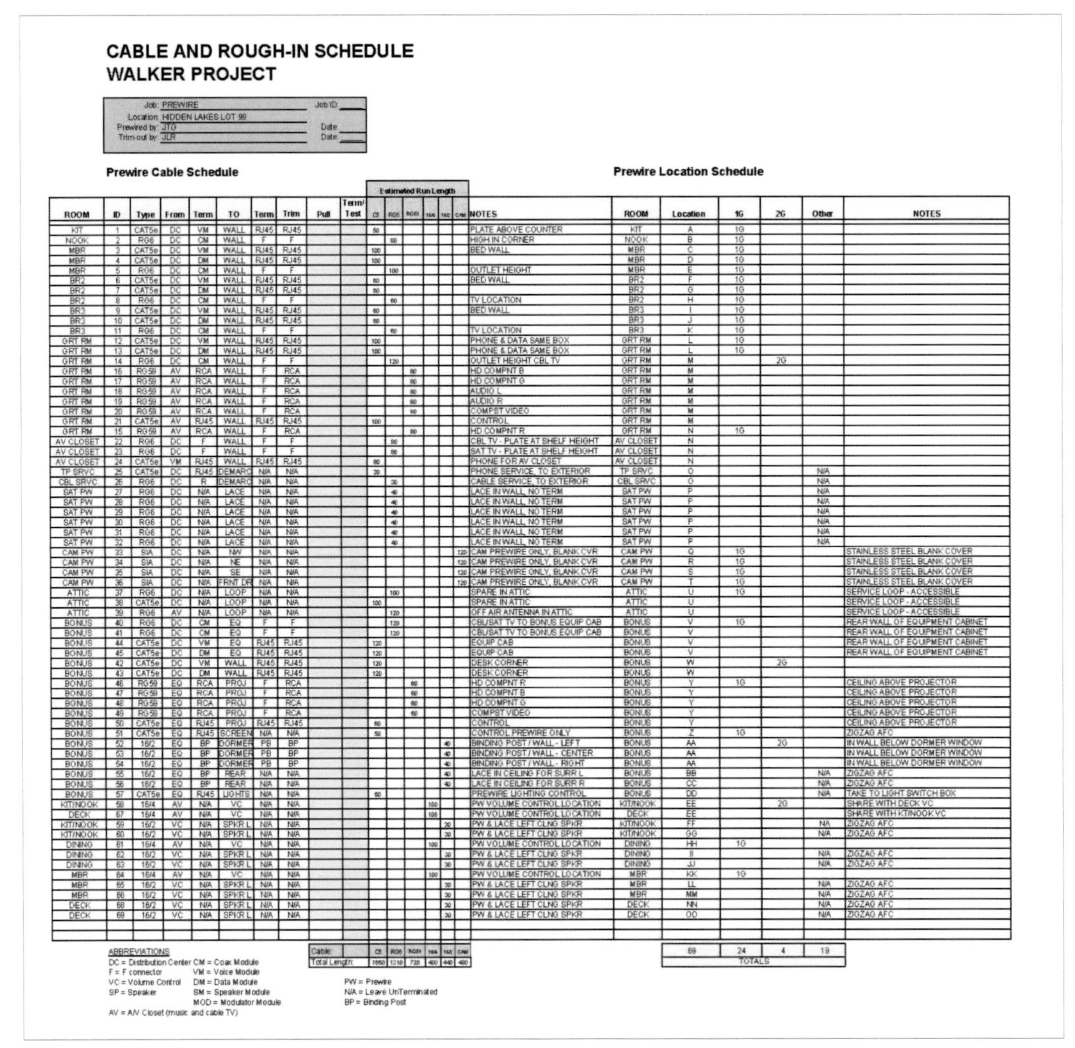

CABLE AND ROUGH-IN SCHEDULE
WALKER PROJECT

Job: PREWIRE Job ID: ____
Location: HIDDEN LAKES LOT 99
Prewired by: JTG Date: ____
Trim-out by: JLR Date: ____

Prewire Cable Schedule | **Prewire Location Schedule**

ROOM	ID	Type	From	Term	TO	Term	Trim	Pull	Term/ Test	Estimated Run Length: C5	RG6	RG59	16/4	16/2	CAM	NOTES	ROOM	Location	1G	2G	Other	NOTES
KIT	1	CAT5e	DC	VM	WALL	RJ45	RJ45			50						PLATE ABOVE COUNTER	KIT	A	1G			
NOOK	2	RG6	DC	CM	WALL	F	F				60					HIGH IN CORNER	NOOK	B	1G			
MBR	3	CAT5e	DC	VM	WALL	RJ45	RJ45			100						BED WALL	MBR	C	1G			
MBR	4	CAT5e	DC	DM	WALL	RJ45	RJ45			100							MBR	D	1G			
MBR	5	RG6	DC	CM	WALL	F	F				100					OUTLET HEIGHT	MBR	E	1G			
BR2	6	CAT5e	DC	VM	WALL	RJ45	RJ45			60						BED WALL	BR2	F	1G			
BR2	7	CAT5e	DC	DM	WALL	RJ45	RJ45			60							BR2	G	1G			
BR2	8	RG6	DC	CM	WALL	F	F				60					TV LOCATION	BR2	H	1G			
BR3	9	CAT5e	DC	VM	WALL	RJ45	RJ45			60						BED WALL	BR3	I	1G			
BR3	10	CAT5e	DC	DM	WALL	RJ45	RJ45			60							BR3	J	1G			
BR3	11	RG6	DC	CM	WALL	F	F				60					TV LOCATION	BR3	K	1G			
GRT RM	12	CAT5e	DC	VM	WALL	RJ45	RJ45			100						PHONE & DATA SAME BOX	GRT RM	L	1G			
GRT RM	13	CAT5e	DC	DM	WALL	RJ45	RJ45			100						PHONE & DATA SAME BOX	GRT RM	L	1G			
GRT RM	14	RG6	DC	CM	WALL	F	F				120					OUTLET HEIGHT CBL TV	GRT RM	M		2G		
GRT RM	16	RG59	AV	RCA	WALL	F	RCA					80				HD COMPNT B	GRT RM	M				
GRT RM	17	RG59	AV	RCA	WALL	F	RCA					80				HD COMPNT G	GRT RM	M				
GRT RM	18	RG59	AV	RCA	WALL	F	RCA					80				AUDIO L	GRT RM	M				
GRT RM	19	RG59	AV	RCA	WALL	F	RCA					80				AUDIO R	GRT RM	M				
GRT RM	20	RG59	AV	RCA	WALL	F	RCA					80				COMPST VIDEO	GRT RM	M				
GRT RM	21	CAT5e	AV	RJ45	WALL	RJ45	RJ45			100						CONTROL	GRT RM	M				
GRT RM	15	RG59	AV	RCA	WALL	F	RCA					80				HD COMPNT R	GRT RM	N	1G			
AV CLOSET	22	RG6	DC	F	WALL	F	F				80					CBL TV - PLATE AT SHELF HEIGHT	AV CLOSET	N				
AV CLOSET	23	RG6	DC	F	WALL	F	F				80					SAT TV - PLATE AT SHELF HEIGHT	AV CLOSET	N				
AV CLOSET	24	CAT5e	VM	RJ45	WALL	RJ45	RJ45			80						PHONE FOR AV CLOSET	AV CLOSET	N				
TP SRVC	25	CAT5e	DC	RJ45	DEMARC	N/A	N/A			30						PHONE SERVICE, TO EXTERIOR	TP SRVC	O			N/A	
CBL SRVC	26	RG6	DC	R	DEMARC	N/A	N/A				30					CABLE SERVICE, TO EXTERIOR	CBL SRVC	O			N/A	
SAT PW	27	RG6	DC	N/A	LACE	N/A	N/A				40					LACE IN WALL, NO TERM	SAT PW	P			N/A	
SAT PW	28	RG6	DC	N/A	LACE	N/A	N/A				40					LACE IN WALL, NO TERM	SAT PW	P			N/A	
SAT PW	29	RG6	DC	N/A	LACE	N/A	N/A				40					LACE IN WALL, NO TERM	SAT PW	P			N/A	
SAT PW	30	RG6	DC	N/A	LACE	N/A	N/A				40					LACE IN WALL, NO TERM	SAT PW	P			N/A	
SAT PW	31	RG6	DC	N/A	LACE	RJ45	N/A				40					LACE IN WALL, NO TERM	SAT PW	P			N/A	
SAT PW	32	RG6	DC	N/A	LACE	N/A	N/A				40					LACE IN WALL, NO TERM	SAT PW	P			N/A	
CAM PW	33	SIA	DC	N/A	NW	N/A	N/A								120	CAM PREWIRE ONLY, BLANK CVR	CAM PW	Q	1G			STAINLESS STEEL BLANK COVER
CAM PW	34	SIA	DC	N/A	NE	N/A	N/A								120	CAM PREWIRE ONLY, BLANK CVR	CAM PW	R	1G			STAINLESS STEEL BLANK COVER
CAM PW	35	SIA	DC	N/A	SE	N/A	N/A								120	CAM PREWIRE ONLY, BLANK CVR	CAM PW	S	1G			STAINLESS STEEL BLANK COVER
CAM PW	36	SIA	DC	N/A	FRNT DR	N/A	N/A								120	CAM PREWIRE ONLY, BLANK CVR	CAM PW	T	1G			STAINLESS STEEL BLANK COVER
ATTIC	37	RG6	DC	N/A	LOOP	N/A	N/A				100					SPARE IN ATTIC	ATTIC	U	1G			SERVICE LOOP - ACCESSIBLE
ATTIC	38	CAT5e	DC	N/A	LOOP	N/A	N/A			100						SPARE IN ATTIC	ATTIC	U				SERVICE LOOP - ACCESSIBLE
ATTIC	39	RG6	AV	N/A	LOOP	N/A	N/A				120					OFF AIR ANTENNA IN ATTIC	ATTIC	U				SERVICE LOOP - ACCESSIBLE
BONUS	40	RG6	DC	CM	EQ	F	F				120					CBL/SAT TV TO BONUS EQUIP CAB	BONUS	V	1G			REAR WALL OF EQUIPMENT CABINET
BONUS	41	RG6	DC	CM	EQ	F	F				120					CBL/SAT TV TO BONUS EQUIP CAB	BONUS	V				REAR WALL OF EQUIPMENT CABINET
BONUS	44	CAT5e	DC	VM	EQ	RJ45	RJ45			120						EQUIP CAB	BONUS	V				REAR WALL OF EQUIPMENT CABINET
BONUS	45	CAT5e	DC	DM	EQ	RJ45	RJ45			120						EQUIP CAB	BONUS	V				REAR WALL OF EQUIPMENT CABINET
BONUS	42	CAT5e	DC	VM	WALL	RJ45	RJ45			120						DESK CORNER	BONUS	W		2G		
BONUS	43	CAT5e	DC	DM	WALL	RJ45	RJ45			120						DESK CORNER	BONUS	W				
BONUS	46	RG59	EQ	RCA	PROJ	F	RCA					60				HD COMPNT R	BONUS	Y	1G			CEILING ABOVE PROJECTOR
BONUS	47	RG59	EQ	RCA	PROJ	F	RCA					60				HD COMPNT B	BONUS	Y				CEILING ABOVE PROJECTOR
BONUS	48	RG59	EQ	RCA	PROJ	F	RCA					60				HD COMPNT G	BONUS	Y				CEILING ABOVE PROJECTOR
BONUS	49	RG59	EQ	RCA	PROJ	F	RCA					60				COMPST VIDEO	BONUS	Y				CEILING ABOVE PROJECTOR
BONUS	50	CAT5e	EQ	RJ45	PROJ	RJ45	RJ45			60						CONTROL	BONUS	Y				CEILING ABOVE PROJECTOR
BONUS	51	CAT5e	EQ	RJ45	SCREEN	N/A	N/A			50						CONTROL PREWIRE ONLY	BONUS	Z	1G			ZIGZAG AFC
BONUS	52	16/2	EQ	BP	DORMER	PB	BP							40		BINDING POST / WALL - LEFT	BONUS	AA		2G		IN WALL BELOW DORMER WINDOW
BONUS	53	16/2	EQ	BP	DORMER	PB	BP							40		BINDING POST / WALL - CENTER	BONUS	AA				IN WALL BELOW DORMER WINDOW
BONUS	54	16/2	EQ	BP	DORMER	PB	BP							40		BINDING POST / WALL - RIGHT	BONUS	AA				IN WALL BELOW DORMER WINDOW
BONUS	55	16/2	EQ	BP	REAR	N/A	N/A							40		LACE IN CEILING FOR SURR L	BONUS	BB			N/A	ZIGZAG AFC
BONUS	56	16/2	EQ	BP	REAR	N/A	N/A							40		LACE IN CEILING FOR SURR R	BONUS	CC			N/A	ZIGZAG AFC
BONUS	57	CAT5e	EQ	RJ45	LIGHTS	N/A	N/A			60						PREWIRE LIGHTING CONTROL	BONUS	DD			N/A	TAKE TO LIGHT SWITCH BOX
KIT/NOOK	58	16/4	AV	N/A	VC	N/A	N/A						100			PW VOLUME CONTROL LOCATION	KIT/NOOK	EE		2G		SHARE WITH DECK VC
DECK	67	16/4	AV	N/A	VC	N/A	N/A						100			PW VOLUME CONTROL LOCATION	DECK	EE				SHARE WITH KIT/NOOK VC
KIT/NOOK	59	16/2	VC	N/A	SPKR L	N/A	N/A							30		PW & LACE LEFT CLNG SPKR	KIT/NOOK	FF			N/A	ZIGZAG AFC
KIT/NOOK	60	16/2	VC	N/A	SPKR R	N/A	N/A							30		PW & LACE LEFT CLNG SPKR	KIT/NOOK	GG			N/A	ZIGZAG AFC
DINING	61	16/4	AV	N/A	VC	N/A	N/A						100			PW VOLUME CONTROL LOCATION	DINING	HH	1G			
DINING	62	16/2	VC	N/A	SPKR L	N/A	N/A							30		PW & LACE LEFT CLNG SPKR	DINING	II			N/A	ZIGZAG AFC
DINING	63	16/2	VC	N/A	SPKR R	N/A	N/A							30		PW & LACE LEFT CLNG SPKR	DINING	JJ			N/A	ZIGZAG AFC
MBR	64	16/4	AV	N/A	VC	N/A	N/A						100			PW VOLUME CONTROL LOCATION	MBR	KK	1G			
MBR	65	16/2	VC	N/A	SPKR L	N/A	N/A							30		PW & LACE LEFT CLNG SPKR	MBR	LL			N/A	ZIGZAG AFC
MBR	66	16/2	VC	N/A	SPKR L	N/A	N/A							30		PW & LACE LEFT CLNG SPKR	MBR	MM			N/A	ZIGZAG AFC
DECK	68	16/2	VC	N/A	SPKR L	N/A	N/A							30		PW & LACE LEFT CLNG SPKR	DECK	NN			N/A	ZIGZAG AFC
DECK	69	16/2	VC	N/A	SPKR L	N/A	N/A							30		PW & LACE LEFT CLNG SPKR	DECK	OO			N/A	ZIGZAG AFC

Cable:	C5	RG6	RG59	16/4	16/2	CAM
Total Length:	1950	1230	720	400	440	480

69	24	4	19
	TOTALS		

ABBREVIATIONS
DC = Distribution Center CM = Coax Module
F = F connector VM = Voice Module
VC = Volume Control DM = Data Module
SP = Speaker SM = Speaker Module
MOD = Modulator Module
AV = A/V Closet (music and cable TV)

PW = Prewire
N/A = Leave UnTerminated
BP = Binding Post

Figure 6-2 Example of Combined Cable Schedule

It is also a good idea to have a checklist to refer as you prepare for a prewire. A pre-wire checklist can be used to make sure you have everything you need prior to leaving the office for the job.

- Make sure you have:
  - » Documentation
  - » Rough-in equipment
  - » Needed cable (enough)
  - » Labels, pens
- Make sure you know the location
- Make sure you have the tools you need including:
  - » Drills, bits, extension cords, charged batteries
  - » Hardhat, glasses, gloves
  - » Hand tools
- Make sure you have the equipment you need including:
  - » Ladders
  - » Clean-up equipment

Others on the Jobsite

During the walk-through, you may have the opportunity to meet several different stakeholders. It is important that you communicate professionally with them and represent your company well:

- Homeowner
- Builder
- Job foreman
- Other trades

Exchange contact information and let them know:

- Who you are
- What you are doing
- How long you plan to be working

Determine Cable Paths

The next step is to determine where everything on the job will be located. Usually the general locations of each device you will install have been marked on a stud, joist or other object in the home when the initial walkthrough was done by the salesperson or project manager.

Each location was given a number or other ID. The ID was later transferred to a set of plans.

The marked plans will use a set of symbols to identify (in general) what is installed at each location. The marked plans will be your primary guide for performing you initial walk through.

Each location may be identified using a label, a tag, or simply marking on the framing member. In all cases this ID should include:

- What type of device is installed
- What type of rough-in hardware is installed
- The exact height or positioning of the device
- The cables that terminate at this location

Almost all residential projects will use a home run cabling, known as a star topology. Normally a location central to the home is chosen as the cable termination point and all low-voltage cables for audio, video, data and control will run from this location out to each device.

In a 2-story framed home you should try to find a common wall on both floors to mount the structured cable distribution center. This will allow drilling a common chase between the floors and attic/basement.

When metal studs are in use, this step is similar. If walls are concrete, the cable paths must be planned out in advance, so that slots can be cut, or pre-located conduit can be installed for low-voltage wiring. Refer to your local best practices and learn the techniques and terminologies which are expected by other stakeholders and contractors.

Plan for Cable Separation

Paths for cable from each device back to the termination point must be determined while keeping in mind the need to keep AC power circuit cables separate from low-voltage cables. A separation from electrical wiring of 12 -18" is recommended. When low-voltage cables must cross electrical cabling, it should be done at a right angle to minimize interference.

Special Needs for Retrofit Installations

Cable routing in new construction is usually straight forward and paths can be easily modified to avoid or overcome obstructions. However, in retrofit installations, cable paths can be more difficult to create and may require special tools and skill in fishing walls, ceilings and floors.

It is often necessary to install conduit or surfaced mounted raceway to overcome obstacles presented by retrofit installations, so these needs should be noted during the walk through to allow time for purchase and delivery to the jobsite.

SETTING BOXES

Various types of rough-in hardware are used to:

- Provide a physical support for the equipment
- Provide an opening for the sheetrock
- Provide a cable termination support
- Interface with conduit

Commonly used box types include:

- Plastic nail-on/screw-on boxes (may be backless depending on code)
- Boxes made for installation in concrete
- Plaster rings
- Multi-gang box extenders
- Divided boxes that support co-location of AC power next to a low-voltage device
- Special backboxes for loudspeakers and control devices

Accurate rough-in hardware placement is critical for a good installation. It is standard practice to mount new outlets consistent with any existing outlets in a room. In addition, always know your local code before doing this type of installation.

Allow for the thickness of the sheetrock and any other finished surface. Careful attention to the drawings should show if there is paneling or ceramic tile being installed. Outlet hardware should be flush or a little recessed from the outer surface of the sheetrock.

Stay consistent in how you mount the outlet or other rough-in hardware. You should measure from the sub-floor to the bottom hole of outlet boxes. You can adopt other standards, but it should be consistent throughout the job.

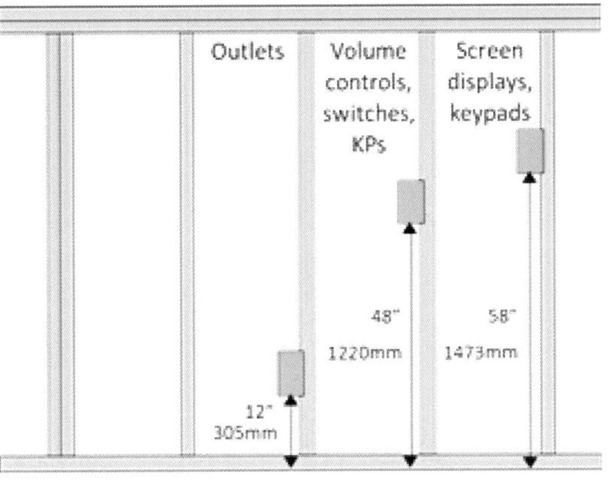

Figure 6-3 Typical Device Heights

Figure 3 shows typical mounting heights for rough-in hardware when there are no existing rough-ins to match. These are not strict standards and heights will vary by region and to accommodate client needs. When measuring or matching device heights, always measure from the sub-floor to the bottom screw hole of each box. Different box types may have different shapes and sizes, but using the bottom hole will always provide accurate on-center location.

You can usually co-locate low-voltage outlets with high-voltage outlets as long as they are separated by a non-metallic or grounded barrier. Check local codes.

You will also be installing the enclosure used for the distribution center. This enclosure may be mounted in-wall or surface mounted. It should be provided with a dedicated AC circuit, so co-ordination with the electrical contractor may be required. Also if the enclosure is metal it may need to be grounded per local electrical code.

CREATING PATHS

Wood Frame Construction

Before learning about cabling we must understand how the frame, or "skeleton", of a home is constructed. Typical "stick-built" homes are framed with "dimensional" lumber, such as 2x4s, 2x6s, 2x8s, etc. Some elements may be pre-fab or factory-built, while custom homes are often framed entirely on-site.

The following diagrams point out some of the standard framing members used in wood framed residential construction. Knowledge of these names will make it easier to communicate with the other trades onsite.

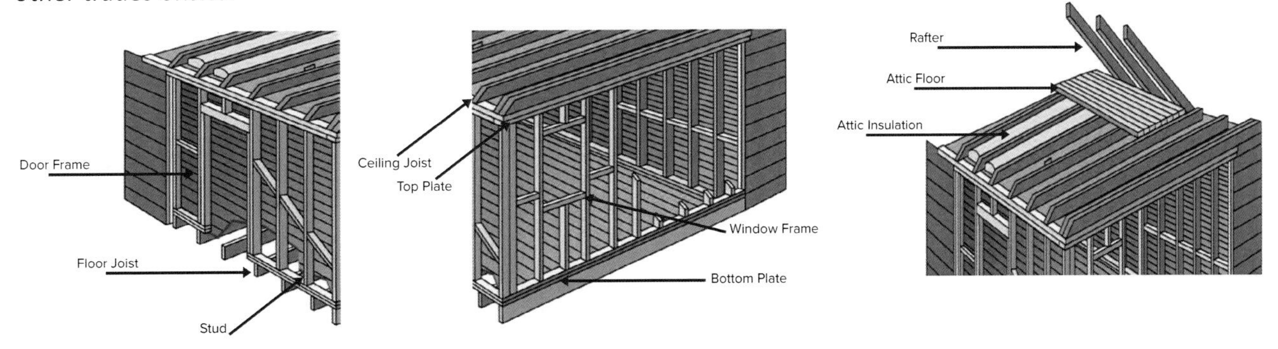

Wall studs in a residential home are usually constructed using 2x4s, with 2x6s being used when additional insulation is to be added. Standard stud spacing is 16" (406mm) on center for interior and exterior residential walls in most areas. Some applications such as garages may allow for 24" (610mm) on center construction.

If a wall has a double top plate it is load-bearing. All exterior/perimeter walls are typically load-bearing because they are supporting a roof or another floor.

When in doubt, assume the wall may be load bearing.

Creating Paths in Wood Frame Construction

In wood frame construction, cable paths will be created by drilling holes in framing members. Specific guidelines must be followed to meet local code. In most regions the suggestions here apply.

The drawing at the bottom left of the page shows the maximum hole sizes in a load bearing and non-load bearing 2X4 wall stud. In load bearing studs, the hole may be no larger than 40% of the depth of the stud. In non-loading bearing studs the maximum hole size can be 60% of the actual depth of the stud.

Remember a “2x4” is really 1 ½” x 3 ½“(38 x 89mm), so the maximum hole size for a non-load bearing wall stud is 2” (51mm) and for a load-bearing stud 1 3/8” (35mm).

When the hole is closer than 1-1/4 inch (32mm) to either outside surface of a wooden stud, a steel guard plate must be secured to the stud to protect any cable in the hole from possible damage from sheetrock screws, nails, etc.

Besides wall studs, it is often necessary to create cable paths by drilling holes in floor and ceiling joists. Like the wall studs, general guidelines have been created to ensure holes drilled into joist will not compromise the integrity of the structure. The general rule is that the hole size be limited to no more than 33% or 1/3rd of the depth of the joist. The holes should always be centered in the joist between the two edges and the holes should be kept away from the horizontal center and the very ends of the joist. Holes should be at least 2” (51mm) from either edge of the joist.

Laminated beams are engineered to support a great deal of weight. Compromising them in any way will be a red flag for the inspector and it is not unusual for an entire project to be halted for a lam beam to be replaced. Find another route.

Engineered beams (look like an I-Beam fabricated from wood products) are OK to drill through in most jurisdictions, and they usually have 2” knockouts built right in, which makes routing easy (if they just happen to all line up).

Steel I-beams are sometimes used in residential construction and should not be drilled. Find another route.

As a general rule, only fill about half of the hole’s area with cable. Cables should be loose and not binding in the hole. Never use holes that contain other services, and do NOT use the holes of other trades!

The drawings below illustrate typical guidelines for hole sizes in studs and joists in wood frame construction

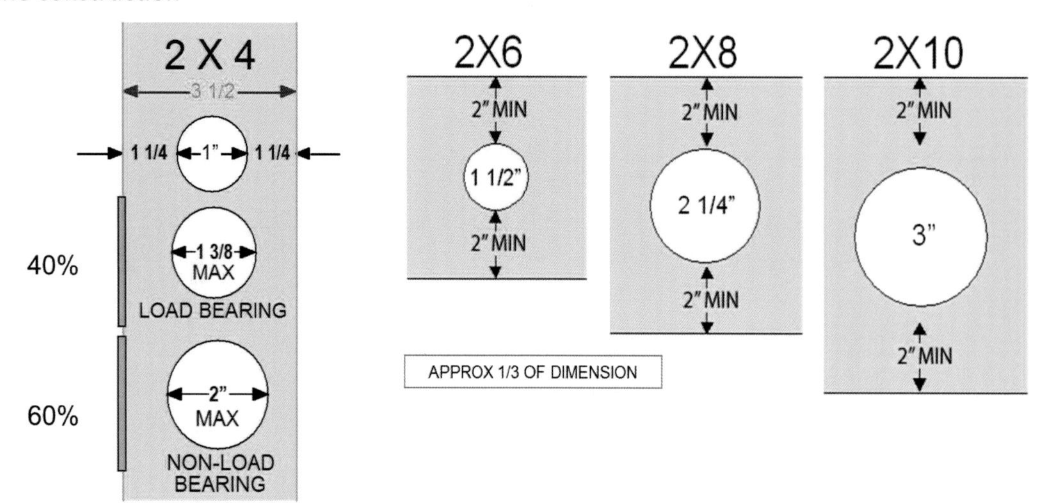

Metal Stud Construction

Prefabricated metal studs are a popular alternative in many parts of the world and in commercial/MDU construction in the US. They are installed using screws rather than nails, and are very lightweight. But installing cable into metal stud construction requires a different skill set.

Most metal studs have circular or elongated holes already punched in them for convenience. A special tool is required to punching holes for cabling where there are none pre-punched.

The metal edges of the holes can be very sharp and caution must be taken to avoid cuts. Plastic grommets, bushings or inserts should be installed in the holes punched in metal studs to protect the cables as they run through the studs.

Cable installed into concrete walls requires paths to be created for conduit, and special boxes to be installed for outlets.

Placing Hooks and Anchors

Cable paths through open spaces, such as ceilings, crawl spaces and attics, will require the use of hooks and anchors to route and secure the cables. The purpose of securing cables is to prevent damage to the cable by other trades (such as the insulation or sheetrock installer) or anything else that will be installed in the home after the pre-wire.

Devices such as J-hooks can be pre-installed along the path to provide a route in which the cable can run.

DO NOT USE electrician's U staples! It is too easy to hit the staple once too often and crush the cable. It is then impossible to remove the staple without damaging the cable. Avoid the T-25 style rounded staples. They can damage almost any cable you install.

Conduit Installation

Many locations require all low voltae cabling to be in conduit, even in wood frame construction. Know your local codes and standards and always observe them.

When conduit is not required it may still be a good idea to "future-proof" primary runs to racks and projectors in order to pull different cable several years after the initial installation.

Metal and plastic conduit and surface mounted raceways can also be used in exposed areas to protect cables.

Obstacles

As you prepare your path for the cable installation, watch out for obstacles that will prevent a successful pull. Watch out for paths used by other trades. It is best if your cabling can be installed after the electrical and mechanical trades have finished their work, but before walls and ceilings are closed up.

CABLE TYPES

Before we move to the cable pulling step of the prewire, we will quickly review the types of cable commonly used in residential infrastructure. Most cable installed will be considered communications cable. There are a few other types which we will also discuss.

A commonly recognized standard for home wiring is TIA570-C. This document serves as a general guide to communications cabling in the home. It states that the three types of communication cables are:

- Twisted Pair (such as Cat5e and Cat6A)
- Coaxial (RG6, Series 6, WF100)
- Optical Fiber

The standard also states:

- All communications cables are to be "home run" to a common location
- Communications cables installed in walls should not exceed 90 meters in length
- Cabling to outside service providers should have a disconnect for troubleshooting (the most common of these is the RJ-31X device, which not only disconnects the security system from the phone line, it allows the security system to override, or "seize" the line in the event of an alarm state which requires dialing a monitoring service.

Twisted Pair

Most twisted pair cable in the home has 8 solid copper conductors and is Unshielded Twisted Pair (UTP). It is configured as 4 individual pairs and uses industry standard color coding. Each pair consists of one wire with solid color insulation and one wire with a striped insulator.
The following list shows the standard telephone line assignment to each pair when used for voice telephone service:

- Blue Line 1
- Orange Line 2
- Green Line 3
- Brown Line 4

Category ratings for Unshielded Twisted Pair cable were established by the TIA (Telecommunications Industry Association) to insure to the buyer that the cable meets minimum performance standards (outlined in TIA-67) for different applications and bandwidths. The following chart details the bandwidth supplied by each category of cable, along with its application.

Rating	Type	Bandwidth	Typical Application
CAT 3	UTP	16MHz	Voice, data
CAT 4	UTP	20MHz	same
CAT 5	UTP	100MHz	10 or 100 MB/s
CAT 5e	UTP	100 MHz	Optimized for 100 Mb/s
CAT 6	UTP	250 MHz	Gigabit data
CAT 6a	UTP	500 MHz	10G/Base T

Figure 6-4 UTP Specifications

According to the TIA 570-C (2012) standard, the current minimum standard for the home is CAT5e, with CAT6 recommended. As shown in the chart, CAT6a provides up to 10 Gb/s performance, much better for the demands of the future. The connector used to terminate the UTP cable is generically known as an RJ-45 connector.

There are two common termination schemes for UTP at the jack; 568-A and 568-B. Both will work fine for data. TIA570-C calls for the use of 568-A, because it is compatible with Line 1 and Line 2 on most telephone systems. You should observe the standard used by your company.

Note; UTP can also be used to carry balanced analog audio and video and signals, using special devices at each end known as Baluns (Balance/Unbalance). UTP is often used to carry control signals as well. Repurposing existing twisted pair cable in a home can be a very efficient way to solve retrofit and upgrade projects.

Coaxial Cable

Coaxial cable (abbreviated "coax" or sometimes "CX") consists of a center conductor and one to four shields separated by an insulator (dielectric). They are all arranged around a common axis, hence the term "coaxial". This cable was designed to be rugged and have a great deal of tensile strength when run between poles and from the pole to the house. Most have outer jackets made of PVC (polyvinyl chloride). Plenum rated and direct burial cable utilize different materials.

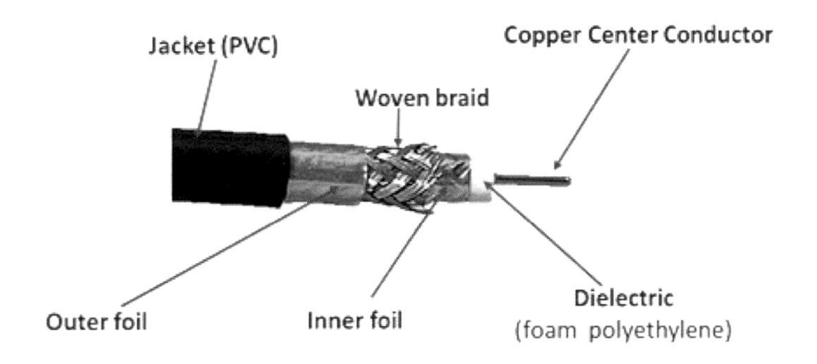

Figure 6-3 Anatomy of Broadband Coaxial Cable

Most broadband coax cable used in the home will have 2 or 4 (quad) shields.

- Solid center conductor is copper or copper clad steel (CCS). Copper is preferred for today's digital signals.
- The dielectric is an insulator which often foam polyethylene. It is used to maintain a set distance between the center conductor and the shield.
- RG-6, also known as Series 6 and WF100, is usually used for RF (Radio Frequency) applications, such as satellite, cable and antenna, and also works for baseband applications. It is the most widely use type of coaxial cable. RG-11 is larger, and used for long runs and provider feeds in the neighborhood.
- RG-59, or Series 59, is smaller and is usually used for baseband video, camera signals, and analog or digital audio.

The signal loss in the coaxial cable is directly related to the diameter in material of the center conductor, as well as distance. The larger the diameter of the inner conductor, the lower the signal loss. Remember loss is higher at higher frequencies and today's digital signals require higher bandwidth.

Most of today's video applications use the HDMI interface for interconnection of high definition video and multi-channel audio as analog video is being phased out. The specification for HDMI provides limited range for standard pre-made HDMI cables. When longer distances are required powered extenders can be used to place HDMI signals unto standard category UTP cables. Like UTP, coaxial cable can also be re-purposed in an existing home. Using a variety of technologies it can be used to carry analog video or HDMI signals.

Fiber Optic Cable

Optical fiber cable is becoming increasingly popular in residential communications installs. With advances in field termination techniques and a sharp decline in costs, fiber optic cabling will likely be an important part of infrastructure in the future.

- Glass strands that provide a path to support the transmission of light pulses (multimode and single-mode).
- The two main elements of the strand are the core and cladding.

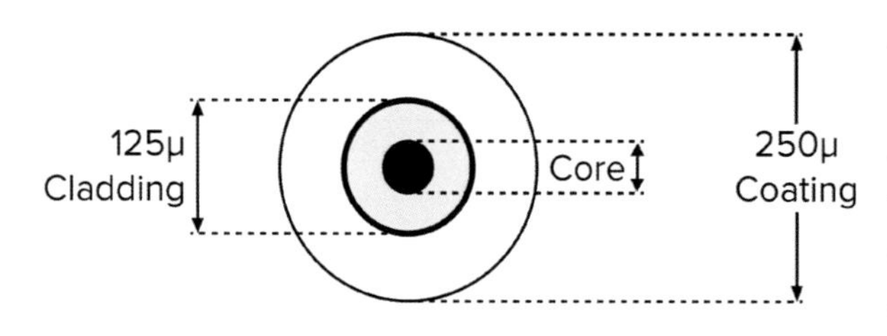

The reference used to define the size of a fiber strand is a micrometer or micron (µm) which is equal to one millionth of a meter. The size of the core and the cladding of optical fiber strands are also stated in µm. The size of the optical core varies with single mode fiber at 9 µm for long distance applications to multimode fiber either 50 µm or 62.5 µm for shorter distances. Today 50 µm fibers are more popular for residential applications. All optical fiber strands have a diameter of 125 µm.

To reinforce the strand and make it easier to handle, an acrylate coating is applied to the outside of the cladding. The application of the acrylate coating increases the original 125 µm diameter of the strand to 250 µm. Each optical fiber cable is then protected with a Kevlar fiber strand for pulling the cable and an outer jacket. The cable jacket provides physical and environmental protection for the fiber strands inside the cable. To reinforce the strand and make it easier to handle, an acrylate coating is applied to the outside of the cladding. The application of the acrylate coating increases the original 125 µm diameter of the strand to 250 µm. Each optical fiber cable is then protected with a Kevlar fiber strand for pulling the cable and an outer jacket. The cable jacket provides physical and environmental protection for the fiber strands inside the cable.

Other Cable Types

There are other cables which are commonly installed in homes that are not considered communications cable, but are important to system integration.

Speaker cable used for in-wall installation is simple and easy to work with. It will have either 2 or 4 conductors (typically 16 - 12 gauge stranded copper). Remember that the larger the conductor, the less resistance it has and the more current it will carry. Proper termination and proper polarity are critical to audio system performance and the following standard should be followed. CEA/CEDIA-2030-A specifies the following color code standard for speaker cable.

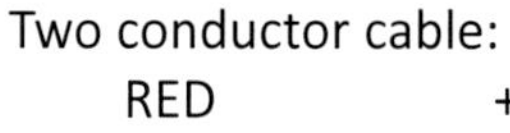

Two conductor cable:

RED	+
BLACK	- (ground)

Four conductor cable:

RED	Right +
BLACK	Right -
GREEN	Left -
WHITE	Left +

The conductors in speaker cable may be paired and twisted, but often are just two or four conductors in a common jacket. The size, or gauge, of the conductors is dependent upon the speaker load and the distance between the speaker and the amplifier.

Control and Security Cable

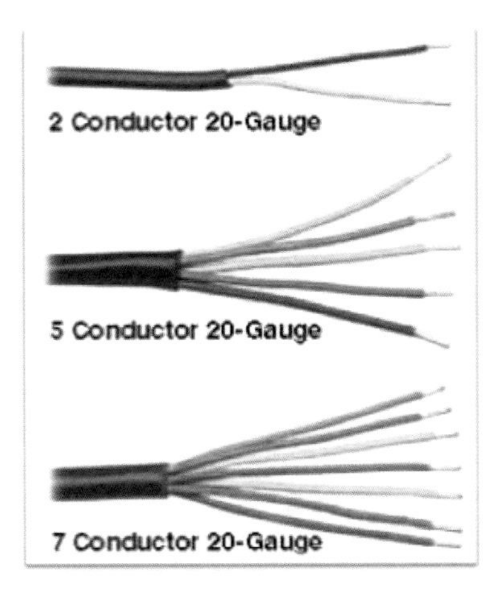

Low-voltage (less than 90 volts) and control cables are used for everything from security systems to lighting control. They are usually untwisted and unpaired. They may or may not be shielded (typically not). They may use stranded (most likely) or solid conductors

The color coding used for the cable varies by application. For example, cable used for connection to a thermostat usually uses a predefined set of colors for each conductor to indicate the function of the conductor (black - common return, red - power, green - fan control, yellow - heat control, etc.)

Do not substitute shielded for unshielded, as they have different performance characteristics. Security system wiring falls into this category and is usually unshielded 22/2 (22 guage, 2 conductor) or 22/4.

Wire Characteristics

In North America, the diameter of a wire conductor is typically given as an American Wire Gauge (AWG) size (instead of a diameter in fractions of an inch.

Wire gauges start at 50 for the smallest wire, and go up to 0000 (quad-zero) for the largest diameter wire.

Most of the wire you will be dealing with used in A/V and communications cable is in the range of 24 to 14 gauge. The smaller the wire, the higher AWG number, the more resistance. The larger the wire, the lower the AWG number, the less resistance.

In other parts of the world conductor size is most often designated by its metric measurement in millimeters (mm) or its cross-sectional area in square millimeters (mm2).

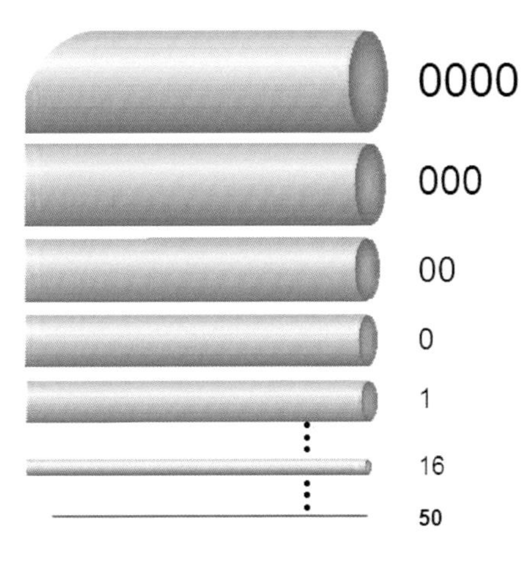

Figure 6-4 AWG (American Wire Gauge)

CABLE PULLING PROCEDURES

Gather Materials

One of the first steps in preparing for the cable installation is to gather all the equipment and material you will need on the jobsite and place them in the appropriate locations.

The cable schedule we mentioned earlier, along with the installation notes made during the initial walk-through should now be reviewed. Check to see you have the amount and types of cable you will need to complete the cable installation. Keep track of cable used to be sure you have enough on the spool to complete each run.

When preparing for the cable pull, you should also have a supply of fasteners to support and secure the cables; and cable labels to mark every cable as it is being installed. If you have to install cable in conduit, you may also have a need for cable lubricant to reduce friction and ease cable installation. Gather all your required tools, including specialty tools like fish tapes, pull ropes, cable reels and ladders.

Setup Location

Although cable can be pulled in either direction, from the distribution center location to the outlet or from the outlet to the distribution center; the preferred location for cable spools is centrally located at the distribution center location. You will still need to pull cables from different locations, such as from volume controls to speakers.

Cable caddies make an excellent way to hold several cable spools. They allow you to pull several cables at a time from several spools.

You Must Label the Cable Before You Pull It!

Remember every cable must be labeled before it is pulled. Make two labels for each cable by referring to the cable schedule. Apply one label to the end you are pulling. When you are done with the pull, apply the identical label to the other end of the cable when you cut it from the cable spool. You may want to make a second set of labels to be used during trim-out to replace any labels damaged or removed during installation. Always re-label if a label is removed. Labels should be located where they can be read by others

Factors to Consider

Cable performance can be greatly compromised if it is not handled properly when installed. The following guidelines should be observed.

Pull Tension

This is the amount of stress placed on the cable when pulled. Cable can be easily damaged by trying to pull it through too many holes at once or over a long distance.

- Coaxial (Solid Copper Conductor): Maximum Pull Tension = 40 lbs (18Kg)
- UTP: Maximum Pull Tension = 25 lbs (11Kg)
- Fiber Optic: Maximum Pull Tension = 50 lbs (22Kg) – Refer to manufacturer specs

Bend Ratio

Cable performance can also be affected when bent too tightly, especially coaxial cable. Coaxial cable should never be pinched or kinked, as this changes its impedance. If you have accidentally pulled a twisted pair or coax cable with a kink, it is probably damaged and should be replaced.

- Coax: Minimum Bend Radius = 10 x the Cable Diameter
- UTP: Minimum Bend Radius = 4 x the Cable Diameter
- Fiber Optic: Minimum Bend Radius = 10 x the Cable Diameter

Flex Life

Although not usually an issue in prewiring, it is important to understand flex life. The number of times a cable can be bent and straighten without damage is known as its flex life. Solid cables have short flex life and can be more easily damaged when repeatedly bent. This comes into play when racks or cabinets are repeatedly moved and when cable is installed in motorized devices like TV lifts.

Other Considerations

NEVER splice a cable. Simply make sure you have enough cable on the spool for the complete run. Do not pull cable through holes used by other trades. While this may not violate any codes (be sure and check local codes), it is bad practice since the other cable (pipe, etc.) may be moved or may move after you have done your install and could damage your cable.

Never use a hole occupied by a water or gas pipe. These pipes can vibrate in the hole, damaging your cable.

Make sure you leave enough slack at each end of the cable to allow enough room to:

- attach connectors easily at outlets
- route cables inside of equipment enclosures

- allow removable equipment enough slack or remove
- allow service loops (for future changes/moves) of equipment

At outlet locations, this should be at least 12-18 inches (305 – 457 mm).

If the cabling is being run to a location that will have a rack, the same thing applies; leave LOTS of extra cabling and wrap the bundle with plastic wrap to protect it from drywall & paint.

Leave "service loops" anywhere you may need to move the equipment or make changes. The length of the loop will depend on where the equipment might be moved. This also allows slack for cutting the cable and reapplying connectors.

In some cases there will be cable installed for future use but not landed in an outlet or other rough-in device. An example of this would be prewiring for speakers in a ceiling or wall when none will be installed during construction of the home. In the case of wood or metal stud construction the cable may be temporarily attached so it can be easily accessed later when the sheetrock is cut.

The location of all non-visible un-terminated cable must be well documented so it can be found later. Take a photo or take accurate measurements from a corner or easily identifiable feature. Keep the notes/photo with the cable schedule.

Securing Cables

Once all cables are installed, they need to be secured.

The purpose of securing cables is to prevent damage to the cable by other trades such as the insulation and sheetrock installer, or anything else that will be installed in the home after the pre-wire.

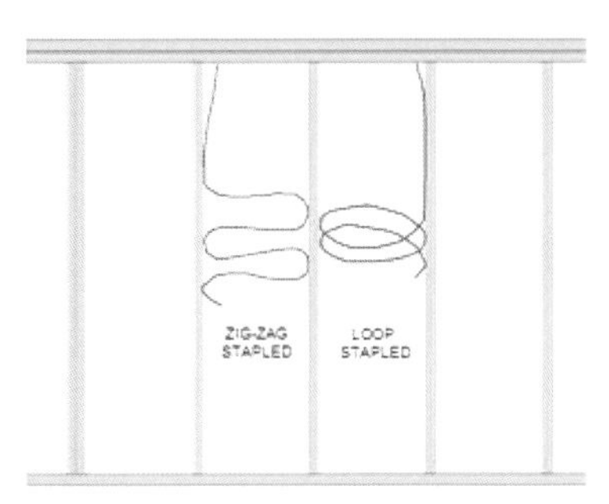

A wide variety of cable securing hardware is available. Choose hardware that does not pinch the cable. Cable should be able to slip through a cable tie down, but not be loose. Use the appropriate sized cable clip, or a small clip with a cable tie through it, then tightened on cable.

Cable should also be secured in attics and crawlspaces to prevent anyone working in the area from damaging the cable.

Use cable ties or Velcro™ strapping to secure cables to joists or other framing members. They may also be attached to joists in attics, but only near the perimeter of the attic, so they are not walked on or tripped over.

Any staple that is not made specifically for communication runs the risk of distorting the cable and affecting performance.

Plastic staples are available with rounded plastic inserts that protect the cable from damage. Again, make sure the staple is large enough so the cable is not pinched. This is the only "staple" that is OK to use. No metal staples, no Romex staples (unless specifically used as an anchor point for a tie strap)!

Cable must also be secured at rough-in hardware. The cable must be secure, but must be easy to retrieve from only the access hole. If you have any doubt, try it before you finish the pre-wire.

If you use cable ties, make sure you can access the tie through the opening!

Protect the pre-wire from other damage such as painting, sheetrock repair, wall texturing, etc. Cover all enclosures with stiff cardboard and tape the cardboard in place. Pre-made covers are available for outlet rough-in hardware and distribution enclosures.

Completing the Pre-wire

After the pre-wire, do a final walkthrough to make sure everything is installed correctly. With the equipment schedule and the cable schedule, go to each location in the home and check the following:

- The proper rough-in hardware is installed
- The correct cables are at the location and the label is correct
- The cables are secured correctly
- All locations are properly protected

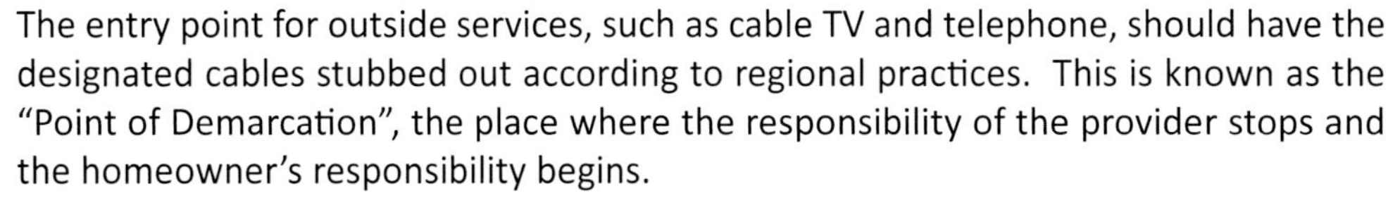

The entry point for outside services, such as cable TV and telephone, should have the designated cables stubbed out according to regional practices. This is known as the "Point of Demarcation", the place where the responsibility of the provider stops and the homeowner's responsibility begins.

For some installations, preliminary testing can be completed at this point. Since this usually requires cables to be terminated, testing often waits until completion of the trim-out phase.

Clean up

- Remove all equipment, cable, unused parts
- Collect and store all tools
- Sweep up and remove trash, wire scraps

Inform job foreman or builder that prewire is complete.

SUMMARY

If your career path includes basic EST responsibilities, this base of knowledge will be very important, both for new construction projects and retrofit/smaller projects. It is also imperative that you are trained in the areas of local codes and safety regulations. Violating either of these can be very costly or dangerous.

More information on cabling and its application in video, audio and system control will be found later in this book, as well as training on other skills which are important to the infrastructure installation process, such as documentation, professionalism, and the sub-systems that the cabling supports.

Questions

1. The one document that contains all critical information about the prewire (cables, loca tions, rough-in devices, etc.) is the:
   a. Floorplan
   b. Cable schedule
   c. Interior elevation
   d. Electrical plan

2. Outlet height should always be:
   a. At 12"
   b. The same in every room of the home
   c. Consistent with electrical outlets already installed on the same wall
   d. "Hammer-height"

3. The wooden framing members which is horizontal and supports a floor or ceiling are called:
   a. Studs
   b. Joists
   c. Rafters
   d. Headers

4. Unshielded Twisted Pair cable can be used for:
   a. Telephone
   b. Data
   c. Control signals
   d. All of the above

5. Observing minimum bend ratio is especially critical with coaxial cable because:
   a. Compressing the dielectric changes the cable's impedance and affects performance
   b. Bending too tight can damage the center conductor
   c. Stretching the jacket can make it susceptible to moisture
   d. The braid can be easily torn

Chapter 7
TRIMOUT AND TESTING

CABLE PREPARATION

Cable Preparation and Termination

The trim-out phase of a new construction project usually takes place once the walls are painted, ceilings are finished and sometimes flooring is completed. During this phase, cables that have been installed during the pre-wire are prepped and terminated on appropriate outlets and equipment. This includes all communications cabling, outlets and specialized platework, architectural speakers and volume controls. Cable termination and testing are skills which all ESTs need to master in a hands-on environment. In this chapter we will provide an overview of the tools and techniques needed.

Using the Proper Tools

Proper tools for cable preparation are essential. Using the right tools will save you time and prevent cable damage.

Cable Cutters

Cable cutters are used to cut cable, and are different than wire cutters (dykes, diagonal cutters, electrician's pliers, etc.) Cable cutters cut using a shearing action (like scissors). This action is less likely to stretch, compress, or deform the conductors.

Jacket Strippers

Use the correct jacket removal tool for each type of cable. Twisted -pair jacket strippers are designed to cut through the jacket without damaging the wire insulation. These jacket strippers can also be used for many other cables up to about .03 inches in diameter including speaker, security, and other control cable.

For larger cables, use appropriate jacket strippers. This type of tool is designed for cables up to 3/4" in O.D. (Outside Diameter)

Bundled cable (cables that combine other cables in a common jacket) jackets can be stripped back using a special tool.

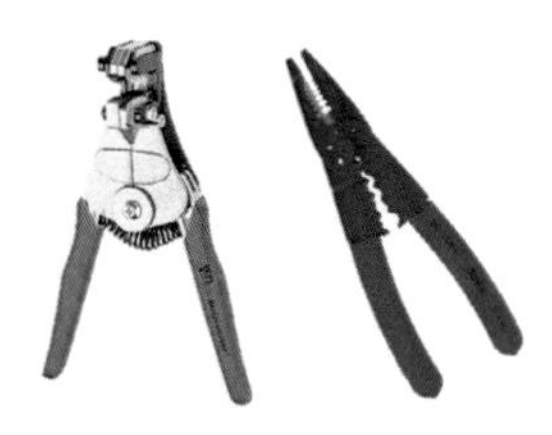

Wire Strippers

Using the appropriate tool for stripping the actual insulation from the wire conductors is also very important. The tools shown here, have jaws that accommodate wire gauges from 28 to 12 AWG (American Wire Gauge). Avoid using diagonal cutters, scissors or knifes when stripping wire, they can easily nick a conductor and affect cable performance. When copper is removed from a conductor its performance characteristics change.

Cable Labelers

You may want to replace some temporary cable labels with permanent labels during the Trim-Out Phase. The TIA 570-C specification recommends the use of machine-generated labels. The original hand-written labels may be adequate for cable that remain in the wall, but cabling which will be seen (in a rack or cabinet, or between components) should be professionally labeled with a label maker.

Outlet labeling is desired but not a requirement (due to aesthetic considerations). Icons or labels may be mounted of the faceplate of outlets to denote the intended application for each outlet.

Cables in the distribution center should be professionally labeled.

CONNECTOR TERMINATION

Phoenix Connectors

A popular type of termination requiring only wire strippers and a small screwdriver for termination is commonly known as the "Phoenix" connector. This connector is used is a variety of different systems.

- Security
- Power supplies
- Some audio components
- Control devices
- Speaker outputs & distribution blocks

Figure 7-1
"Phoenix" Connectors

Pig Tail Terminations Using Crimp-on Splices

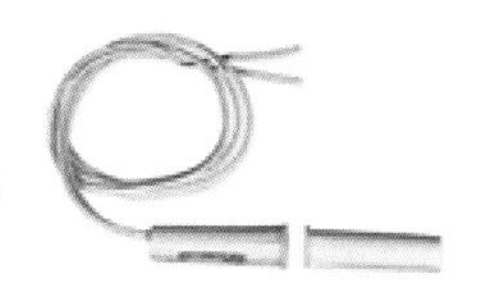

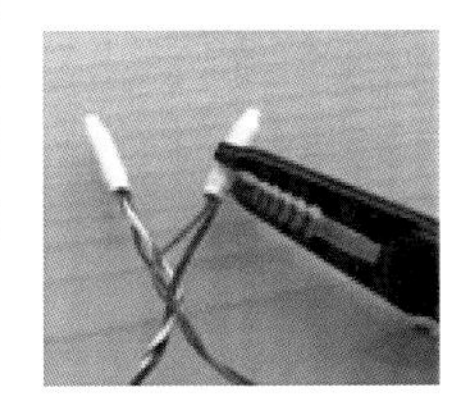

Many devices, like security sensors and volume controls have wires pre-attached. These wires are called "pig tails" and are terminated to the installed cable using a crimped-on splice. Small crimp connectors, known as "beanie" or "dolphin" connectors are used for small gauge (up to 18 gauge) solid and stranded conductors. Each connector will accept 2 (possibly 3) wires. Remove 1/4 - 3/8" of insulation, twist conductors, apply connector over wires and crimp in place. Some connectors of this type are called Insulation Displacement Connections (IDC) and do not require stripping the wire first.

The "F" Connector

Almost all video applications installed in the home for CCTV or CATV use the "F" connector for termination of the coax cable. The standard coaxial cable "F" connector consists of a barrel where the connector makes contact with the shield, and a threaded nut is used to secure the male connector to the female connector. There are many different types of F connectors, but they all perform essentially the same task.

However, the quality of the connection can vary greatly:

- The **screw-on type "F" connector** is not a preferred connector because it does not provide a dependable connection. It allows oxidation and can "literally" fall off the cable for no reason.

- The **crimp-type connector** was widely used for many years and is somewhat better but it is still not very strong, and allows moisture and air to get to the conductors via the gaps between the shell and the cable jacket.

- The **compression-type connector** is the preferred method. It is very strong in the way it attaches to the cable jacket, and provides a nearly waterproof seal to prevent oxidation of the conductors.

Compression-Type Connector Termination

When terminating coax cable, you need to select the appropriate coax stripping tool and crimping tool or compression tool for the brand of coax and connector you are using.
Different connectors require different types of cable stripping tools and tool settings. The major difference is how many shields are stripped.

Figure 7-2 Coax Prep Tools

Figure 7-3 Coax Compression Tools

Let's take a quick walk through the steps required to terminate a coax cable with the preferred compression-type "F" connector.

Step 1	Place the quad RG-6 coax flush with the stop of the jacket stripper tool and twist the tool 3 times around in each direction until it stops cutting the braid.	
Step 2	Remove the jacket and insulation/braids cut by the tool. Your cut should look like the cable pictured here.	
Step 3	Fold back the outer three shields (braid, foil, braid) over the jacket. DO NOT DISTURB the inner foil shield. It should be attached to the dielectric insulation.	

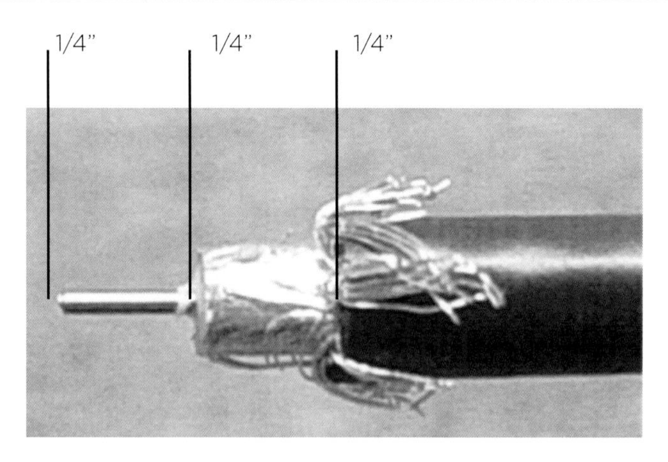

Ensure that any braid extending longer than 1/4" is trimmed to the proper length

This is what the prepared cable should look like prior to installing the connector. Make sure there are no stray strands and that the foil on the dielectric is intact.

Step 4	Install the connector over the coax. Push the connector onto the coax so the insulation and inner foil go inside the inner barrel, and the inner barrel is forced under the inner braid shield.	
Step 5	Push the connector onto the coax until the dielec-tric insulation is flush with the inside barrel of the connector (A). Some manufacturers have the captive plastic sleeve which guides the center con-ductor (B). The important thing is to make sure the dielectric is in far enough to full the opening. The center conductor will extend just slightly beyond the rim of the connector (C).	A B C

Step 6	Insert the connector and cable into the compression tool as shown.	
Step 7	Squeeze the tool to force the plastic seal firmly into the connector, which also forms a tight attachment to the cable jacket.	

Figure 7-4 The Finished Connector

There are a variety of compression type connectors available for coax, including RCA and BNC type connectors. The technique for termination is similar for all, but be sure to follow the manufacturer's instructions as there are some variations. Some of these connector types are also available for "Mini 59" cable, a small version of RG-59 bundled with five separate coaxial cables in one jacket.

UTP Termination

There are three common ways to terminate Unshielded Twisted Pair (UTP) cable.

- 110 style punch down connectors (other types are available, but the 110 is used most often in residential networks)
- Modular jacks (this is the female connector in a wall outlet)
- Modular plugs (this is the male connector on the end of a patch cable)

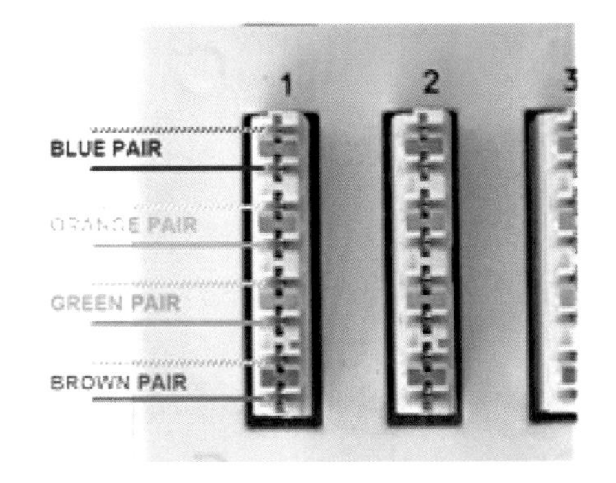

Terminating UTP cable to a 110 style punch-down connector is fairly straightforward. The diagram shows the proper pair color coding. This remains the same regardless of what pin configuration is used at the other end.

The 110 style connector uses insulation displacement contacts (IDC). The contact cuts through the wire insulation and makes a tight contact with the copper conductor and since the penetration of the insulation is so small, air and moisture are somewhat deterred from getting to the copper.

You will need the appropriate jacket stripping tool and a UTP punch-down tool. In this case you will only be stripping the cable jacket, not individual conductors.

Termination on a 110 Connector

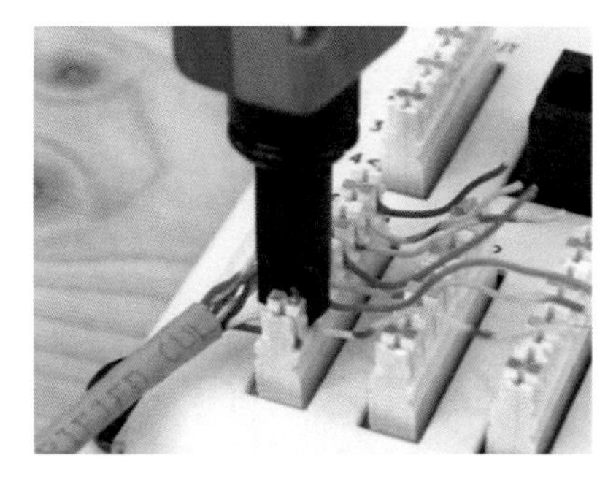

Start at whichever end of the 110 connector is more convenient to route the UTP cable. In the example at the right, the cable is routed from the bottom; therefore the brown pair is connected first.

Untwist each pair as little as necessary to insert each wire into the corresponding contact.

Use your fingers to firmly press each wire into its contact.

The wire should stay in place until it is punched.

The punch-down tool blade has a cutting and non-cutting side. Make sure the cutting side is on the side you want to cut off!

Use a quick, straight-down motion to force the wire into its contact. When the proper pressure is reached, the tool will release its spring loaded plunger and drive the blade down until the excess wire is cut off. A second punch will ensure good contact.

If the blade does not cut off the excess wire, do not keep punching-down the wire. Move the wire back and forth, or cut off the excess with diagonal cutters.

The finished installation should look like the example on the right.

Note that the pair untwisting is kept to a minimum.

Common mistakes made with this type of termination are:

- Untwisting the cable pair too much.
- Removing too much of the cable jacket.
- Cables crossing over each other

The errors can impact cable performance and present a general impression of sloppy workmanship.

Termination on a Modular Connector

UTP cable uses standard 6 and 8 pin modular connectors developed by AT&T. You will only be installing 8-pin connectors.

The modular jack is on the left, the modular plug is on the right.

TIA-570 specifies only 8-pin connectors for all UTP cable terminations in the home, both data and phone.

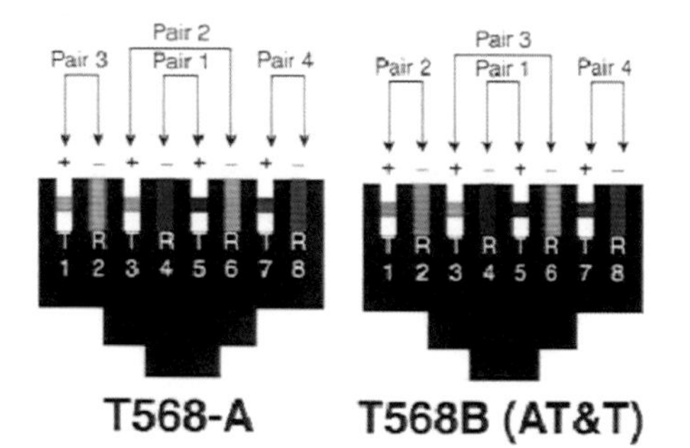

	+	-
PAIR 1	BLUE/WHITE	BLUE
PAIR 2	ORANGE/WHITE	ORANGE
PAIR 3	GREEN/WHITE	GREEN
PAIR 4	BROWN/WHITE	BROWN

The diagram to the left shows the wiring configuration (which wires in which pair are connected to which contacts) for the TIA-T568A wiring configuration. This is the recommended configuration for all residential installations, because it is compatible with the traditional locations of Line 1 and Line 2 for telephone use. However, TIA-T568B is still widely used in many regions, so you must know your company's policy. A color image of this diagram can also be found on the back cover of this book.

Modular Jack Details

The contacts on the modular jack are numbered left to right looking into the connector. Notice that Line 1 is the most inside pair, and Line 2 is the two conductors just outside that pair. This is consistent with the legacy RJ-11 connectors used for standard two-line telephone systems.

There are a variety of modular RJ-45 jacks on the market which utilize the IDC type of termination. The one thing you will find on all of them, in one form or another, is color coding for both T568A and T568B termination. You must look carefully to make sure you are following their color code correctly.

TIA-570 dictates that ALL 8 position modular connectors in residential installation are to be terminated to the T568A protocol.

The tools for terminating cable to modular jacks are the same as those used for the 110 style connectors; the UTP cable stripper and the UTP punch down tool.

This shows how UTP cable is typically installed on modular jacks which have the two rows of 4 contacts (4 on each side). This is the most popular contact configuration. You have to follow the wire color code on the connector.

Step 1	Remove about 2" of jacket from the UTP by scoring, bending, and pulling jacket off. Fold back all but the pair which is to be punched down in the slots farthest from the modular jack snap-in. Place the connector in the holder. The holder holds the connector stable but always punch on a solid surface. The spring loading of the punch tool will be less effective if you hold the connector/holder in your hand.	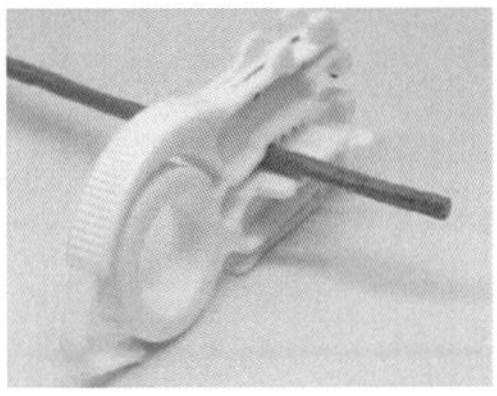
Step 2	Untwist the first pair down to the jacket. Insert the wires into the rear contacts as shown, using your fingers to firmly seat the wires into the contracts. You may punch one pair at a time, then move to the next, or set all 8 and punch all of them. Just be careful never to damage other wires when punching.	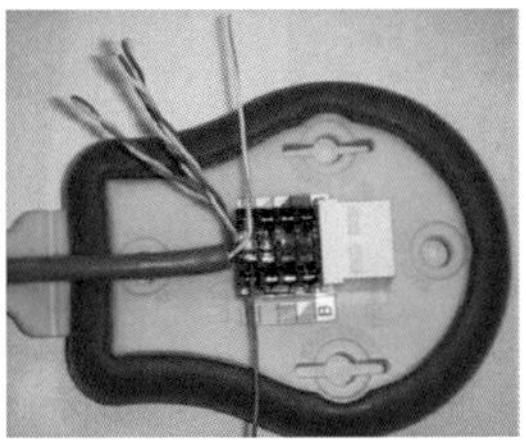
Step 3	Use the punch-down tool to seat the wire in each contact, keeping the cutting edge on the outside of the connector. Remove excess wires. Push until the spring loading "pops".	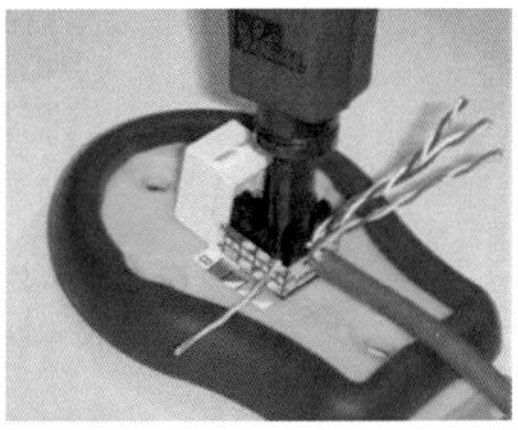
Step 4	Work your way through all pairs, following the T568A or B protocol as specified. (T568A is recommended for all residential networks)	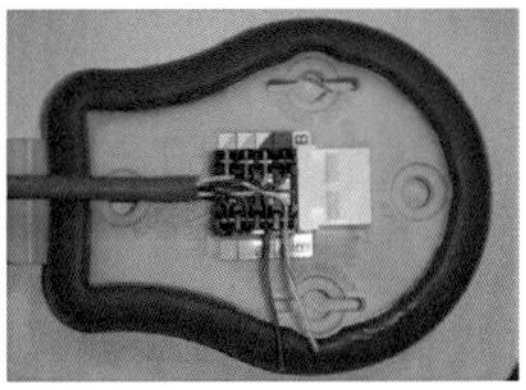
Step 5	Place the plastic cover over the contacts and snap it in place.	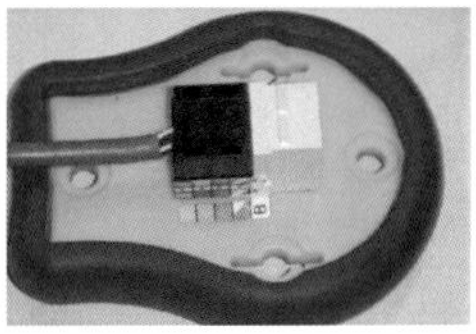

Common mistakes made when terminating UTP jacks:

- Too much jacket removal
- Too much pair untwisting

Modular Plug

To install a modular plug, you will need the same cable jacket stripping tool, plus a plug crimping tool. Make sure the crimping tool was designed to crimp the style of connector you are using. RJ-45 EZ style modular plugs are widely used because the conductors go all the way through the connecter and are cut off by the crimping tool.

Steps for plug installation using the RJ-45 EZ plug:

Step 1	Remove about 2" of insulation and spread the pairs out (green left, blue center, brown right, and orange up).	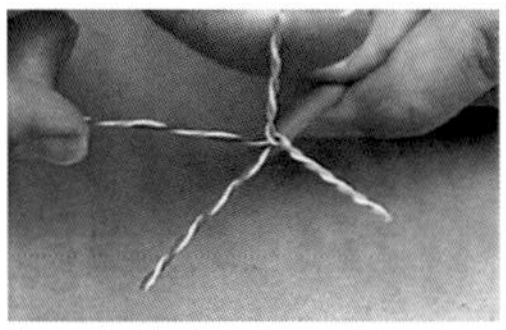
Step 2	Untwist each pair and while holding wires between your thumb and index finger, order the wires in each pair in the order shown from left to right.	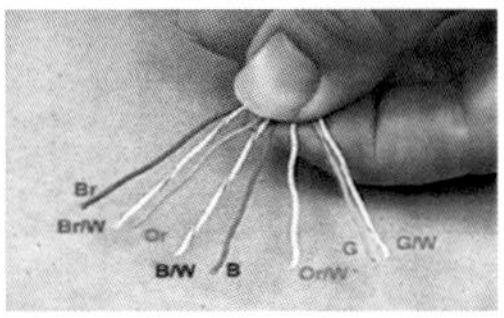 Br, Br/W, Or, B/W, B, G, Or/W, G/W
Step 3	Carefully bring the wires together and work out the twisting by bending the flat wires back and forth between your fingers.	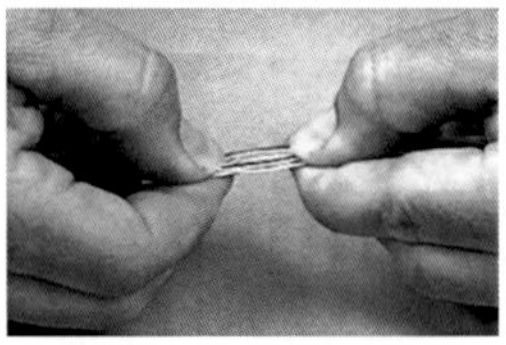
Step 4	Insert the wires into the plug body with green/white wire on the left and the plug tab down. The wires should slide into the channels for each wire. Make sure there is no overlap of wires across channels. Force the cable all the way into the plug until the jacket is at least ¼" into the plug body.	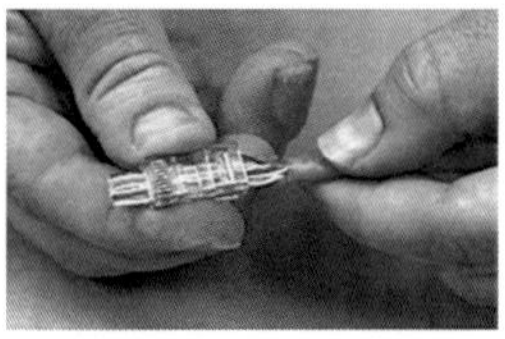
Step 5	Make sure the jacket is under the strain relief clamp and all wires are protruding out the front of the plug.	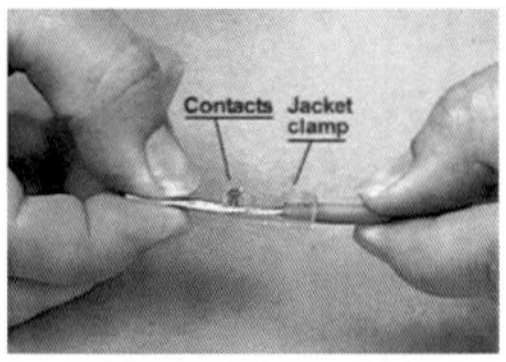 Contacts, Jacket clamp
Step 6	Place the plug into the crimping tool until it snaps in place and crimp the tool through it's complete ratcheting cycle. This should cut off all excess wire.	
Step 7	Inspect the plug to make sure the plastic strain relief clamp is crimping the jacket and none of the contacts are bent.	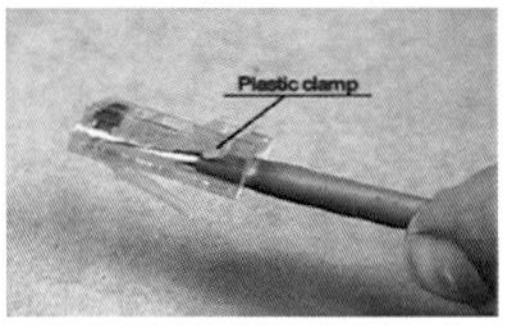 Plastic clamp

Common mistakes made when terminating UTP Plugs:

- Jacket not fully inside plug under jacket strain relief
- Pairs are twisted inside plug, preventing wires from being fully inserted.

Fiber Connector Termination

Pictured here are examples of the common types of fiber optic connectors.

FIBER OPTIC CONNECTORS

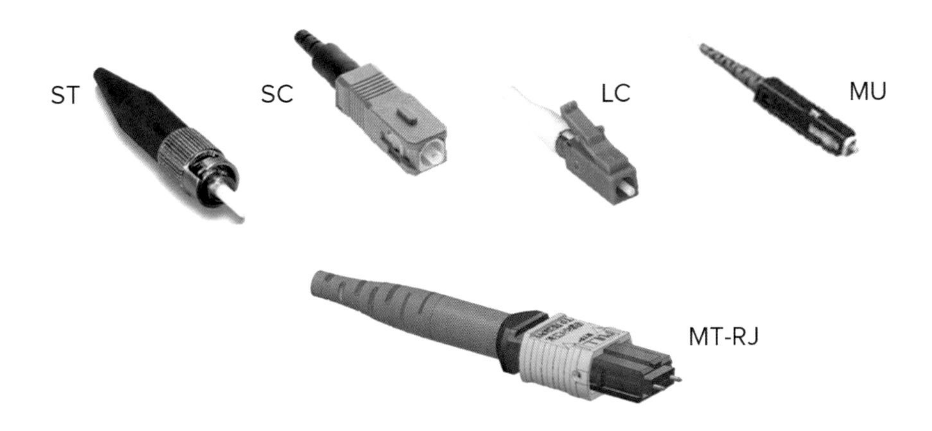

NOTE: TIA-568-C recommends duplex SC connectors be used

Note: The MTRJ is a duplex connector.

There are a number of different methods for terminating the various systems of fiber connectors. Some involve the use of epoxy glue, while other using a crimping method of attachment. The tools required are specific to the style of connector and method of termination. Technicians should refer to the manufacturer specifications for the tools and supplies needed to support the termination of choice. Many connectors require specialized training before attempting to terminate and test fiber optic cables.

One popular method on fiber termination is a method that employs a pre-polished connector and a mechanical splice to attach the quick termination to the unterminated fiber.

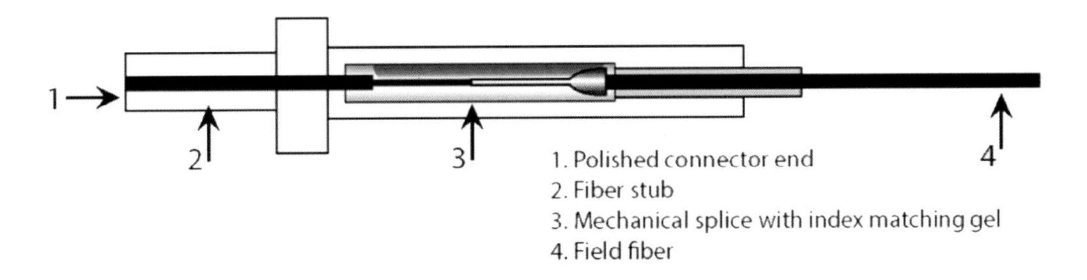

In order to facilitate the coupling and reduce reflections of light inside the mechanical splice an index-matching gel is used to conduct the light between the fiber ends. The gel reduces the need for stringent mechanical tolerances on cleaving and polishing, expensive fusion equipment, and extensive technician training.

Speaker Cable Termination

There are many types of connectors used for terminating speaker cable. These connectors are typically used for connection directly to speakers and amplifier contacts on interconnect cables.

They crimp, solder, or screw to the copper conductor of the cable. In most cases using some kind of connector rather than just the bare conductor ensures a more robust and long lasting connection. Banana plugs on the cable usually makes a great connection in binding posts.

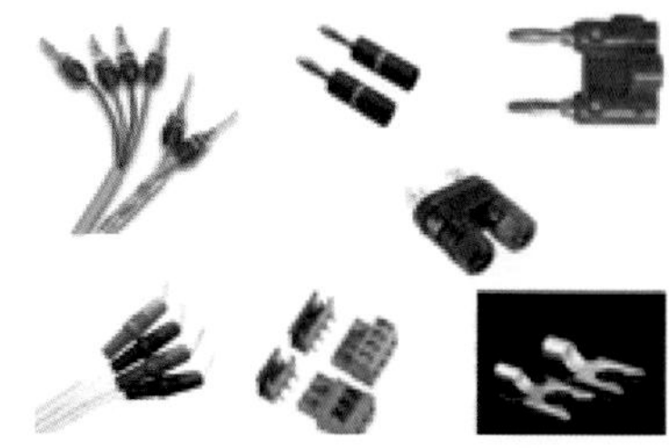

Many speaker connections on distribution blocks and amplifiers are done with Phoenix type connectors. The connector, shown in the photo to the right, is quite common. Each conductor is attached in its own port using either a screw terminal or a spring type connection.

- Remove about 2" of jacket using a jacket stripping tool
- Remove about ¼" of insulation from each wire
- Carefully twist the copper strands together to prevent loose strands from sticking out of the connector.
- Insert each wire into the appropriate contact and screw down the contact to make a tight connection.

You should not see any of the bare copper when properly installed.

A common mistake when make this type of termination is to strip too much of the individual conductor and to have bare wire showing. This may lead to an arc, short circuit and amplifier failure.

While soldering of cable wires may still be needed from time to time, most of the connectors used today will either be installed by a crimp or screw down type of termination.

The binding post is a simple way to connect a stripped wire conductor to a jackplate, amplifier, or speaker. This post often has a hole which will accept a "banana plug". Two posts, + and -, are required for a single loudspeaker connection. Red is always +.

The banana plug is a standardized connector typically used for speaker level audio signals. Many binding posts will accept either single or double banana plugs. Most models allow for the wire conductor to be attached inside a cylinder by tightening a small set screw.

An interesting feature of this connector is that two speaker lines can be attached in parallel by simply plugging one connector into the top of another one which is already plugged into the amplifier. The side with the tab on a double connector is typically the ground or negative side.

EQUIPMENT INSTALLATION

Also during the trim-out phase, built-in devices such as the distribution center components are installed, along with supporting equipment like keypads, controllers, sensor, thermostats, volume controls, antennas and speakers. Any equipment that would be considered part of the house, and that cannot be pilfered.

Outlet Hardware

Outlet plate systems vary by region, but all do the same thing; they allow the system wiring of the home to be accessed in the rooms in an efficient and attractive manner. The most flexible systems accommodate a wide variety of connector types and even allow the connector modules to be changed as needed.

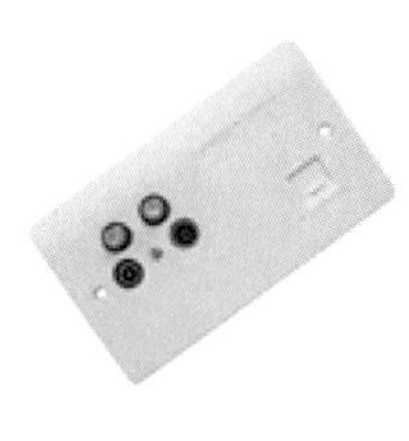

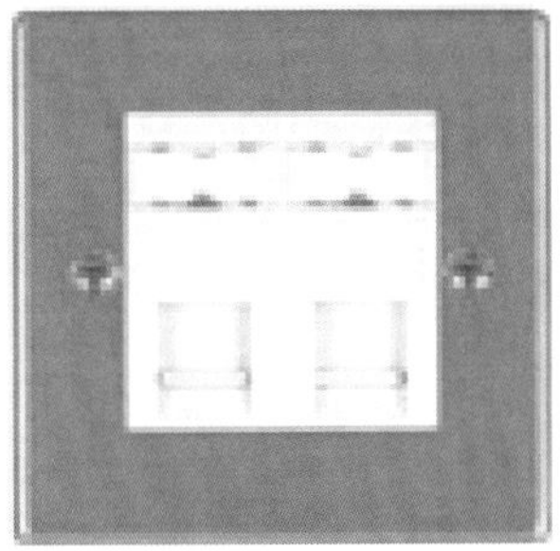

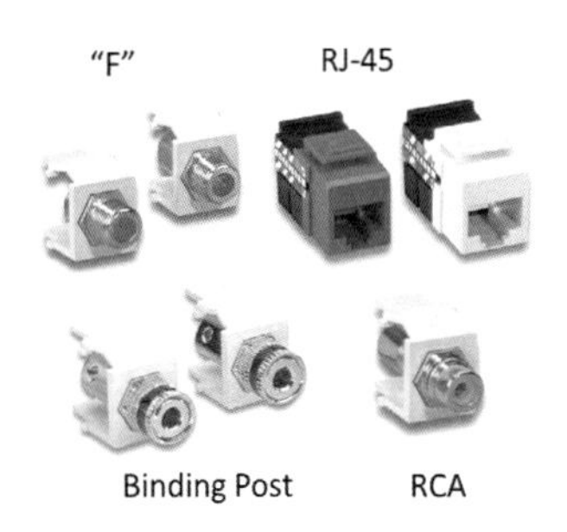

Platework should match the electrical outlets in style and color on a room by room basis.

Company policy will dictate any specific color coding of the inserts such as RJ-45s. In many cases, the jack used for telephone will match the plate, but the network jacks may be of a different color.

These modular systems appear to all be the same, however parts and pieces from different manufacturers may not be compatible.

Speaker Installation

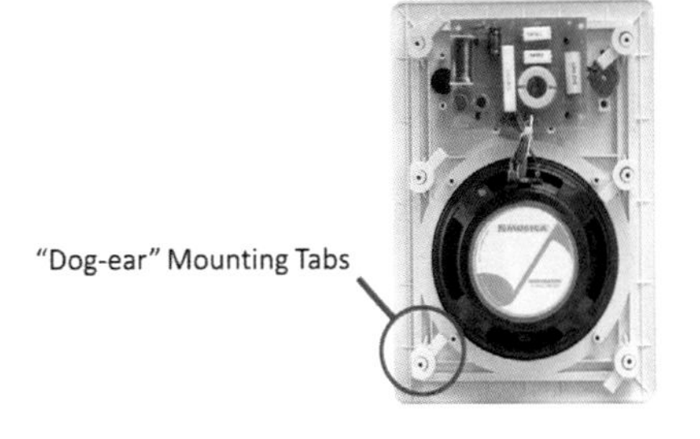

In-wall and in-ceiling speakers will either use a rough-in bracket or mount directly to the sheetrock at trim-out. Cable may be "zig-zagged" and left unterminated during pre-wire. Then at trim-out, the proper hole is cut and the speaker installed.

The speaker has "dog-ear" tabs that swing out and grab the sheetrock (or bracket behind the sheetrock) when tightened with a screwdriver

Volume Controls

Connection of speaker cable to volume controls also often uses Phoenix connectors, with either screw terminals or tension levers. These connectors can be removed from the unit to make wire connections easier. Make sure you don't get input and output confused.

Volume controls typically have an INPUT and an OUTPUT connector.

- INPUT is for the cables from the amplifier
- OUTPUT is for the cable to the speakers

Traditional passive volume controls may also have a way (such as the jumpers shown) to set the impedance seen by the amplifier when more than one volume control is used on the output (connected in parallel). You will need to consult the manufacturer's installation instructions for the model you are using.

The volume control shown uses removable jumpers to set the impedance. Some have a simple slide switch to select x1, x2, x4, x8. The correct jumper positions are show on the back of the PC board.

It is during the Trim-out Phase that final testing of the installed cable is completed. While all cables should be tested, the level of testing depends on the type of system specified.

Types of Testing

Testing is divided into three methods.

- Verification testing
- Qualification testing
- Certification testing

Verification Testing

This is the simplest (and minimum) testing required for all cable. For most cables this is simply an end-to-end test of continuity to verify a signal path existing for each conductor. For UTP cable, verification testing may also include wire-mapping; a test that show pin-to-pin continuity and correct termination.

Simple static verification testers are available to test coax cable as well. These testers check for opens and shorts in the center conductor and shields.

Dynamic testers provide some additional features which can be useful. They have the ability to determine the length of the cable, and if there is an open or short, indicate the distance to fault.

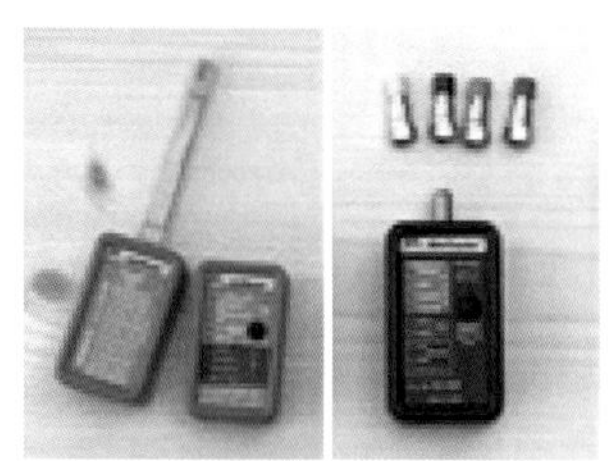

Qualification Testing

This type of testing uses test equipment that passes actual signals on the cable and monitors the results at the remote unit. This test can find faults that are missed by verification testers. This type of testing will tell you if a network line is performing at the required speed for the application.

Certification Testing

Certification test equipment also measures the actual performance of the cable to TIA specifications (such as cross-talk, attenuation, etc.) Certification testers can provide a database and printout of the test results for each cable. This level of testing is not common in residential applications; but is frequently required for institutional and industrial installations where a high level of network performance is necessary.

While commonly used only for UTP cables, qualification and certification testers can check coax cable for attenuation and proper impedance throughout the length of the cable. This will determine if the cable has any damage that might affect picture quality.

It is important to document all testing on the cable schedule as this will let the final technician know that all runs have been checked and are fully functional. Simply go down the list of cables and test each cable on the schedule. As the cable is tested, check off or initial the cable.

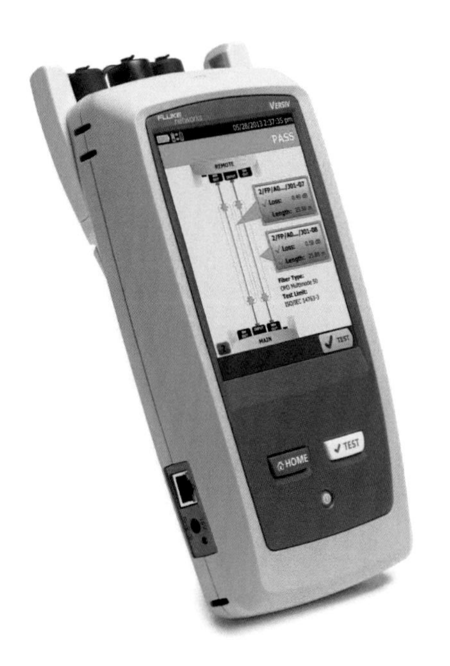

Unidentified Cable Testing

The tone generator and tone detector can be used to help locate cables which are not properly identified or located behind a wall or ceiling. The tone generator places an audio AC signal on one or more of the conductors of a cable under test.

The tone detector picks up the AC signal (using a high gain amplifier) and converts it to an audible sound.

It is good practice to fix any faults that you find while testing when you find them. Faults such as mislabeled cables, if not fixed when they are found, will cause serious problems later.

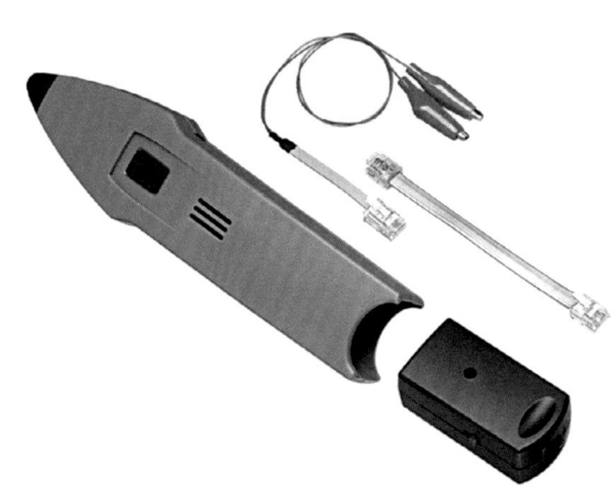

DOCUMENTATION

The main documentation needed during the trim-out phase is updating the cable schedule with as-built information such as test results. Also collect any equipment user manuals, installation instructions and warranty cards that come packed with equipment. These can be used later when creating the final documentation for the owner.

Check off all installed equipment against the equipment list and gather serial numbers of main components in order to validate warranty information.

Questions

1. The connector which is universally used to terminate coaxial cable for television signal distribution (antenna, cable, satellite) in the home is the _____ connector.
   a. BNC
   b. F
   c. RCA
   d. RG59

2. The RJ-45 jack configuration which is compatible with both data application and telephone lines 1 and 2 is:
   a. TIA T568B
   b. TIA 570
   c. TIA T568A
   d. BICSI

3. The highest bandwidth and throughput is found in _______.
   a. Fiber optic cable
   b. Coaxial cable
   c. Unshielded Twisted Pair cable
   d. Shielded Twisted Pair cable

4. The simplest type of communication cable testing (continuity and/or mapping) is called __________.
   a. Certification
   b. Verification
   c. Qualification
   d. Clarification

5. The crimp-style F-connector is not recommended because:
   a. It is not compatible with all F-type female jacks
   b. It doesn't have adequate bandwidth for digital signals
   c. It allows air and moisture to oxidize the conductors inside
   d. It has a left-handed thread

Chapter 8
EQUIPMENT INSTALLATION

INTRODUCTION

Whether in a new home, or a retrofit project, the way equipment of all kinds is installed impacts the performance and serviceability of the system. But just as important are the many ways the installation affects the client, including ease of use, safety, and aesthetics.

In a well-planned and executed project, the systems designer provides documentation which spells out in detail the location, wiring, and configuration of every component. And the actual installation is carried out by an qualified EST (Electronic Systems Technician) in the field. Because both of these positions require several years of training and experience, the more advanced topics are beyond the scope of this book, however we will provide a solid introduction to the fundamentals, including terminology and generally accepted best practices.

The equipment we will be discussing here is generally installed during the final phase of a project, either new construction or retrofit. The infrastructure (cabling between locations) and platework has been tested. Cabinets, millwork, etc. is already in place, and verified to be correct. The importance of planning cannot be overemphasized. When every detail of an installation has been thought out in advanced, and properly documented, the installation will be smooth and there will be little or no troubleshooting needed.

Main System Components

The basic electronic components of an A/V system are usually grouped and housed together to allow ease of hookup, short interconnects, shared power, and convenient use and servicing. This grouping of components may be for a specific subsystem such as a media room or theater, or they may be in a "head end" location where several systems are centrally located, which serve the entire home. In all cases, the configuration of this equipment should be completely thought out in advance. The most common components (all of which are discussed in more detail elsewhere in the book) are:

1. **Sources** – Blu-ray players, VCRs, Cable/Satellite boxes, DVRs, servers, etc.
    - These devices typically require some client interaction
2. A**udio/Video Receivers (AVRs), and Preamp/Processors (Pre/Pros)**
    - These should also be in sight and reach of the client
3. **Power Amplifiers**
    - These are often heavy and produce a good deal of heat
    - Typically not adjusted by the client
4. **Power Conditioners**
    - Also heavy and do not require much client interaction
5. **Other peripheral equipment** – includes switchers, control processors, and smaller system specific interfaces
    - Typically require no user interaction but must be located for easy access and servicing
    - Sometimes mounted out of view

Design Considerations

This equipment will need to be housed in some sort of rack, cabinet or shelving. As mentioned earlier, the configuration of the equipment will usually be designed in advance, unless the project is very small. We will discuss some of the primary considerations that play a part in this process.

User Interaction

The first consideration is always the client themselves. Who will be using the system? How tall are they? Are there children in the home? Any special situations such as wheelchairs? These will all influence decisions regarding equipment location.

In general the following guidelines should be considered:

1. Equipment with no user interaction – location is not critical, can be based on things like weight, heat, and relationship to other equipment
2. Insertion components (equipment that the user puts media into (Blu-ray, game consoles, etc.) – should be at a convenient height. Those which are "top-loaders" should be low enough to be conveniently loaded/unloaded

3. Components with displays (set-top boxes, tuners, etc. – should be at a height which is easy to see
4. Equipment which uses external inputs from games, cameras, USB devices, etc. – should be accessible

Technician Interaction

Some equipment may never be directly used by the client, but require easy access by the technician. Any component which may need updates, downloads, adjustment, rewiring, calibration, or rewiring should be located so that no other cabling or equipment needs to be moved out of the way. Some equipment of this type is actually better suited for locations NOT accessible to the client. This may prevent problems and service calls related to untrained hands making adjustments. Keeping unnecessary units out of view also makes the installation cleaner.

RACKS AND CABINETS

Equipment Racks

As the residential systems industry matured, one of the concepts which became universal is the use of standard EIA (Electronics Industry Alliance) style equipment racks. This standard began in the telecommunications industry and carried over to virtually any application where electronic equipment needs to be mounted, managed, wired, serviced, and accessed by users. A wide variety of types are available as well as various parts and pieces to customize them to virtually any application. These include, but are not limited to:

Figure 8-1 Equipment Racks, Front and Rear

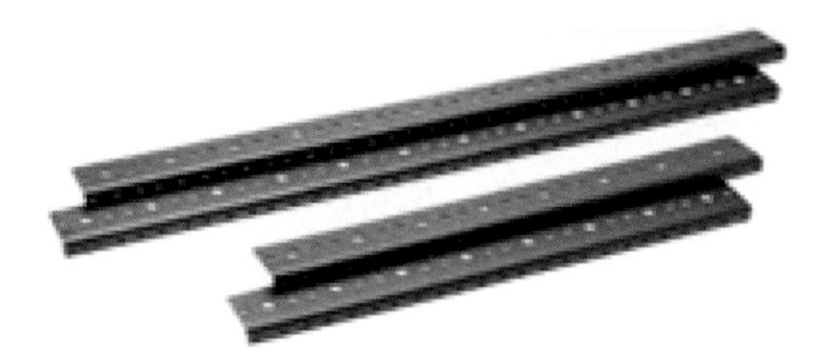

Figure 8-2 Rack Rail

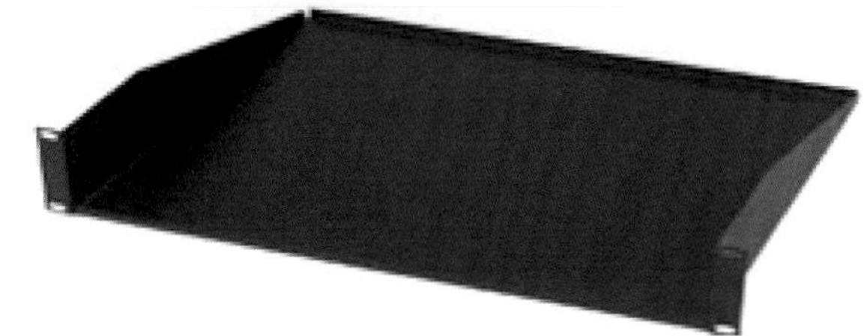

Figure 8-3 Rack Shelf

Figure 8-4 Rack Drawer

Figure 8-5 Blank Panel

1. **Rack Rails** – vertical metal rails which are pre-drilled to exact specifications and support various other elements of the system. In some situations, rack rails can be mounted securely to a frame fabricated in advance, without any other outer shell.
2. **Enclosures** – the outer shell which holds the rails and surrounds the equipment. May be just a frame, or a fully enclosed box made of metal or wood. May be floor standing, built-in, or wall mounted.
3. **Shelves** – mount to the front rail and sometimes other points and provide a resting place for equipment. May be fixed, or moveable.
4. **Panels** – blank covers for unused space, may be solid or vented
5. Drawers, trays, fan modules, etc.

Although the racking system is strictly standardized, it offers a great deal of flexibility and other benefits. Equipment locations can be configured in advance, but easily changes when updates or replacements are made. Installing equipment is simple and dependable. The rack can be pre-loaded, wired, and tested offsite and then delivered all at once, reducing the time on the jobsite. A metal rack provides a common electrical ground. If necessary, a sealed "active" ventilation system can be implemented. Custom faceplates and blanks present a very professional appearance. In addition, there are computer programs available which allow the designer to easily create an elevation drawing which shows all equipment, panels, and accessories to scale. There are some drawbacks to this system which should be noted; racks can be too industrial looking for a typical residential space, and some projects do not justify the expense of a professional style rack.

Rack System Details

The rack system is based on very strict and standardized specifications. The vertical height of panels, shelves, and rack-mountable equipment is always measured in "rack units" or "RU". A rack unit is 1.75 inches in height. One RU is 1.75", two RU is 3.5", etc. so if a unit is 4 RU, we know it is 7" high.

Rack mountable equipment and accessories are 19" wide. This is the width established by the EIA for A/V equipment. The holes are threaded to 10-32, and that is the bolt type that must be used.

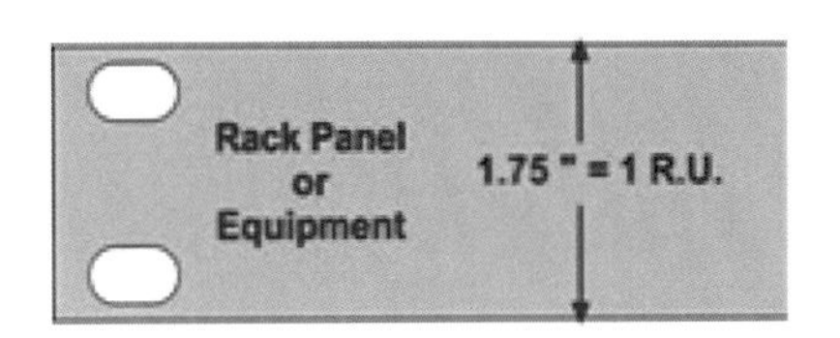

Figure 8-6 Rack Unit Height

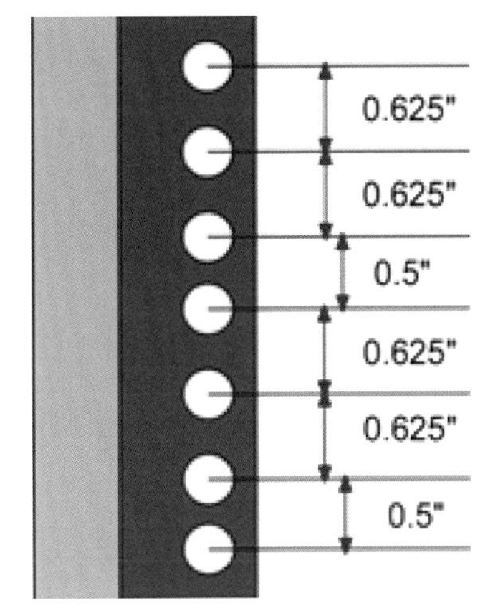

Figure 8-7 Hole Spacing

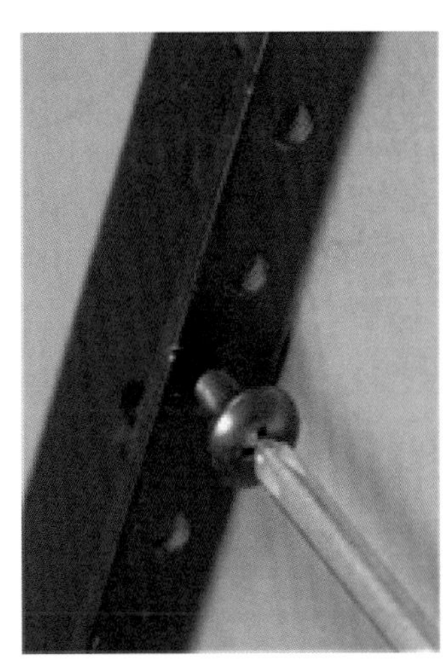

Figure 8-8 Rack Screw

Cabinets

Cabinets may be free-standing prefabricated pieces of furniture, or custom made and built in to the home. In many cases, especially local systems like media rooms, there are only a few electronic components and this is a convenient and attractive solution. Cabinets can be very attractive, chosen or fabricated to work well in the aesthetics of the room, and not just the electronic components, but provide storage and even house loudspeakers in a concealed fashion. On the other hand, there are challenges with using a cabinet rather than a professional rack. Sizes may vary, wiring can be less than ideal, and ventilation can be a huge issue. Reconfiguring, upgrading, and servicing is usually much more difficult, especially if the only access to equipment is from the front.

Once again, however, the key to success is extensive advance planning. If the designer is part of the conversation from the beginning, obstacles can be overcome.

The cabinet must meet several requirements for the installation to be smooth and provide ease of access and long term performance.

- Adequate space for equipment and wiring
- Thermal management; exhaust fans and fresh air intakes
- Flexible shelf configurations, moveable and/or slide out
- Planning for remote control operation (RF or IR routing)

Wall Shelves

In very small systems, components may simply be placed on shelves. While this allows for good ventilation and easy access, it can be difficult to conceal wiring and make the installation look professional.

Thermal Management

Any time an electronic component is placed in an enclosed space, there is the potential for the heat they generate to raise the operating temperature too high. Most equipment failures are the result of excessive heat. Allowing equipment to run hot over a period of time may shorten their operational life by 50%. The general guideline to prevent failure is to keep the operating temperature no higher than 85° F (29.4°C).

Some types of equipment generate more heat than others. Power amplifiers use a lot of power and create a great deal of heat. Some smaller components like cable and satellite set-top boxes have powerful processor chips similar to what is found in a computer, which run hot.

In a rack system, there are two approaches to managing this heat; a "passive" design, which depends on natural convection (heat rises) and adequate ventilation around components, and an "active" design, which involves a sealed system with designated entry points for ambient air, and fans to remove the heated air. This is an advance topic which system designers need to master before designing complex systems.

CABLING PRACTICES

Certain best practices should be observed in the wiring of any combination of equipment in a rack or cabinet. Experience ESTs can turn a complex configuration of equipment into work of art. In fact, a well-dressed rack says a lot about your company's attention to detail. Conversely, a messy job shows the client (and their friends) that you were concerned with a quick installation than doing a professional job.

Pre-fab vs. Onsite Fabrication

Different situations call for different approaches regarding interconnects. Smaller installations work just fine with pre-fabricated interconnects of the appropriate length; however there will always be extra cable to deal with. Fabricating interconnects to exact lengths, especially in larger systems, allows the technician to make interconnects exactly the length needed and route them exactly as desired. Of course, some cable types cannot be made onsite, such as TosLink and HDMI.

Best Practices

Some simple guidelines for wiring in racks and cabinets include:

- Route all signal cabling down one side of the rack, all power on the other side
- Route all cabling so that the back of equipment is visible and not too obstructed
- Use Velcro® to secure interconnects, rather than nylon cable ties
- Leave adequate slack in cabling to remove components if there is only front access
- If rack is moveable, allow for adequate cabling to allow full extension
- Pre-load and wire the rack at the shop whenever possible. Do as much testing and setup as possible. If the rack is too heavy to manage safely, remove heavier components like power amps and power conditioning and replace them onsite.

INSTALLING DISPLAYS AND PROJECTORS

Installing equipment in locations other than a rack or cabinet involves attaching directly to physical boundaries, such as walls and ceilings. This requires a working knowledge of task-specific hardware and a variety of fasteners.

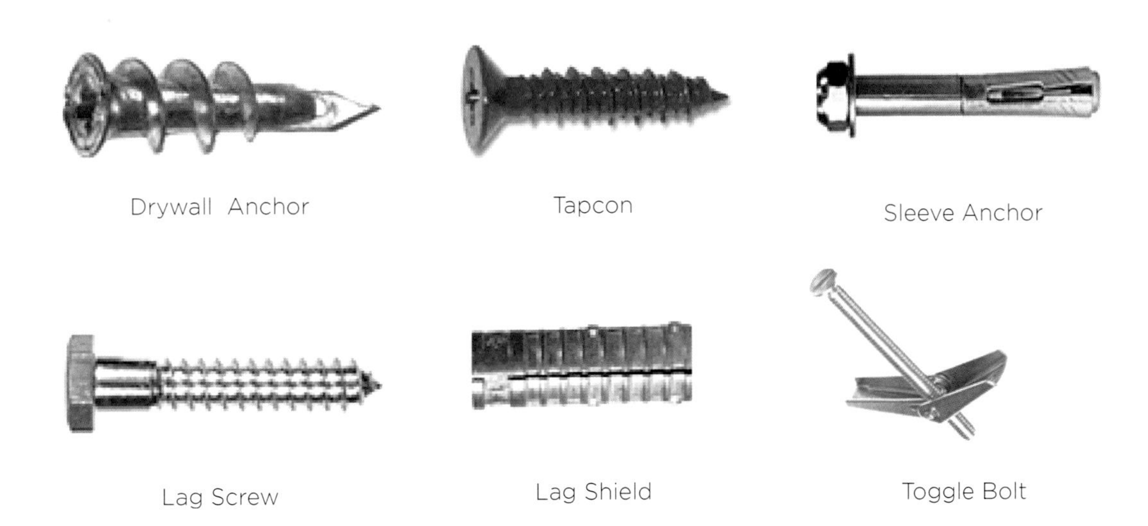

Fasteners and Techniques

The mounting of video displays and projectors must be done with safety as a prime consideration. These installations are only as solid and safe as their weakest link. A heavy-duty mounting bracket will not do the job if attached to the wall incorrectly. Different construction methods require different fasteners and mounting techniques. In all cases, the mounting must be extremely secure and solid.

The recommended guidelines for an entire mounting system (wall/ceiling, hardware, and fasteners) are based on the UL 1678 standard, written by Underwriters Laboratories.

- Static loads (those which are not moveable) should be supported by a mounting system capable of holding 4 times that load.
- Dynamic loads (those which are moveable, especially away from the mounting surface) should be supported by a mounting system capable of holding a load equal to 8 times the actual installed load.

The entire mounting system must be extremely solid and stable. In order to ensure this, the right fasteners must be utilized and they must be attached to the right part of the wall or ceiling. In wood frame structures, this means attached to solid studs or joists. In retrofit situations, it may be necessary to open up the wall and install blocking to provide solid support. In concrete, the anchors must be the correct type and size and properly installed in multiple locations.

You should become familiar with the construction methods and materials in your region, and the recommended fasteners which provide the best results.

Power

The location of the power outlet (and signal outlet) in the wall or ceiling should be planned well in advance. When possible the power to a display or projector should conditioned along with the other equipment in the system

Mounting Flat Panel Displays

The transition to lightweight flat panel televisions and monitors has opened up a number of mounting options which were not feasible with heavy CRT (Cathode Ray Tube) TVs. These thin profile displays provide a clean, contemporary appearance when mounted close to the wall or in a shallow "shadow box" and work very well in motorized lifts and drop-downs (discussed in Chapter 15).

As with other topics in this chapter, the decisions regarding mounting location, sightlines, and type of mount should be made by the designer or sales person well before the actual installation takes place. We will focus on an overview of mounting options and the issues directly related to installation in the field.

Mounting Options

There are several mounting methods you should be familiar with:

- Factory-provided pedestal – most displays are shipped with a pedestal which can easily be attached and provides solid support for the display to rest on a table, desk, or shelf. If this piece is not used, make sure it is stored in a safe place in case the display eventually is uninstalled and taken to a different room or a new home.
- Wall mount – these are widely used for displays of all sizes and several types are available:
- Flat mounts hold the display flat on the wall, no variation
- Tilt mounts offer a few degrees of vertical adjustment, up or down
- Swivel mounts provide some form of movement left or right and sometimes include a small amount of tilt
- Fully articulating mounts have multiple pivot points which allow the display to be pulled away from the wall, and angled in a variety of ways.
- Ceiling mounts may have a proprietary bracket or utilize a pole or pipe which drops down from the ceiling (more common in commercial applications)

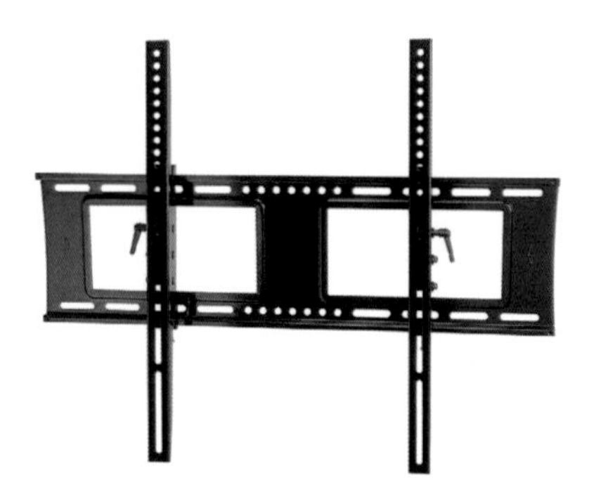

Figure 8-9 Flat Mount

Figure 8-10 Tilt Mount

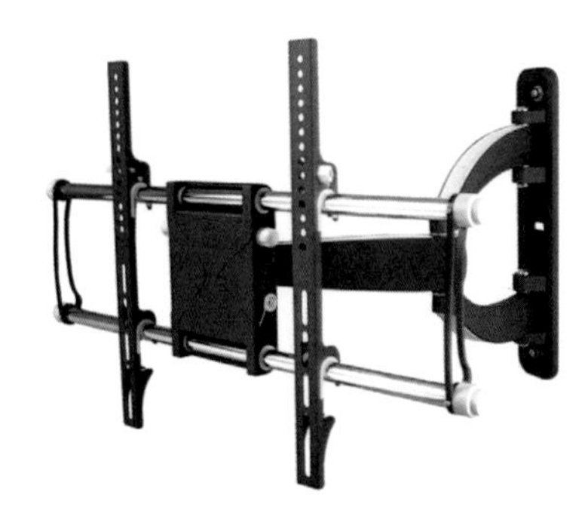

Figure 8-11 Articulating Mount

Standards

Many flat panel displays have pre-configured, threaded holes which conform to an international standard. This is known as the VESA FDMI (Video Electronics Standards Association Flat Display Mounting Interface.) The standard defines mounting interfaces, hole patterns, and associated cable/power supply locations for LCD monitors, plasma displays and other flat panel devices. If a display does not conform to the standard, a universal mounting bracket must be used which allows fastener locations to be changed as needed.

Installing Projectors

The same guidelines apply to projectors as to flat panels. The location of the signal cabling and power must be exactly right and the mounting system must be solid and strong. Cables need to be dressed and secured for a neat and professional appearance. If the projector is installed using any type of moveable or motorized mount, the cabling must be dressed in such a way as to accommodate this movement. In this case all conductors should be stranded, rather than solid, to avoid flex life issues due to repeated bending of the cable.

In some cases the projector mount will allow for fine adjustment of pitch, yaw, and roll. A good mount will allow them to be easily adjusted to fine tune the location of the image on the screen.

Figure 8-12
Universal Projector Mount

- **Pitch** – vertical tilt up and down
- **Yaw** – movement left and right on a horizontal plane
- **Roll** – tilting the image to the left or right with the center unchanged

For details on motorized projector mounting options, see Chapter 15.

Remember that working above floor level requires attention to your safety and the safety of others. Most projector mounting systems are configured so that half can be attached to the ceiling and the other half to the projector. Then when the projector is lifted into position the two can be paired easily. Lifting a projector into position should be done with two people and with great care.

THE WORK AREA

When working in a finished home it is imperative that technicians do everything possible to protect all surfaces such as walls and floors from damage. Drop cloths should be thick and padded. Ladders should have non-scratching feet. Locations should be verified with certainty before any holes are drilled. When the work is done the entire area must be completely cleared of all debris and tools and left exactly as it was found.

The work area should be checked after the work is done to make sure that floor, cabinets, furniture, walls and ceiling are cleaned. All wires should be concealed and no wire debris should be left behind.

CONCLUSION

It is the responsibility of the designer to specify the equipment, cabling, and location of all installed equipment. But the EST is the one who must ensure that the installation is accurate and professional looking – and done in a way which is safe for the technicians and the end users. Even if a system performs perfectly, any shortcomings in the way the installation looks can reflect negatively on the company. Knowing and following best practices will help make every installation safe, attractive and functional.

Questions

1. A professional equipment rack has what advantage over a typical cabinet?
   a. Available accessories
   b. Ventilation management
   c. Accessibility
   d. All of the above

2. The recommended method of securing interconnects in an equipment rack is:
   a. Velcro™
   b. Nylon wire ties
   c. Electrical or gaffers tape
   d. None, leave unsecured for ease of access

3. According to UL 1678, when a piece of equipment is mounted so that it is moveable (dynamic load), the entire mounting system must be able to support;
   a. The weight of the device
   b. 2x the weight of the device
   c. 4x the weight of the device
   d. 8x the weight of the device

4. Most failures of electronic equipment are due to:
   a. Faulty manufacturing
   b. Excessive heat over time
   c. Design flaws
   d. User error

5. On a fully adjustable projector mount, the "yaw" adjustment refers to:
   a. Vertical tilt up and down
   b. Left and right tilt with the same center point
   c. Left and right movement on a horizontal plane
   d. Movement toward, and away from, the screen

Chapter 9
FUNDAMENTALS OF SOUND AND AUDIO

INTRODUCTION TO SOUND

What is Sound?

In this chapter we will cover all of the fundamentals of sound; what its, and how it is recorded and reproduced. This foundation of knowledge will allow you to better understand later discussions of distributed audio and home theater.

In order to understand how to properly design and install an audio system with today's technology, we must first understand sound itself. How is it created? How is it measured? What are the attributes ascribed to it? We will begin by exploring what sound is, then defining several terms associated with sound (and audio).

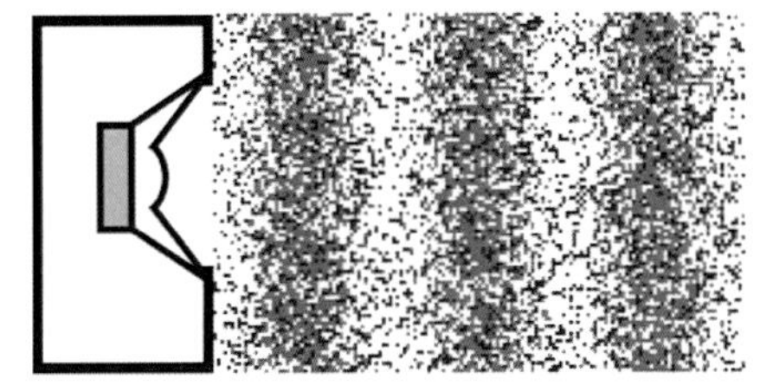

Sound Wave Basics

This is an oversimplification, but it shows what happens when a simple sound wave travels through the air. There are alternating zones of higher and lower pressure, just like the waves in a pool of water when an object is dropped into it. When these variations in pressure reach our eardrum the sympathetic vibration is converted to neurological signals which travel to the brain to be interpreted. In nature the source of a sound is obviously not a loudspeaker, and the waves are not this simple. In fact the wave shown is actually a sine wave, which is only produced by a tuning fork or an audio generator. Music, speech, and cinema sound are all much more complex than this.

The same theory holds true for all types of sound. When an object vibrates, it causes alternating positive and negative pressure zones in the air (or other material). These waves travel to another location and can be heard as sound.

Frequency

Sound propagates at about 1,130 ft. /sec (344 m/sec) at sea level, regardless of its pitch. The higher the pitch, the shorter the wavelength, so more cycles pass a point in space per second. A full cycle starts at zero, goes to the positive (higher pressure), back past zero, into the negative (lower pressure) and back to zero (pressure equal to the air at rest). The frequency is described in terms of cycles per second (cps) or as Hertz (Hz), in honor of the German physicist who studied this relationship of frequency and wavelength.

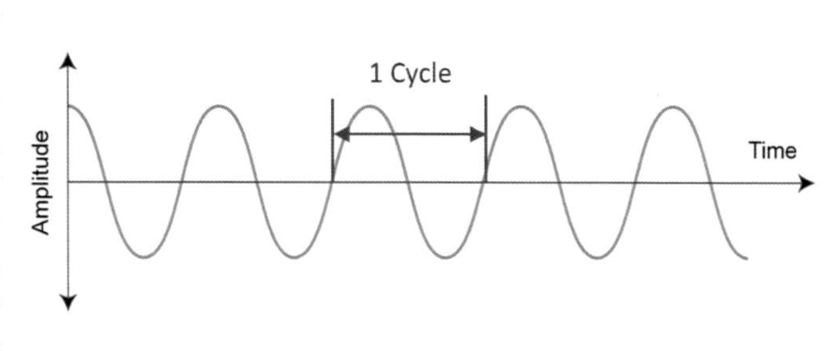

Wavelength

Frequency and wavelength always work hand in hand. As frequency increases, wavelength decreases. It is important to be able to calculate one or the other when predicting how a room or acoustical treatment will behave. Wavelength in the audio range (20Hz – 20kHz) is expressed in feet or inches, or in metric measurements of length (meters, centimeters). In formulas, wavelength is often indicated by the Greek letter Lambda (λ).

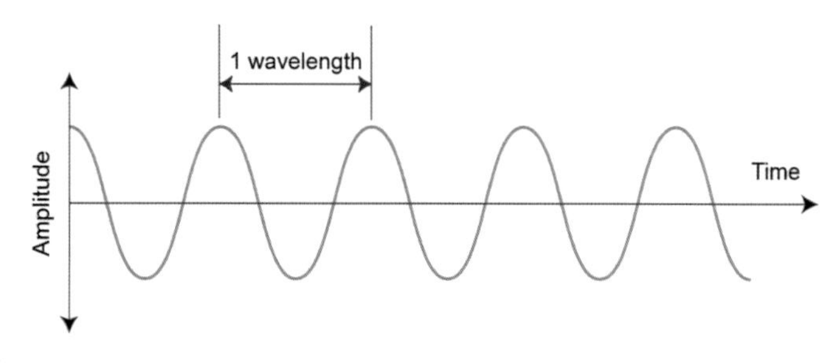

Octave

We are all familiar with a typical major musical scale in Western music; Do, Re, Mi, Fa, So, La, Ti, Do. This would represent a scale on the piano from, say, C2 to C3. Both notes called "Do" are C notes, but they are one octave (8 notes of the scale) apart.

In audio and acoustics, an octave is also the interval between two frequencies that are half or double the frequency of the other. For instance 2,000Hz is one octave up from 1,000Hz. This is important to know because this also means the wavelength of 2KHz is half that of 1KHz.

Frequency, Wavelength, and the Speed of Sound

We now know the mathematical relationship between Frequency (F), Wavelength (λ) and The Speed of Sound (V). If you know the wavelength, you can find the frequency, and vice versa. This assumes the speed of sound to be a constant.

Calculations can be done in Imperial/US measurements or Metric. Here is a summary:

- Frequency (F) is the number of cycles per second (Hz)
- Wavelength (λ) is the length of one complete wave (in feet, inches, cm, etc.)
- Velocity (V) is the directional speed of sound (use 1,130 ft/sec or 344 m/sec)

The calculations can easily be made:

$$\lambda \times F = V$$

or

$$V / F = \lambda$$

or

$$V / \lambda = F$$

Examples:

We want to find the wavelength for a sound at 1,000 Hz.
Divide 1,000 into the Speed of Sound (1,130 ft/sec) and get 1.13 ft for the wavelength.

We want to find what frequency has a wavelength of 2".
We first need to convert 2" into feet because the constant Speed of Sound is in feet/second.
0.167 feet divided into 1,130 = 6,766 Hz

Amplitude

Amplitude can be seen here as the height of the wave or how far it varies from the zero line (both negative and positive). This is the same whether the sound is in the air or in the form of an analog electronic signal. An amplifier simply takes a wave with little amplitude and makes it bigger, or amplifies it.

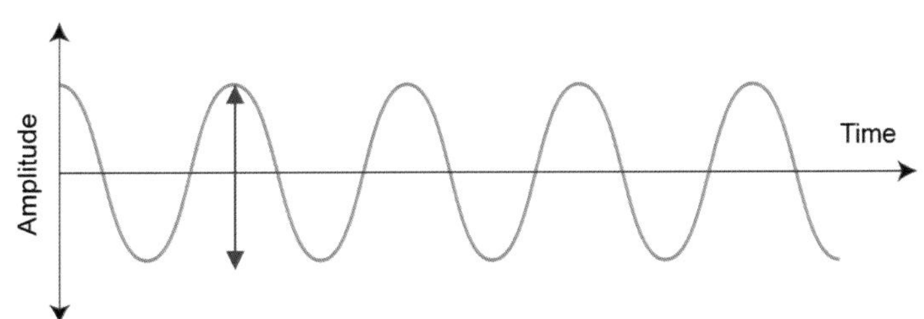

Amplitude is the absolute value (combination of both positive and negative variations from zero) of the wave; the total variation in pressure. When measured in the air, this called Sound Pressure Level (SPL), and is expressed in decibels. Bels and decibels will be discussed in more detail later in the course.

Loudness

At different sound pressure levels the human ear exhibits a different sensitivity curve. In general, as overall levels drop humans become less and less able to hear very low frequencies and some very high frequencies. This means these frequencies must be at a much louder level in order to be perceived as equal volume to the mid-range sounds. On early stereo receivers, it was not unusual to see a button called "Loudness", which typically boosted low and high frequencies for a more smooth sound when listening to music at lower levels.

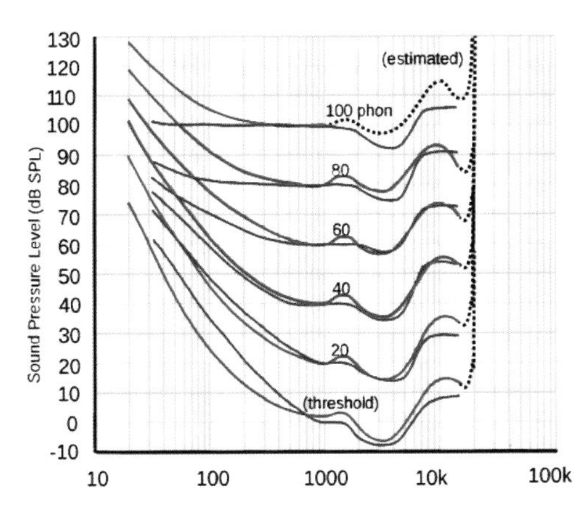

The graph shows the Fletcher-Munson Curve, which was developed it 1933. Each line indicates the relative SPL needed for all frequencies to "sound" like they are equally loud. As overall volume goes down the variation between mid-range and other frequencies becomes more pronounced. This explains why it is difficult to hear highs and lows at lower listening levels. It also explains why so much power is needed to make subwoofers seem evenly matched with the full range speakers in a system.

Measuring Sound Pressure Level

An SPL meter is a device that all technicians should own and understand. It is used for theater audio calibration, measuring the coverage of distributed audio systems, and evaluating the sensitivity of loudspeakers.

Digital and analog SPL meters work in the same way, but display the level differently. An analog meter has a needle which responds to sound picked up by the microphone. A digital meter shows the SPL level numerically.

Figure 9-1 SPL Meters

Both have settings that allow compensation for loudness curves and control how fast the measurements change. A slower response allows easier reading of the level as averaged over a few seconds, rather than jumping around rapidly.

SPL Meter Weighting

When measuring low sound levels, such as background noise, A-weighting is generally the best choice. It has built in compensation that accounts for the way our ears work at lower levels. For louder measurements, such as theater calibration at 85db, C-weighting should be used. Its response is flatter, like our ears are at higher levels.

Types of Sounds

These artificially generated waveforms are not found in nature but can be combined and modified by an electronic instrument known as an analog synthesizer to create sounds that imitate those of real musical instruments.

The top waveform is a sine wave, and it has the most pure sound. We often use this waveform for system and component testing. The others have varying amounts of "harshness".

Sine
Square
Triangle
Sawtooth

It should be noted that when an amplifier or input stage is overdriven, it may "clip" the signal. This means it chops off the top and bottom of the wave, making it appear more like the second wave shown – a square wave. Any time a waveform is modified from its original form this is known as distortion. Some distortion is very minor, but in this case it is easy to hear, and unpleasant……and often results in the overheating of a loudspeaker, since a loudspeaker cools itself when in motion and builds up heat when at rest. With the square wave, the speaker is at rest more than in motion, leading to increased heat and possible failure. This explains why speakers are more often damaged when there is inadequate amplifier power than by too much power.

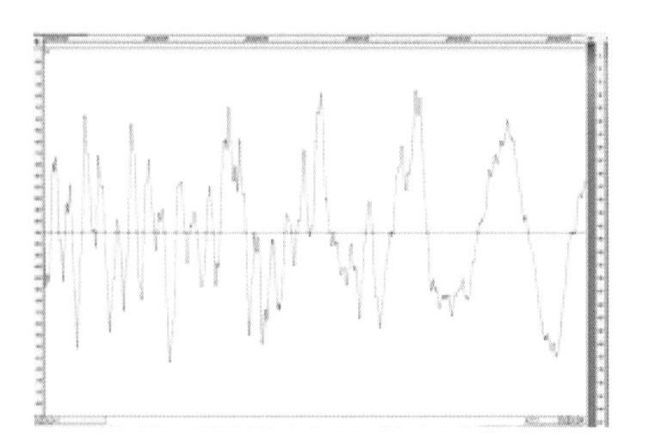

A sound source like a voice or instrument, creates a complex mix of different frequencies.This is why the same note played by both a violin and a saxophone sound very different. Each is playing the same note (or fundamental frequency), however, its voice, or "timbre", is also made up of a mixture of other sounds known as partials, harmonics, and overtones. These smaller components of the overall sound give the instrument its unique sound. When multiple sounds or instruments or voices are combined the waveform gets even more complex, as shown here. Although the wave is more complex, the same rules apply; shorter wavelengths represent higher frequencies, and amplitude is seen as the overall excursion above and below the zero line.

Noise

There are a couple of different definitions of noise. The first is "any unwanted sound". Examples of undesirable noise include:

- Air handler noise
- Traffic
- Tape hiss or line noise
- EMI (Electro-magnetic Interference) caused by proximity to electrical wiring
- RFI (Radio Frequency Interference) like a CB radio heard in the system
- Plumbing (rattling or groaning pipes, drain noise, etc.)
- Footfalls from room above

However, certain kinds of randomized noise are intentionally created for testing purposes. This noise has no particular note or pitch but sounds like a "shhhhh" sound. We can get noise test signals from signal generators, CD's, or DVDs.

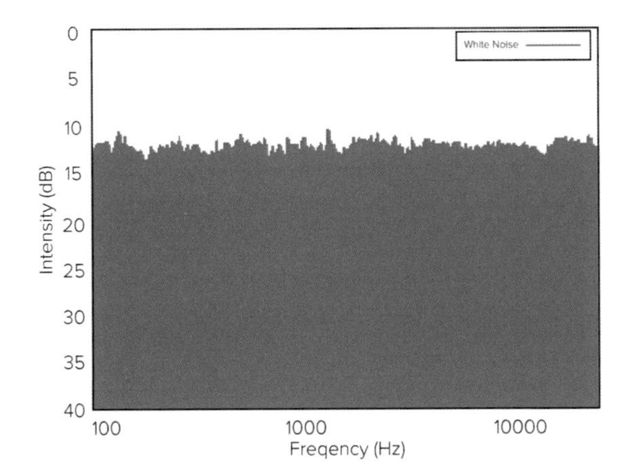

Figure 9-2 White Noise

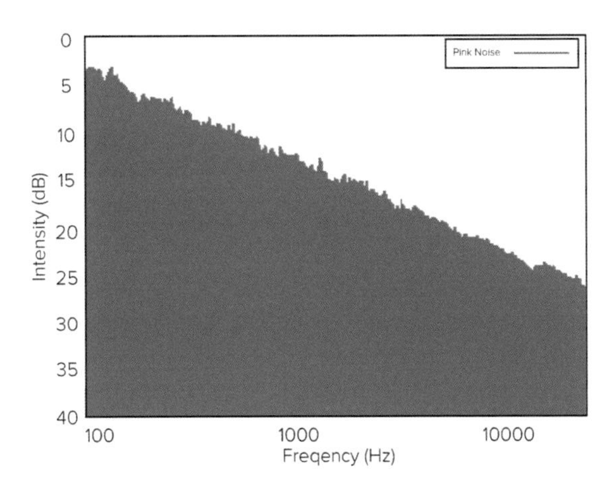

Figure 9-3 Pink Noise

White Noise

White noise has equal energy at all frequencies. If we were looking at this display in real time it would be constantly fluctuating but maintaining the same general appearance. You can see approximately equal levels at all frequencies in the audible range, 20Hz – 20KHz. This noise sounds bright and thin to our ears because we hear relative levels between octaves.

Pink Noise

Pink noise, however, is measured and shown according to its energy at in each octave. This noise sounds less bright and "hissy" to our ears. In fact, noise generate according this response curve will actually sound like flat response to our ears – equal energy in all octaves. Most audio calibrations are done with pink noise and devices which read this curve as flat response. So pink noise, measured on a Real Time Analyzer, will look more like a flat line.

Measuring Noise

The two measuring devices used to analyze broadband sound or noise are the spectrum analyzer and the real time analyzer. Spectrum analyzers are useful for a wide variety of applications outside of the audio spectrum because they are equally sensitive at all frequencies. But for measuring audio, the RTA is generally used because it is calibrated to see pink noise (equal energy in each octave) which is the way our ears work.

More about the RTA

Real Time Analyzers may display the frequency response in increments of octaves, 1/3 octaves or even 1/12 octaves. They also have settings to control the range and response time of the display. Many RTAs have built-in pink noise generators and come with calibrated microphones.

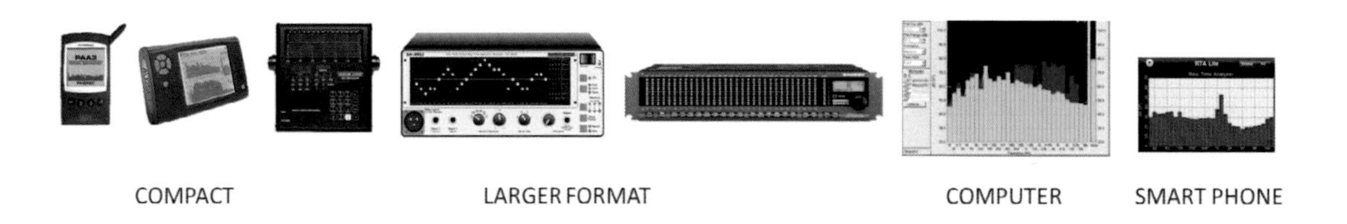

Figure 9-4 Real Time Analyzers (RTAs)

RTA Types

Real Time Analyzers are available in many sizes, shapes, price ranges, and degrees of accuracy. If you choose to use a software based system, be sure you have a high quality calibration microphone and external microphone preamp. Your measurements are only as accurate as the weakest link in the system. Compact and smart phone RTAs may not be accurate enough for critical calibration but can measure relative levels and general evaluation in the field.

Sound Perception

Next, let's talk about how we hear. We use the range of 20 – 20,000 Hz as the audible sound bandwidth for humans with normal hearing. In reality, most people cannot hear frequencies at the two extremes of this range. Older folks typically have measurable hearing loss above 16 KHz.
Referring back to the loudness curves discussed earlier, we know that our ears are more sensitive to mid-range frequencies, especially when overall amplitude is reduced.

In the grand scheme of nature, our hearing range is a relatively narrow band of frequencies. Some very large animals can hear frequencies below 20 Hz (Infrasound), and other animals, especially smaller ones, can hear well above our upper limit (Ultrasound). Whales can hear extremely low frequencies which are transmitted through the water. Remember these low frequency sounds have very long wavelengths. Higher frequencies have extremely short wavelengths, a fraction of an inch. These frequencies can be heard by very small animals and insects, mostly because their sensing organs are very small and respond to short wavelengths.

Most sound reaches the ear by travelling through the air. When the sound waves enter the ear they travel through the ear canal and strike the eardrum, which is a tympanic membrane – like a real drum. The eardrum vibrates and passes the energy through a complex mechanical system into the fluid in the inner ear where there are millions of tiny fibers or hairs, each sensitive to specific frequencies. When these fibers are vibrated they create electrical signals, which go to the brain, where they are interpreted. We could say that the ear is just a microphone, and that the real processing takes place in the brain, the computer of the human body.

It is important to protect your ears from high sound levels or loud sound over a long period of time. Once fibers are damaged they do not recover, so hearing at specific frequencies can be lost forever.

Decibels

A decibel is 1/10 of a Bel. The Bel was developed by Bell Labs in the early days of telephone (named after Alexander Graham Bell). Since it is such a large measurement, we find the decibel to be a more practical increment for most applications.

When working with loudspeakers, it is important to understand the relationship between power and perceived volume. In order to achieve a 3dB increase in output (3dB is considered to be the minimum easily perceptible increase humans can hear), you must DOUBLE the amount of power going to the loudspeaker!

Also note; in an anechoic space, SPL drops by 6dB when the distance from the sound source to the ear is doubled. In a room, where reflections play an important role, that dropoff over distance can be much smaller.

AUDIO RECORDING

In this section we will talk about early audio technologies and the recording / playback process today.

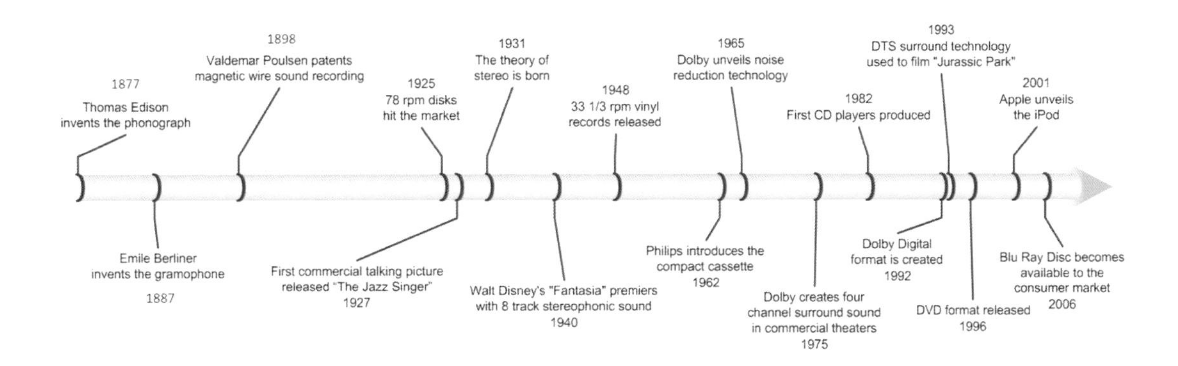

History of Audio

Here is a simplified timeline of the history of audio reproduction. Thomas Edison's cylinder phonograph is shown as the starting point. In 1898 a system was invented which allowed sound to be converted to an electrical signal and recorded magnetically on a piece of wire. Motion pictures became "talkies" in 1927 when amplified sound enabled the recorded voice to be heard throughout a commercial theater.

From there the technology advanced using both vinyl records and magnetic tape in a variety of formats. In 1975 Dolby created the first surround sound system for theaters. In the 80's the advent of digital recording brought us the Compact Disc and in the 90's this technology allowed for more sophisticated multi-channel formats and brought surround sound to the home theater.

Helpful Hint: Optical media such as CDs and Blu-rays are called "discs". Magnetic media such as hard drives are called "disks".

Phonograph

History tells us that the very first recording and playback of the human voice was Thomas Edison saying "Mary had a little lamb, whose fleece was white as snow, and everywhere that Mary went, the lamb was sure to go." However, it is speculated by some that his chief assistant and head machinist (tasked with building the first prototype) likely tested the machine before presenting it to Edison himself. Therefore, it would be one of their voices (not Edison's) that was actually the first to be recorded!

The first recordings were made on a cylinder of foil. Each cylinder was a unique recording. Later, wax cylinders were used, which allowed multiple cylinders to be made from one plaster "master" mold. These recordings could be played over 100 times. It was also possible to smooth the surface of the cylinder and re-record it.

The frequency range of these devices was very limited, but since most speech is in the mid-range octaves, they worked well for the spoken word.

Early phonograph records were played at 78 rpm (revolutions per minute). Later the 45 rpm "single" (one song per side) became popular, and finally the long playing 33 1/3 rpm "LP" was the standard, before finally being replaced by digital recording distributed as the Compact Disc (CD) beginning in 1982.

Tape Recording

Magnetic recording started with the wire recorder, and evolved into several formats of magnetic tape. First the open reel deck, then the popular Compact Cassette and the 8-track tape cartridge. Multi-track analog tape recorders are still sometimes used in studio recording.

The Compact Cassette used 1/8" tape in a small case with two tiny reels that the tape travelled between. The tape head recorded and played back 4 tracks on the tape. In one direction the head played tracks 1 and 2, and in the other direction, 3 and 4. This delivered stereo sound, Left and Right. This configuration also allowed stereo recordings to be played on monaural machines, which summed the adjacent tracks to mono. Some machines reversed direction when the tape got to the end; others required ejecting the cartridge and flipping it over.

The 8-track tape was used widely in automobiles, especially in the US. It used ¼" tape and faster speed, for better fidelity. The tape was configured as a continuous loop in the cartridge. There was a moving playback head which played two tracks until the "splice point", then switched to two different tracks, then two more and two more – a total of 8 tracks of music or 4 segments of stereo programming.

The 8-track tape disappeared in the late 80's. Cassettes are still in limited use today, although the term "cassette tape" was dropped from the Oxford Dictionary in 2011. Digital streaming, MP3 recordings, etc. have replaced cassettes for everyday music listening.

Before Electronics

Before there was any form of audio recording, it was known that horn shaped devices, even cupped hands, could be used to collect sound and focus it into the ears to make it appear louder.

This concept was used to record the first phonograph cylinders and records, before electricity or electronics were in use. These crude recordings were then played back by directing the vibrations of the needle which tracked the groove through another horn, this time to control the dispersion of the sound and aim it at the listener. You see horn type components on all of the early phonographs and even Victrolas.

Early Days of Amplified Sound

When the development of early electronics and amplifiers allowed live and recorded sound to be reproduced by loudspeakers, the power was so limited that horn loading continued to see use in order to make the signal louder.

Even with today's high powered amplifiers, many speaker designs for a variety of applications still utilize horn devices to couple the driver efficiently to the air.

The introduction of electronic sound reproduction made a lot of things possible; playing back music through loudspeakers, speaking and performing live for hundreds, even thousands of people at once, and more extensive sharing of musical traditions around the globe.

Capturing Sound Electronically

So how do we start with real sound waves and end up with something that can be amplified and played back over loudspeakers? Before any sound can be manipulated, stored, or amplified – it must first be captured and converted from acoustic energy to an electrical signal which is the same, or "analogous" to the acoustical energy. This is done with a microphone. There are many different types of microphones in use. Shown in Figure 5 is the most widely used technology, the dynamic microphone. The dynamic "mic" is a passive device which is nearly the same mechanically as a dynamic loudspeaker. A diaphragm moves a coil near a magnet, which produces an electrical current which is analogous to the sound wave.

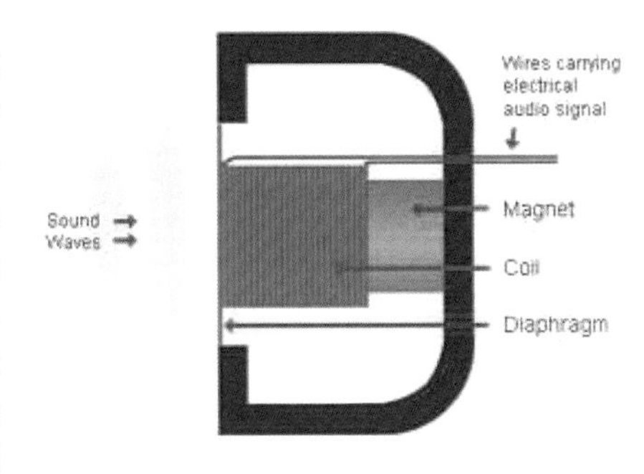

Figure 9-5 Dynamic Microphone

The Recording/Reproduction Path

Today's recording/reproduction path is more diverse and flexible than in the days of the early phonograph. Here is an example:

- Sources may include live instruments recorded using microphones, and direct sources such as electronic keyboards and sampling technology.
- Multiple sources are then recorded digitally on multiple discrete audio tracks in a Digital Audio Workstation which is computer based.
- The mixing and processing of these tracks is done in the digital domain.
- The final product, in either stereo or surround, is then distributed digitally; streamed, downloaded as a file, or on physical medium like CD or Blu-ray
- A receiver or other component converts the digital content back to analog so that an amplifier can provide an analog speaker signal strong enough to drive loudspeakers, which in turn convert electrical energy to acoustical energy, in the form of audible sound

AUDIO EQUIPMENT FUNDAMENTALS

We will next look at the typical audio equipment you will deal with in the residential systems industry, its operation, and typical signal interconnection.

Sources

We will start with sources. Some are audio only, others are audio and video. All provide either analog or digital audio content.

Disc Players

An optical disc stores digital audio in a binary format that can be read by a laser. The shorter the wavelength of the laser beam, the smaller the "pits" it can track and read. The Blu-ray disc utilizes a blue laser, which allows for over 25 times the data compared to a CD.

CD players are usually very simple, having L/R analog outputs and a digital output. Remember, Coaxial and Optical outputs carry identical signals, one over copper cable and one over optical fiber. This analog output is unbalanced, utilizing RCA connectors.

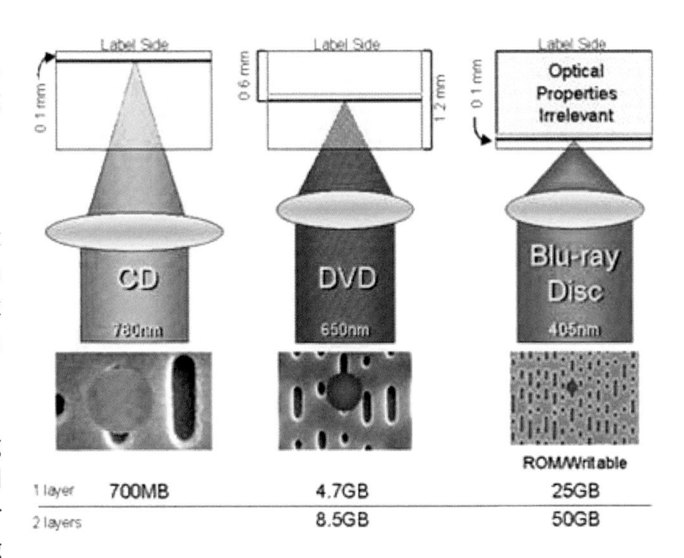

Figure 9-6 Optical Disc Characteristics

Some CD players may have IR remote controls. CD players may be single or multiple disc, or even a changer holding as many as 300 discs. Note that high-end CD players may also have RS-232 or other control options. SACD players provide multi-channel audio via discrete analog outputs or HDMI.

DVD and Blu-ray players deliver motion pictures and other video content with surround sound capabilities. Audio outputs may include Stereo Analog outputs, Digital Coax or Optical, and HDMI. HDMI is the only output capable of carrying full uncompressed surround sound.

Legacy Sources

In addition to the equipment listed previously, many customers will request that some of their "legacy" devices be integrated for use with their new system. Turntables are gaining in popularity among audiophiles. Many clients may still have video tapes or Laserdiscs they want to play. Most will want the ability to use game consoles, not just for games but for streaming media. It is also important to have auxiliary inputs that are available for external devices such as MP3 players, mobile devices, etc.

Remember, phono cartridges have a very low signal, and require a phono preamp (sometimes built into receivers and pre-pros). Most other devices have line level (RCA) analog audio and are easy to integrate. However, controlling these components may be a challenge, and some create noise and heat which may present a problem.

Digital Servers & Portable Sources

A digital music source can be any device that contains and organizes music digitally onto a hard drive or streams it from a remote location. Many manufacturers produce chassis style units that may be placed with other traditional audio equipment and hooked up as a permanent source component. These types of servers typically contain hard drives with Terabytes of storage space, and allow access and browsing through a Graphical User Interface (GUI) displayed on the TV screen. The same type of capabilities can be achieved using a personal computer and software interface.

Cable and satellite set-top boxes can be a good source for commercial free music, as can a variety of portable devices including MP3 players, phones, and tablets. Mobile devices are often linked to installed systems via WiFi or Bluetooth, allowing the individual to bring their own music collection with them from room to room or elsewhere. Remember, streaming audio or video requires a dependable and robust network.

Receivers

Receivers are the "Swiss-army knife" of A/V gear. The surround receiver is the heart of a multichannel audio/video system.

In the early days of stereo, the receiver was simply a source selector, AM/FM tuner, tone control, and stereo amplifier. Today's A/V receiver does much more, taking advantage of digital technology and microprocessors. Receivers are typically able to switch input source audio (analog & digital and audio from HDMI). They can also switch video sources associated (or not) with an audio source (HDMI, component or composite).

Surround receivers include multiple amplifier channels, and often have "Zone 2" capability (to drive an additional room with similar or different content).

Most high-end receivers can be remote controlled via IR, RS-232 or IP and have network access to internet radio, streaming video, etc. This array of inputs and outputs may look confusing but once you understand what everything is, it is pretty straightforward.

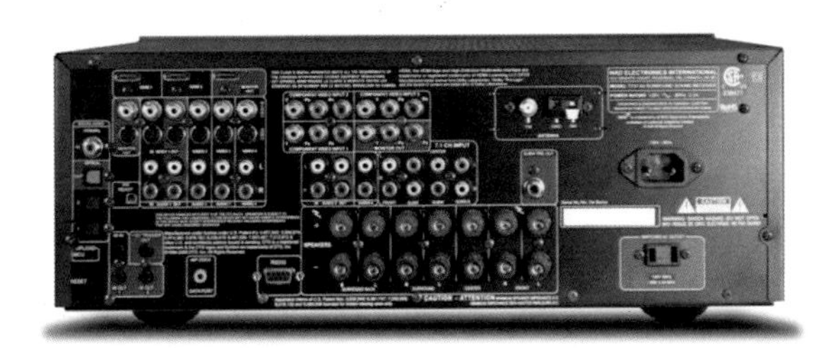

A/V Receiver

Preamplifier/Processors

A preamp/processor (or "pre-pro) may contain all the functions of a receiver except the output amplifiers, and radio tuners. Recently, many are being designed for switching and processing video as well as audio, but some may still be strictly geared towards audio. They will sometimes utilize balanced audio (via XLR) inputs/output. Balanced interconnects should always be used when possible.

Power Amplifiers

Stand-alone power amplifiers will usually contain from 2 to 8 separate amplifier channels, each dedicated to a specific loudspeaker. They accept the input from a preamplifier or sound processor such as an equalizer and increase the amplitude of that signal to a level adequate to drive a loudspeaker and create sound. This requires a great deal of power, which is drawn from the wall socket and managed by a power supply. A high quality amplifier will increase signal amplitude without altering the signal in any other way. Each amplifier channel may have a gain adjustment to compensate for differences in audio input levels.

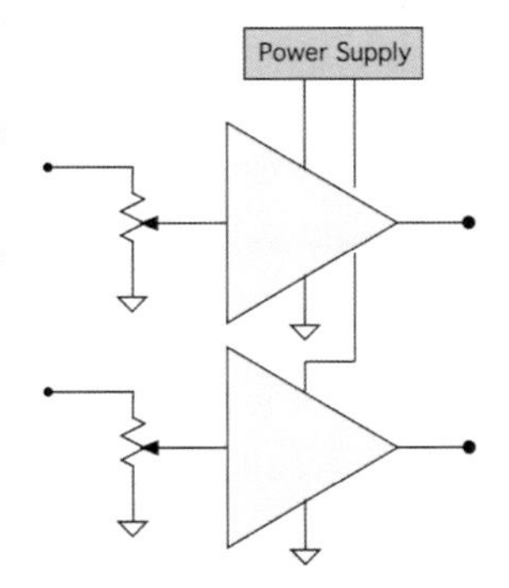

Basic 2-Channel Amp

Amplifier Specifications

It is important to understand amplifier specifications. Amplifier specs can be very misleading especially for low-end equipment that tends to "stretch the truth" of provided specs without any qualification.

The main amplifier specs to understand are:

- **Maximum Power** (RMS and peak): Per channel. You must also know how the amplifier channel was tested – with one or two channels driven or all channels driven. Driving multiple channels tests the limitations of the power supply. RMS (Root Mean Square) power is about .7 of peak power and about .35 of Peak to Peak power.
- **Load Impedance:** When load impedance goes down, current and therefore wattage goes up. An amplifier not rated at 4 ohms should not be driven into a 4 ohm load.
- **Total Harmonic Distortion (THD)**: This is essentially a measurement of the difference between the input signal and output signal. The difference should be extremely small. The lower this number is, the more accurate the sound reproduction. 0% THD would be an exact reproduction of the original. In other words the amplifier increased the amplitude of the waveform, without changing anything else.
- **Bandwidth**: This is the frequency range reproduced, from lowest to highest, at approximately flat response (within a given range, usually +/- 1dB). A rating of 20Hz to 20KHz is ideal, and the smaller the variation (+/-) is, the flatter the response is – in other words the response is more even from one frequency to another.
- **Dynamic range**: not always stated, but higher quality devices will handle a wider dynamic range (from softest to loudest) more accurately.

**140 watts RMS/channel X 7 channels,
all channels driven simultaneously into 8Ω,
less than 0.09% THD from 20 Hz to 20 kHz,
+/- 1dB**

Amplifier Specifications Example

The above is an example of good amplifier specifications. All the necessary information is provided.

- 140 watts of real RMS power (not peak) per channel
- All channels driven continuously—140 watt rating was tested with all channels driven to maximum power at same time, putting the power supply to a real test.
- Into 8 ohms—an 8Ω resistive load (not speaker) was used. If the amplifier also shows a rating at 4Ω, then it will work well at that load. It will show increased power output at the lower impedance.
- Less than 0.09% THD—the distortion that the amp is producing at maximum power level. This figure is very good. The lower the better.
- From 20 Hz to 20 kHz—maximum power bandwidth (for THD and RMS power level). Should state +/- dB variation. +/- 1 dB typically used in audio industry. A rating of +/- 3 dB means more variation in frequency response is tolerated (an overall variance of 6 dB possible)

Audio Equalizers

Equalizers are used to adjust the relative amplitude of "bands" of frequencies in the typical 20 Hz-20 kHz audio frequency spectrum. Remember, an octave is double or half a given frequency. For instance 400 Hz is one octave higher than 200 Hz, 100 Hz is one octave below 200 Hz.

Primarily used to adjust (or flatten) the frequency response at the listening position, to compensate for speaker characteristics, placement, and room acoustics, with the ideal goal being "smooth" response, with no extreme peaks or nulls. When a frequency range is augmented at the listening position, reducing the relative energy reproduced in that range will bring the audible response closer to smooth and accurate.

Audio equalizers come in two "flavors": graphic EQs and parametric EQs.

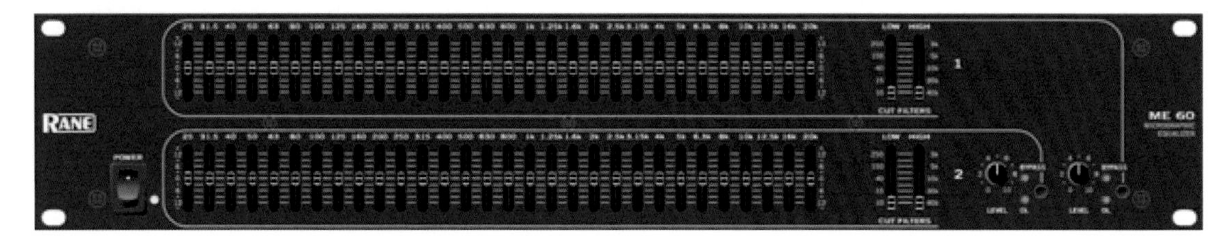

Figure 9-7 Two Channel Graphic Equalizer (EQ)

In a **graphic EQ**, the bands (and their width) are fixed and the interface is designed to allow easy adjustment of each band. Each band typically has a vertical slider which goes above or below the "0" indent, to either boost or cut the level in that frequency range. The range could be an octave or 1/3 of an octave. This is the most widely understood and used type of EQ, and provides a graphic indication of what is being done to the signal. Shown is a two-channel graphic EQ, with a row of faders for each channel.

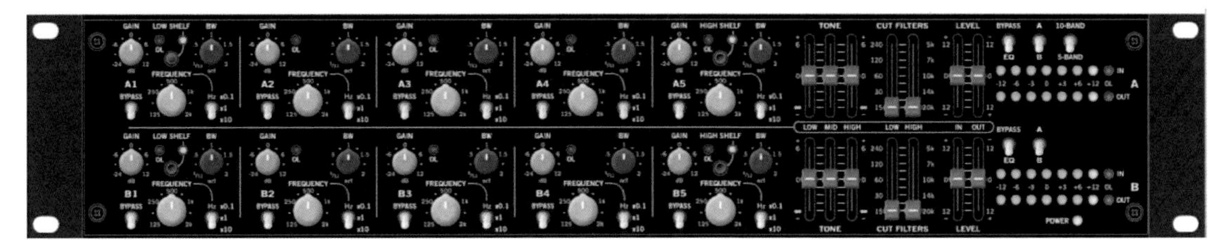

Figure 9-8 Two Channel Parametric Equalizer (EQ)

In a **parametric EQ**, for each "band" the center frequency, gain and bandwidth are all adjustable. The user can fine tune the frequency to be affected, determine how "wide" a range to affect (1/3rd octave = narrow. 1 octave = wide), then choose to either boost or cut to any desired degree, at that frequency. Thus, only a few bands are needed to solve most room problems, rather than 10 or 31. Shown is a two-channel parametric EQ, each channel with four fully adjustable bands. Many modern surround receivers have equalization built into their Digital Signal Processing (DSP), accessible via the internal menu. This means if you understand how graphic and parametric EQs actually work, you can take full advantage of this feature to calibrate the system without needing several channels of outboard equalization.

AUDIO SIGNALS AND CONNECTIVITY

Audio signals used in the home can be divided into two general categories:

- Analog signals
- Digital signals

We will cover the major types in each category.

Analog Audio Signals

An analog audio signal is complex AC waveform, with energy alternating between positive and negative, exactly the way a sound wave behaves in the air. This electrical signal is then carried over cable to its destination. This method is simple and dependable, however there are a couple of disadvantages:

- Cable has resistance and over long distances this results in line loss
- Any noise that gets into the signal becomes part of the waveform and cannot be removed
- Therefore when the signal is amplified, so is the noise

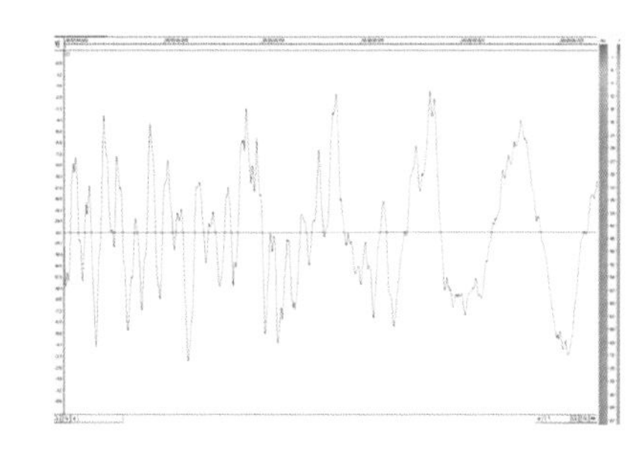

Figure 9-9 Complex Audio Waveform

Line Level Audio Signals

Low amplitude analog signals which travel between components, such as from a source to a receiver, are known as line level signals. A line level signal is not strong enough to drive a loudspeaker or even headphones.

Line Level Analog audio signals can be subdivided into:

- Unbalanced signals
- Balanced audio signals

Line-level unbalanced signals are typically used for consumer audio/video equipment. Balanced audio signals are typically used for professional audio/video equipment and very high end audiophile applications.

Analog Line-Level Audio: Unbalanced

Unbalanced line-level is the predominant form of audio signal interconnections between traditional A/V devices. It uses two-conductor RCA style phono connectors (or ¼" phone plug in pro audio and music industries). Two cables are required for stereo, and they are sometimes manufactured as a pair.

The cable is a coaxial style with a braid and and a stranded center conductor. The stranded center provides more flexibility for interconnects. RG59 or RG6 may be used to fabricate interconnects in the field or carry this type of signal in a permanent installation.

It is referred to as "unbalanced" because the outer shield does "double-duty" as a shield and ground connection, as well as being one of the audio signal conductors. Since current in the two conductors is unbalanced with respect to the electrical ground, it is referred to as unbalanced. Any noise that gets into the shield becomes part of the signal.

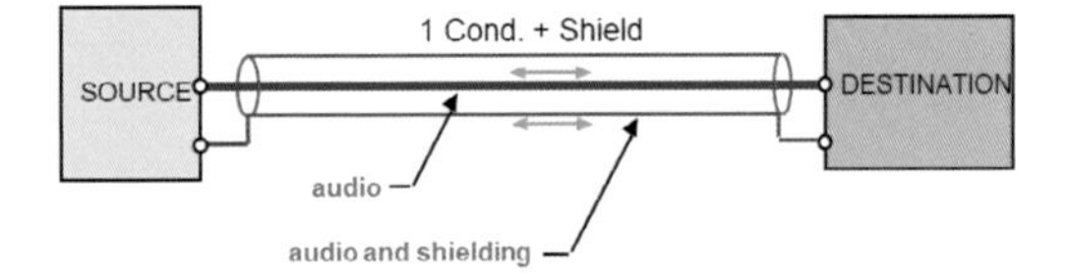

Line-level audio signals are typically in the range of 0.1 to 1.5 volts peak-to-peak (from the peak of the positive part of a cycle to the peak of the negative part), depending on the volume level. The volume level is directly expressed as the signal amplitude.

Line-level inputs are high-impedance, therefore very low-level signals, including induced noise on the cable, are easily amplified. This is just one of the reasons that unbalanced line-level connections are very susceptible to noise.

Analog Line-Level Audio: Balanced

A balanced audio signal uses 3 conductors; +, -, and ground.

The conductors carrying the signal are isolated from the ground and carry only the audio. The signals in each conductor are equal and opposite and both have the same potential to ground as measured in ohms. Therefore, it is referred to as a balanced circuit.

If external noise gets into the cable, it is usually caught by the shield and sent to ground. If it is strong enough to get through the braided shield, it will be equally present on both the + and - conductors. Since these two conductors are always 180° out of phase, that noise is cancelled out when the signal reaches its destination. This is known as Common Mode Rejection, or CMR.

Balanced audio signals almost always use the XLR style connector, which has 3 pins.

Audio inputs use the female jack (and male plug); audio outputs use the male jack (and female plug).

The pins of an XLR connector are numbered as seen in the diagram. The ground conductor is always connected to the cable shield and typically to the connector's metal shell, which is in contact with the chassis of both source and destination equipment.

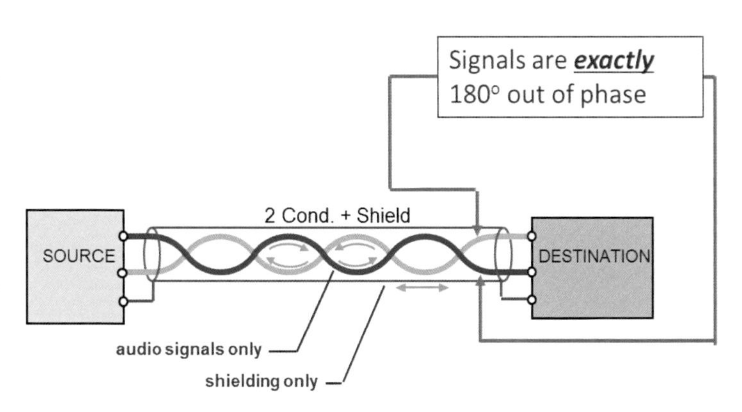

This connector is used for pro audio and microphone cables, as well as high end audiophile and theater connections. Some audio equipment may use a 3-conductor ¼" phone jack for balanced audio, instead of an XLR. Either connector will work, as will simple screw terminals.

It is also possible to send audio over a balanced line using UTP cable and special transformer devices known as baluns. Baluns are devices which balance an unbalanced signal at one end of the run, and then unbalance it at the other end. Thus, the name "bal"..."un". Baluns can also be used to send analog video signals over twisted pair cable.

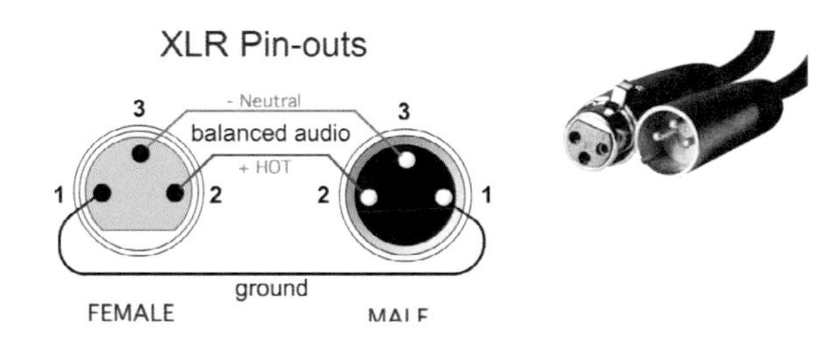

Speaker Level Audio

A speaker level audio signal has low-voltage and high current to drive the voice coil of a speaker. A signal that is delivering about 10 watts to an 8-ohm speaker will be from 0-12 volts peak and 1.5 amps peak. A signal delivering around 25 watts will be in the 0-32 volt range at 4 amps peak.

Therefore, the cable used to connect a speaker to an amplifier must be a large enough gauge to handle relatively high currents without significant voltage drop. The signal is so strong that external noise rarely is an issue.

Digital Audio Signals

Digital audio signals represent the original analog information as a series of numbers derived by sampling the analog source thousands of times per second. The numbers are transmitted as a series of on/off "bits" represented as the presence or absence of some predetermined voltage level.

This technology opened up a whole new world of audio storage, transmission, and processing. Unlike analog recordings, digital audio files can be copied over and over with no degradation of the signal. Since the numeric values are all that count, background noise is ignored so noise does not increase on each generation like it does with analog recordings.

It is important to understand the encoding/decoding process, sampling rates and sample size, as well as the various digital audio formats.

DIGITAL AUDIO TECHNOLOGIES

What is Digital Audio?

Sound waves in the air are analog. Analog recordings represent these sound waves in a similar fashion, with infinitely variable amplitude. This type of "analogous" signal was used for phonograph records and analog tape recordings. Today most audio is recorded and stored digitally. "Digital audio" is just another way of representing the analog signal, using numbers. Most digital audio is derived from an original analog source.

The variable voltage of the analog signal is fed into an analog-to-digital converter (A/D). The converter samples (measures) the instantaneous voltage of the analog signal, and converts that voltage (say 0-1 volt) to a binary number, usually in the range of 0-255 (an 8-bit number). Therefore each number represents 1/255th of a volt. The larger the number of numeric choices (bits), the better the sampling accuracy. The binary numbers are transmitted at a rate necessary to keep up with the sampling speed which is the number of "snapshots" taken per second.

So, two things determine the accuracy and quality of the digitized signal, and thus the ability to recreate the original waveform:

- The sampling rate (how often the waveform is sampled), and
- The sampling resolution (how many bits are used to represent the voltage).
- These two factors are combined to tell us something about the quality of the recording:
- CD audio is 16/44.1, which means 16 bit samples taken 44,100 times per second.
- Studio recordings are sometimes 24/96, or 24 bit samples, at 96,000 samples/second.

The total amount of throughput needed depends on these specifications, plus the number of channels. For instance, the coax and optical outputs of a CD player carry two channels of digital audio. Although CD audio is at a sampling rate of 44.1, the outputs are capable of a little more: 48k x 16bps x 2 channels = about 1.5 Mbps

Analog to Digital and Back

Since the original sound is analog, and the sound created by loudspeakers is analog, digital recording and playback includes two processes. The process of converting analog to digital is called encoding. This is done by an Analog to Digital (A/D) Converter. When the digital information is converted back to analog, that is called decoding, and it is done by a Digital to Analog (D/A) converter.

Encoding

An analog signal is converted and "encoded" into digital. The waveform is sampled, converted to numbers and it is now in the digital domain. When the source contains multiple audio channels (stereo or surround), there is an A/D converter for each channel. The output from each converter is then interleaved or multiplexed into one data stream, multiplying the bandwidth required by the number of channels, but allowing multiple channels to be carried over one signal path. This is a total departure from analog, where two combined signals are combined forever. With digital audio one cable can carry many from the same source.

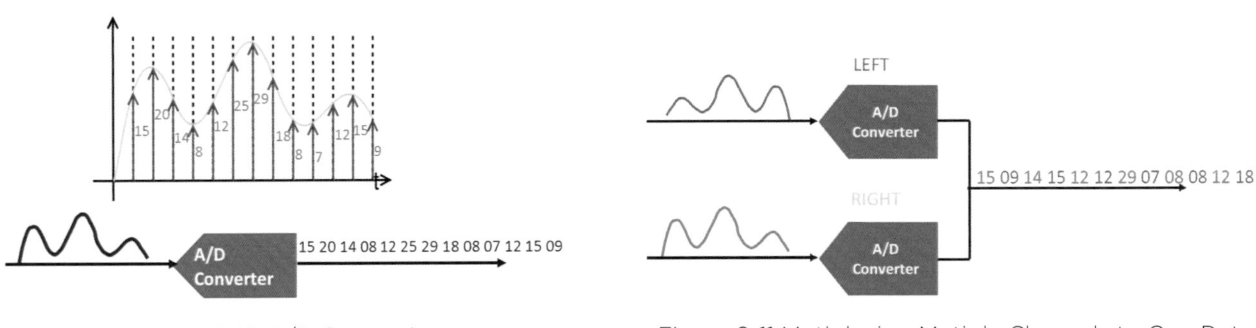

Figure 9-10 A/D Conversion

Figure 9-11 Mutiplexing Mutiple Channels to One Data Stream

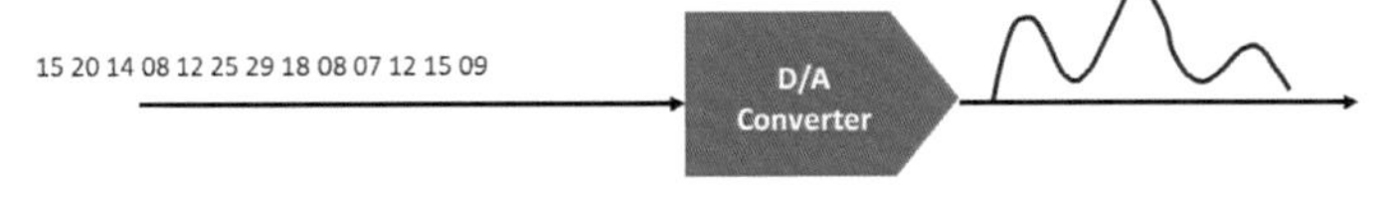

Figure 9-12 D/A Conversion

Decoding

To convert the digital signal back to an analog form, the inverse process is used: a digital-to-analog converter (D/A).

This device interprets the numeric values, and recreates the analog waveform. The output is then filtered to "smooth-out" the discrete voltages into a smooth analog signal, reproducing the original analog signal as closely as possible. You can see why resolution and sample rate are important to recreating an accurate analog waveform: fewer samples per second or fewer points of resolution requires more "guessing" on the part of the converter to fill in the unknown gaps between sample points. This is especially critical at high frequencies.

Note that the type and format of the A/D and D/A converters must be exactly the same on both ends!

Digital Audio Connections

Stereo PCM (Pulse Code Modulation) audio is uncompressed and is found on the "coaxial digital" RCA connector, or the optical TosLink output of a CD player, where the content is just stereo. These outputs, as we saw earlier, have the throughput necessary to carry two channels of uncompressed PCM audio. In the case of the coaxial, the signal consists of a 1.0 volt peak-to-peak signal representing digital bits. The polarity and position of the "bits" in the data stream determine whether it is a 0 or 1 bit. The data stream on the optical output is identical to coax in quality, but uses light bursts instead of electrical signals. Note: As long as the equipment is on, it outputs a data stream of alternating 0s and 1s, even if nothing is playing.

An HDMI connection may also be used to carry digital audio, and is always found on DVD and Blu-ray players as well as SAT and CATV set top boxes. HDMI has much more bandwidth. This allows full 7.1 digital surround sound information to be sent uncompressed, resulting in a high quality audio experience.

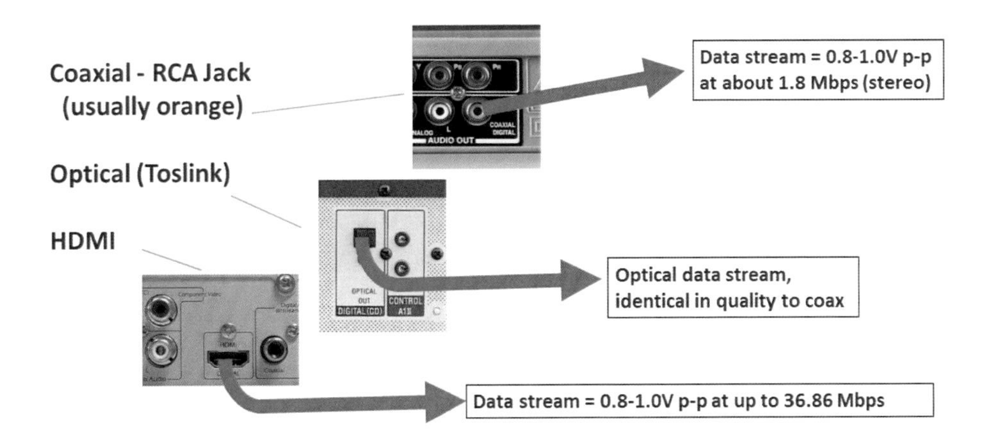

Audio Encoding: Compressed

However, to get more than two channels of audio over Coax or Toslink, the content must be compressed. Compressed formats require a Codec be implemented after the A/D converter. A codec is simply a computer program that manipulates the digital information. It is not a piece of hardware. Codec stand for "Code/Decode" or "Compress/ Decompress".

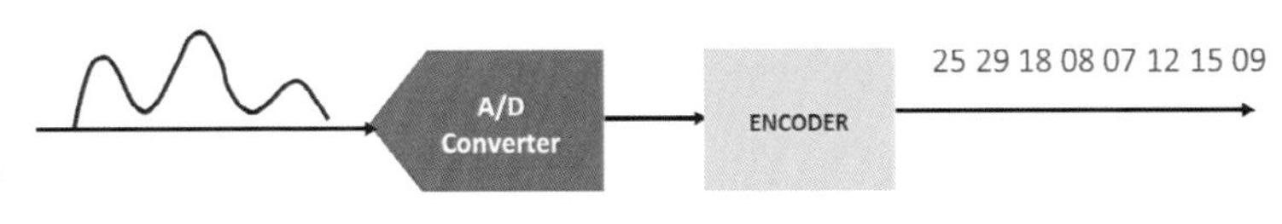

This stage can do a variety of things to the digital signal.

- Compress the data by removing redundant sampled values.
- Multiplex several channels into one data stream.
- Perform proprietary signal processing such as noise reduction and dynamic range compression.

Digital Audio Formats: Compressed

Most digital audio is compressed. The data stream is analyzed and redundant information is removed, lowering the bandwidth and allowing more channels to be transmitted in the same bandwidth. Quality ranges from good to poor, depending on the format. The problem is that there are dozens of compression algorithms, some non-proprietary, and some proprietary.

The most common non-proprietary compression algorithms are MPEG (Moving Picture Experts Group) and AAC (Advanced Audio Coding – iPod/iTunes)

An Mp3 recording requires about 1/10 the storage space needed for an uncompressed wav file. Mp3 really refers to MPEG 1.3, not to be confused with MPEG 2 or 4.

SURROUND SOUND TECHNOLOGIES

In this section we will discuss the technologies used to deliver digital surround sound. For information on home theater design, including speaker configuration, see Chapter 12.

History of Surround Sound

Surround sound is often viewed as a fairly recent technological development in the world of audio. In truth, the idea of surround sound has existed for decades! The first real world experimentation with a surround sound format came in 1940 with Walt Disney's "Fantasia". After talking with the film's conductor (Leopold Stokowski) about the musical score, Disney became excited about the idea that in the movie, during the performance of "Flight of the Bumble Bee", the audience could experience the musical sound of the bees flying all around them instead of just from the front of the theater near the screen. Disney charged his engineers with the task of accomplishing this type of sound environment and what resulted was essentially a six channel surround setup including front left, right, and center channels as well as "house" (surround) left, right and center channels that were reproduced by a total of 54 speakers throughout the theater. Unfortunately, due to the extremely high equipment costs and the onset of WWII, the format was simply too expensive for most theaters to adopt.

The next big step in the evolution of surround sound came 35 years later with Dolby Laboratories introduction of "Dolby Stereo" in 1975. This format was a simple upgrade to the existing optical sound prints the movie industry was already using. By encoding additional tracks within the same stereo print size, the new format allowed for front left, front right, center, and rear speaker locations. It was not an immediate success, but in 1977 with the release of "Star Wars" and "Close Encounters of the Third Kind" (both formatted in Dolby Stereo) the format quickly became the industry standard.

The first application of surround technology to enter the consumer market came in 1982 with the release of Dolby Surround, which enabled home systems to decode the surround channel of a Dolby mix. Dolby Surround was swiftly followed by Dolby Surround Pro Logic, a format that made it possible to decode the center channel along with the other three.

These first surround formats were the industry standard until the mid-1990's, when surround sound entered the digital age with the launch of Dolby AC-3, which we know today as Dolby Digital.

Digital Audio Formats

Earlier, we discussed the process of converting analog audio to digital. This technology not only allowed for low noise, identical copies, and improved storage and distribution, it made it possible to pack multiple channels of audio into less storage space, and was a key element in the development of multichannel surround sound. As a background, we will first discuss the different types of digital audio formats, and what they are used for.

As discussed earlier, digital audio can be stored and delivered as either uncompressed or compressed. Compressed audio requires less bandwidth and less storage space. In the case of music, an uncompressed file such as a .wav file is about ten times the size of a compressed file such as an MP3. In the world of multi-channel audio for surround sound, compressing the multiple channels allowed them to be carried over coaxial or optical interconnects before HDMI was introduced.

Digital audio can be divided into two categories:

Standard Surround Formats: Compressed

Dolby AC-3 (Dolby Digital) was introduced in 1992. This technology was the first to compress sound information in such a way that it would enable 6 discrete channels to be encoded in the space typically reserved for one, while also maintaining an extremely high level of sound quality. The technology entered the home in 1995 via laser disc, but really took off when established as the audio standard for the DVD format.

We know this popular format as Dolby X.1, meaning X channels of surround with 1 channel of LFE (Low Frequency Effects). The most common of these is 5.1 (5 channels of surround [C, LF, RF, LS, and RS] and one channel of low frequencies for a subwoofer).

Digital Theater Systems (DTS) was unveiled in 1993 with the release of "Jurassic Park". Since then, many movie studios have decided to release their films utilizing the DTS format for their soundtracks. Most high quality receivers will decode both Dolby and DTS. In some cases the user must select DTS or the receiver may default to Dolby Digital.

Extended Surround Formats: Compressed (6.1, 7.1)

There are also formats that go beyond the basic 5.1 configuration. These are referred to as Dolby 6.1 or DTS Matrix or Discrete 6.1. Some A/V equipment uses proprietary decoders to generate two separate back channels. They may or may not alter the back channels by adding in slight amounts of other surround channels. Most receivers in use today are either 5.1 or 7.1 in their configuration. When only installing a 5.1 speaker package, the setup will include designating exactly what speakers are present in the menu driven setup of the receiver.

Extended Surround Formats: Uncompressed

Finally, some HD sources now have multichannel audio which is NOT compressed but is essentially the original pure audio translated from analog to digital. With the introduction of HD video sources, both Dolby and DTS introduced a number of uncompressed (but proprietary) formats. These formats take advantage of the much higher bandwidths available from HD media and provide the highest quality digital audio. The only way to experience these uncompressed formats is with a source that is actually delivering it, and the higher resolution requires an HDMI interconnect.

In this section of the chapter we will cover the basics of loudspeakers, speaker design, speaker specifications, and various types of speakers.

Audio drivers (speakers) are always imperfect transducers. There is always a compromise between size, cost, dynamic range and application. The loudspeakers are usually the weak link in the system's ability to reproduce sound accurately. This is partly because the quality of the other electronics is so high, and partly because the loudspeakers are operating in an air space that is almost always acoustically flawed in some way.

There are too many speaker designs to cover in this chapter. Some are very serious and deliver excellent sound reproduction. Many are just gimmicks and are never seriously considered for theater or distributed audio. Although excellent results can often be achieved with such designs as electrostatic and magnetic plane, for the purposes of this course we will focus on typical dynamic type drivers and crossovers.

Typical Cone Speaker Design

Almost all dynamic drivers work the same way. The amplifier drives a coil of wire placed inside a stationary magnetic field. The direction of the current through the wire causes the coil to move in or out, and the amount of current determines the travel distance. The coil is stabilized by a "spider" assembly. The coil is attached to the driver cone. The cone movement moves air and thus creates sound waves. In this manner, electrical energy is converted to acoustical energy. Speakers can be made more efficient and more accurate by making the cone lighter and more rigid, making the voice coil gap smaller, etc.

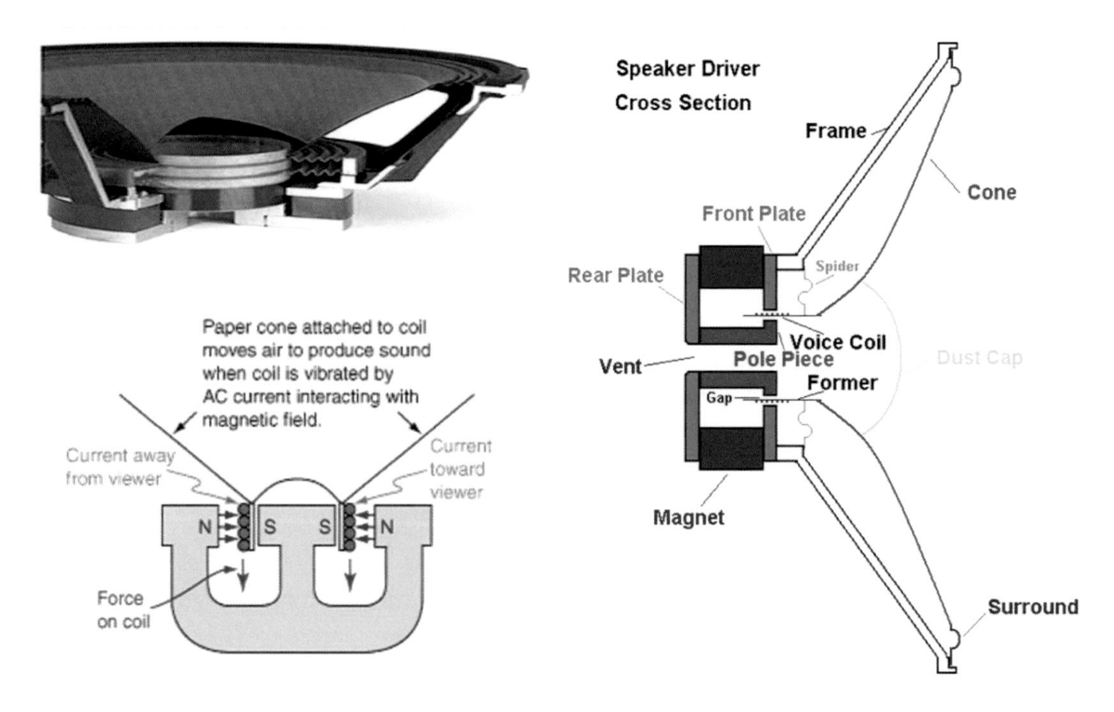

Figure 9-13 Dynamic Speaker Characteristics

Historical note: Back in the "old days" when strong permanent magnets were hard to make, the magnet was actually an electro-magnet. A coil of wire around the "magnet" was connected to a power supply to create the magnetic field.

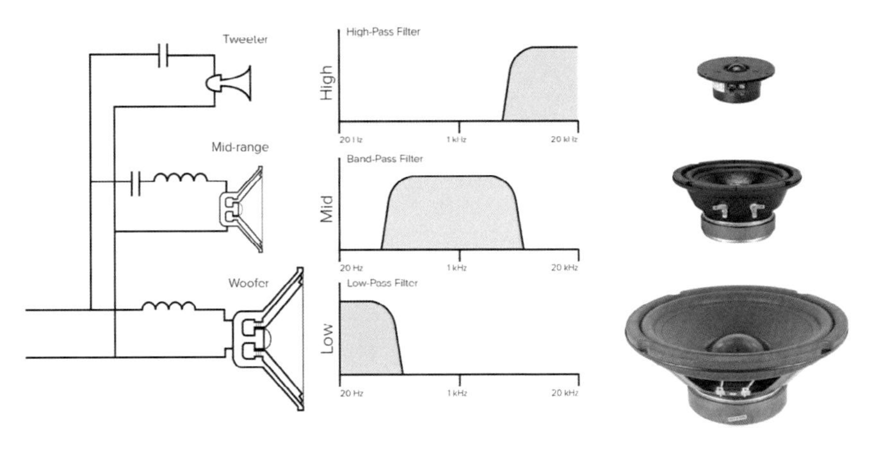

Figure 9-14 Three-way Crossover Network

Crossover Networks

As we have learned earlier, different frequencies have vastly different wavelengths. This makes it difficult for any one transducer to work well across the entire audible range of 20Hz-20K Hz. This problem can be addressed by using different transducers for different frequency ranges. In order to have these multiple drivers, each designed for a specific frequency range, crossover networks are used to divide the full range signal into separate, slightly overlapping, ranges. In a two-way design the frequency dividing network (x-over) sends all audio above a specific frequency to the tweeter and all audio below that frequency to the woofer. In a three-way design there are two crossover points and a mid-range driver receives the audio between those points.

Cross-over networks are usually placed in the speaker enclosure (passive), but can also be placed before the amplifiers (active), which is more efficient but more costly because a separate amp channel is needed for each driver.

The crossover network for a separate subwoofer may be in the audio processor (receiver). There may also be a crossover in the subwoofer electronics too, so care must be taken to make sure these settings are not conflicting with each other.

Basic Speaker Design

Now let's put all of these devices together to make a typical 3-way loudspeaker design. This diagram illustrates the terminology used for the different parts of a typical speaker:

- The enclosure
- The crossover network
- The drivers: three are shown for the three output bands of the cross- over (woofer, mid-range, and tweeter).
- A port to allow the sound waves created inside the enclosure (primarily by the woofer) to exit the enclosure in-phase with the outside generated sound wave. Not all speakers utilize a ported design.

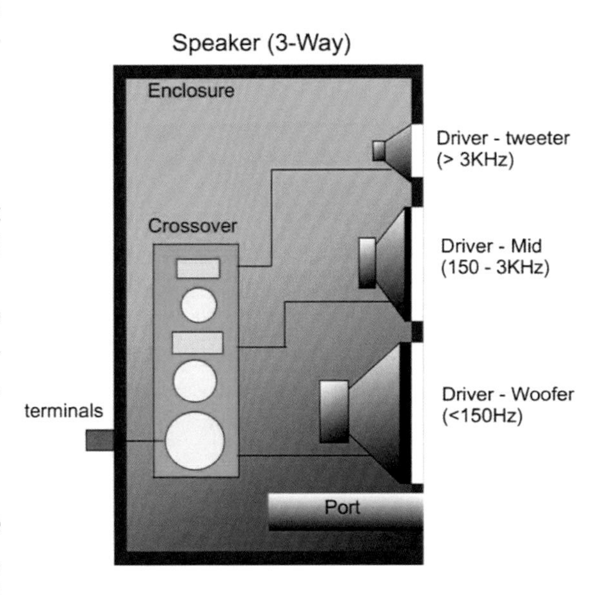

Speaker Sensitivity

Loudspeakers are inherently inefficient devices. Much of the electrical power used is not converted to acoustical power, but wasted as heat. The efficiency of speakers is measured as "sensitivity", which establishes a standardized way to

Watts	SPL in dB	Typical sound
	140+	Jet aircraft
	120+	Threshold of pain/Rock concert
320	110 dB	
160	107	
80	104	Orchestra climax
40	101	
20	98	
10	95	Comfortable music listening
4	91	
2	88	
1	85	Average speaker efficiency
	65-70	Normal speaking voice
	40-60	Home background noise
	15-25	Whisper
	0	Threshold of hearing (1 kHz)

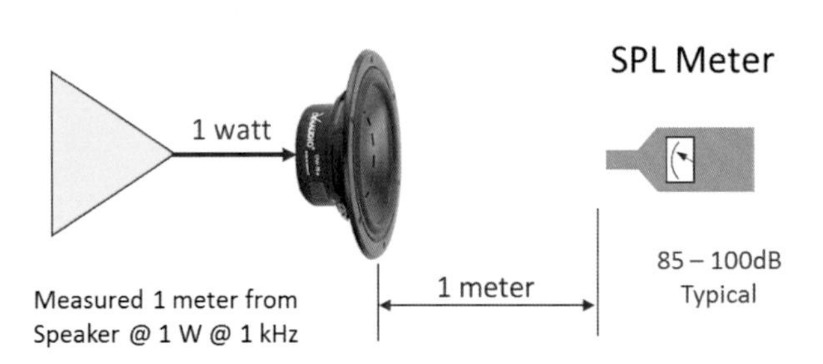

compare the performance of different products. Speaker sensitivity (the ability of a speaker to output a given sound pressure level for a given input power) is measured as shown with an SPL meter. High efficiency speakers will have efficiencies of 90dB and above, sometimes over 100dB. Low efficiency speakers will be below 85dB.

Keep in mind, manufacturers may use different methods to arrive at their ratings. In general they use a signal of 2.83 volts into a speaker with a nominal impedance of 8Ω. But there are other variables, like the surroundings, which affect the measurement.

SPL vs. Speaker Power

O dB is considered the threshold of hearing. 120 dB is the threshold of pain. Note the relationship between the power it takes to drive a speaker and the SPL output in dB. Doubling the power to a loudspeaker increases its output by just 3 dB. (a noticeable change in level). A 10 dB change sounds to our ears to be twice as loud, yet it requires 10 times the power to get that increase.
So to achieve adequate SPL at the listening position, it is clear you need efficient speakers AND lots of power.

Speakers: Basic Types

There are several general categories of "free-standing" (not in-wall/ceiling) speakers used for home A/V applications :

- Floor standing - larger/heavier speakers such as stereo or FL/FR speakers in a home theater
- Bookshelf – usually small lightweight speakers. This designation, created in the 1960s was actually for small speakers to be mounted in bookshelves! The name stuck. We will also see some specialized theater speakers in this smaller category
- Satellite - smaller, good for mids and highs, used in small surround applications
- Subwoofer - specialized speakers used to reproduce low frequencies typically <80 Hz.

Speakers: In-Wall/In-Ceiling

In-ceiling and In-wall speakers, or "architectural speakers" are often used for distributed audio, or in theater situations where there is a desire to make the speakers inconspicuous. This type of speaker is designed to be installed flush with the wall or ceiling surface and can sound very good. Tiltable transducers are preferred because they allow the sound to be aimed, putting the listener on-axis with the speaker.

Speakers: Home Theater Types

Regardless of the type, speakers used in a home theater should ideally be from the same manufacturer and should be the same model class so they will sound basically the same (have the same tonal quality). Some speakers are designed to play a specific role in a home theater system. It is not often practical to have all full range speakers identical, so special speakers are offered to be used for the center channel and surround/rear channels.

Center Channel Speakers are designed to have similar tonality to other speakers in the system, but are often configured horizontally, making them ideal for placement directly above or below the display.

Surround Speakers often just monopole two-ways, but sometimes more specialized designs are chosen for a more enveloping surround experience. They are usually in-wall or on-wall types with transducers aimed in different directions. If these transducers are wired out of phase, the design is known as di-polar. If in phase, bi-polar. Some are switchable.

Subwoofers can be utilized as an enhancement to a two-channel system or even a distributed audio zone. But the most widespread use is in home theaters and media rooms, where deep bass plays such a vital role in the experience. Subwoofer types include free-standing self-powered types, and in-wall designs which require the amplifier to be housed with the rest of the equipment. Self-powered subs require a coaxial feed for analog audio, and an outlet for AC power. Passive subs like in-walls are fed with speaker cable from the remote amplifier. Remember to use large gauge speaker cable. For information on subwoofer placement see Chapter 12.

SUMMARY

In this chapter we have covered the fundamental nature of sound and how it works, the evolution of the technologies that allow us to record and reproduce sound, and devices used for this process. These fundamentals will be applied when you learn about distributed audio and home theater.

Questions

1. Which of these converts acoustical energy to an electrical signal?
   a. A/D convertor
   b. D/A convertor
   c. Microphone
   d. Loudspeaker

2. A high-efficiency loudspeaker is one with a sensitivity of:
   a. 90dB and above
   b. 85dB and below
   c. 75dB and above
   d. 75dB and below

3. Pink noise has;
   a. Equal energy in each octave
   b. Equal energy at all frequencies
   c. Equal energy in all parts of the room
   d. Frequencies adjacent to infra-red signals

4. In surround sound nomenclature, the ".1" in "5.1" stands for:
   a. One center channel output
   b. One subwoofer
   c. One Low Frequency Effects channel output
   d. One source input

5. High frequency sound has:
   a. Higher amplitude
   b. Lower amplitude
   c. Shorter wavelengths
   d. The ability to pass through walls easily

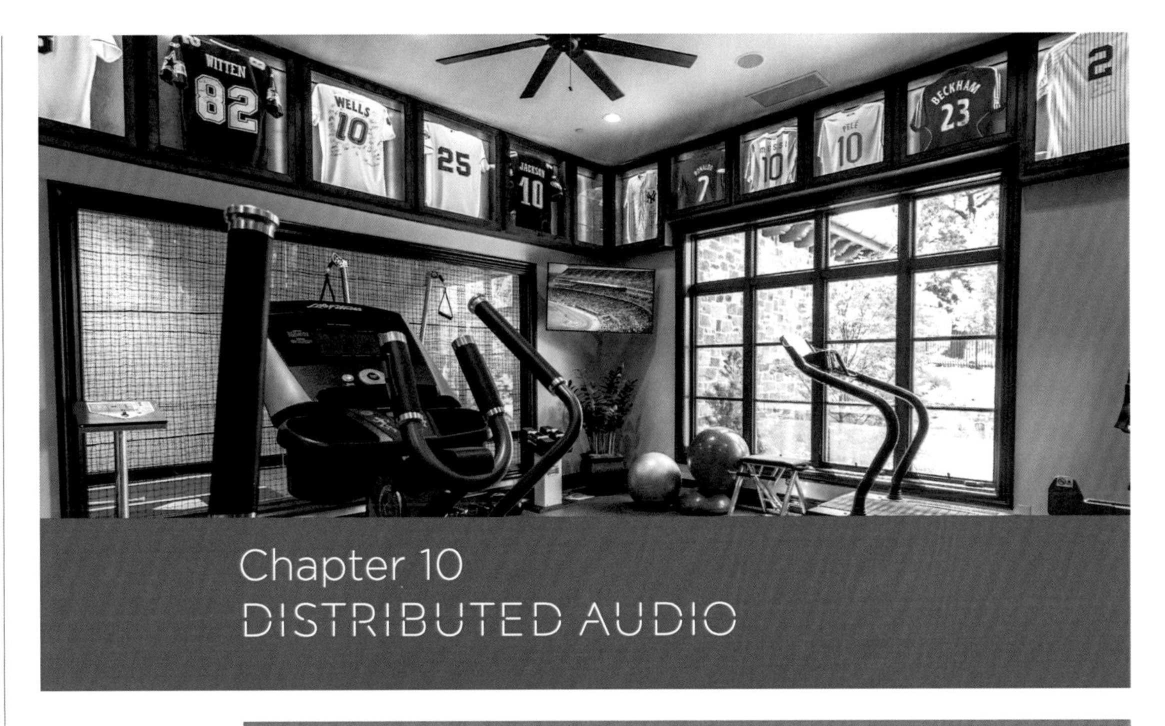

Chapter 10
DISTRIBUTED AUDIO

INTRODUCTION

Definition Of Distributed Audio

Distributed audio (multi-room audio) refers to the delivery of one or more audio sources to two or more areas/zones in a building or house, and this subsystem has been a staple of the industry for many years. It essentially means placing speakers in different rooms of the home, and often outside on a patio or deck. In most cases architectural style speakers are installed in the ceiling or wall and are intended to be as inconspicuous as possible, sometimes completely invisible. The intended use is usually for background music but a properly designed system with the right components can deliver very high quality sound. In recent years the way these systems are installed and used has changed dramatically, as mobile devices and streaming media have augmented (or replaced) centrally located servers and other sources. We will cover all of these options in this chapter.

Goals of Distributed Audio

The primary goal of distributed audio is to provide uniform coverage to a specified area. The placement of the listener is generally not well-defined, and there is often not a "sweet spot." Instead, homeowners should experience a consistent volume level as they move through a space. Ideally, the frequency response will remain constant, and listeners will not experience "hot spots" where the volume level is significantly higher or "nulls" where the volume is too low. This can be accomplished by specifying the right speakers and placing them properly.

Typical Client Uses

Before delving further into the different types of distributed audio systems, we will first discuss how clients typically use distributed audio. When designing distributed audio systems, you must understand how your clients intend to use their system. Different client needs, expectations and finances require totally different designs. In order to meet these varying demands, you should create a needs analysis that analyzes what the client wants, what the client is willing to pay, and what system you can design in order to meet those specific client needs. This section focuses on some of the typical ways clients use distributed audio systems.

Background Music

Background music is the most common use of a multi-room music system. Here people often partake in other activities while using their system. Most commonly, the volume levels used to play background music are modest. Whether listening to a news channel while getting ready for work, or light-jazz while preparing a meal, this type of multi-room music system makes it possible to add a soundtrack to your client's lifestyle. For the most part, plan your speaker placement so that there aren't any hot spots or alternately, null zones (areas in which there is little or no coverage).

Entertainment

Some clients will want a system that they can use while entertaining guests when they host a party. This may require a louder volume level along with the ability to select music that will play continuously. While in-wall or in-ceiling speakers might provide enough sound for a specific area, depending on the size of the area and the volume at which the customer wants to play the music, larger speakers and more power amplification should be considered. Remember that as you fill a space with more bodies, you will need to produce greater volume. Bodies, like any obstacle will dampen the apparent sound level. In addition, you will also need to overcome the base level of noise generated by having that many people talking at the same time. The use of better speakers and proper amplification are important because as you push a system harder, distortion becomes an issue. In some situations, like a bar area, subwoofers may also be implemented to provide deep bass without overdriving the full range speakers.

Critical Listening

As a designer, you may be asked to create a critical listening environment. There are still those clients who take the time to really just listen to music, typically in a 2-channel stereo format (like the type of sound you hear with a car stereo as opposed to the sound in a multi-channel, or surround-sound system).

For these clients, some basic in-wall or in-ceiling speakers are typically not good enough. Most often, 2-channel listeners are more discriminating when it comes to audio clarity and will require, at a minimum, something from nice free-standing bookshelf speakers to tower speakers. In this type of scenario, you can create a "sweet spot" that will provide the best listening environment in one area for the listener. You would create this spot where the clients plan to sit and enjoy the music most of the time. When using smaller speakers, plan on placing them on stands as a discerning audiophile will not want to simply situate them on a bookcase. In some cases, high quality in-wall speakers may deliver similarly good audio, but remember that their location will be fixed once they are installed. As with left/right speakers in a surround system, they will likely be installed at ear level for the most accurate sound.

In addition to providing speakers, you also have the opportunity to provide more sophisticated electronics as well. In many "critical listening" rooms within a multi-room system, the local hi-fi system is completely stand-alone and simply has audio cables that permit it to play the same source component playing on the multi-room system. Here, you can opt to have the local hi-fi system connect to a fixed (vs. variable) line-level audio output from another zone or if you have extra zones available on your multi-room system, make this room its own zone.

History Of Distributed Audio

Distributed audio can trace its history back to the first "high fidelity" systems, when amplifier and speaker technology evolved to the point where fairly accurate music reproduction became possible. We will discuss this evolution in a little depth to establish a solid background for the topic of modern multi-room audio.

From Hi-Fi to Stereophonic Sound

In the '20's and '30's the radio was the center of entertainment in the home. If you wanted to listen to the radio in a particular room, you bought a radio for that room. Then in the 1950's as recording techniques, records and amplifier quality all improved, the term "hi-fi" (high fidelity) became widely used to describe a system that delivered high quality sound. This trend continued with the advent of stereophonic records and FM broadcasts which could deliver a much more enjoyable field of sound which worked with the room to create a more natural representation of music. Still, if you wanted music in the Living Room and the Master Bedroom, you installed a stereo in each room.

In the Beginning

The first stereo systems were quite simple: a receiver that selected AM/FM or other external sources, amplified the signal and powered a pair of speakers. This type of stereo system was very popular and delivered great sound reproduction for a variety of musical styles. The local hi-fi shop was THE place to go and check out the newest audio equipment in the 60's and 70's.

Then a Second Pair of Speakers

In the '70's it became popular to use a single stereo receiver, with ample power, to drive two pairs of speakers (two "rooms") at the same time. The amplifier needed to be robust enough to handle the increased current created when the load impedance was reduced (two 8Ω speakers in parallel on an amp channel presented a 4Ω load). Some receivers provided a switch to turn each "room" off or on as desired. Although this system had two rooms, it was a single "zone" because both rooms always received the same audio.

Then More Rooms

In order to feed more rooms with this single amp configuration, it was necessary to provide impedance matching so that the amplifier was presented with a load it could handle comfortably. The first method used to accomplish this was a centrally located impedance matching network which the amp fed directly. Then its output went to the speakers. This device added impedance to the amplifier load as needed to compensate for the drop caused by adding speakers in parallel. This design was the beginning of multi-room audio in the home. We will cover this in more detail later in the class.

DISTRIBUTED AUDIO TODAY

Most Common System Types

Let's look at the range of distributed audio systems being installed in homes today, how they work, and when to apply different systems based on a client's expectations and needs. For instance, a retired couple in a condo may not need to listen to different sources in different rooms. They may just want music or talk radio smoothly distributed to a couple of areas of the home, and on the deck. But larger homes and larger families may require more complex systems which allow for different sources in different areas, the distribution of audio from a TV or theater area, or local override of the system by a teenager's portable music device.

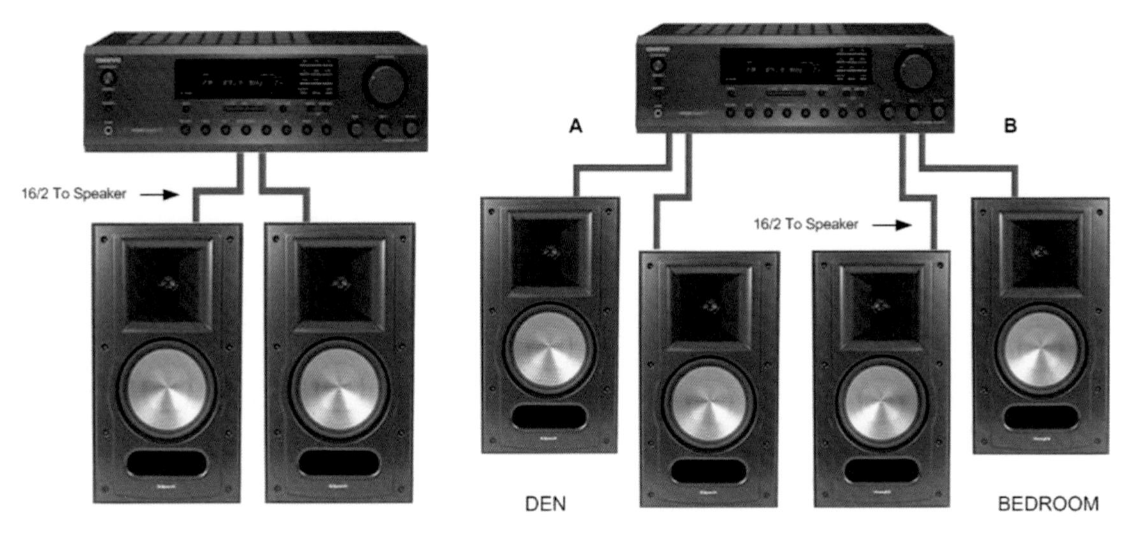

Single Zone, One Room

Single Zone, Two Rooms

Single-Zone Systems

There are still simple applications which call for nothing more than a stereo receiver and a pair of speakers, either free-standing or architectural. Remember this receiver can select from a variety of sources, including radio and other devices which are connected to its inputs. Some receivers may include iPod docks or tuners for HD or Satellite radio in addition to simple stereo RCA inputs. Just like a few decades ago, it is not uncommon to simply add a second pair of speakers (a second "room") to a basic stereo receiver. The speakers may be connected to separate outputs, known as A and B, which would allow the user to selectively turn each "room" on and off.

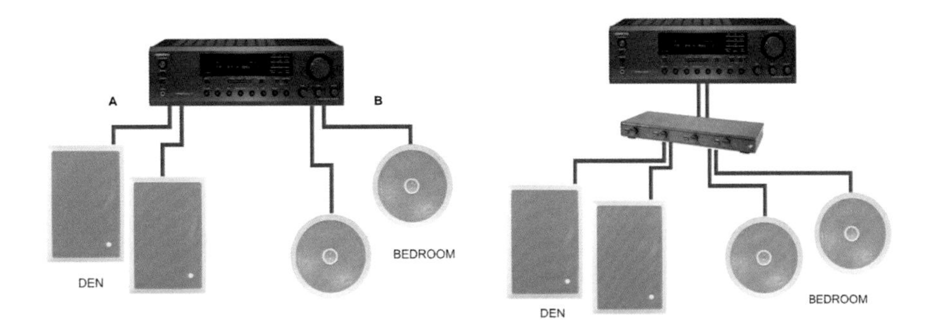

For a less cluttered look, architectural speakers may be used. Shown here are a pair of in-wall speakers and a pair of in-ceiling speakers. This type of speaker requires wiring to be installed during construction

Several years ago, some manufacturers came out with impedance matching multi-room switch boxes, which allowed up to 4 rooms to be driven from a single stereo amplifier (Single-Zone) and each room to be turned on or off as desired. Then volume controls were added so the relative SPL could be balanced between rooms. This is certainly not as convenient as having control in each individual room, but a step in the right direction.

When a stereo speaker feed (2 conductors for Left, 2 for Right) is run from an amplifier, it is convenient to use a 4-conductor cable rather than two 2-conductor cables. This "4-con" is usually 16/4 (four 16 gauge conductors). Then of course the cable to each individual speaker can be 2-conductor. Shown here is the industry standard color code for 4-conductor speaker cable when it is carrying Left and Right speaker audio. This is specified in ANSI/CEA/CEDIA-2030-A. There are a couple of ways to remember this standard; "RED HOT / WHITE HOT" and "RED IS RIGHT, LEFT IS WHITE".

However, don't assume that a previously installed system adheres to this standard. Many companies have used other color codes, and some still use non-standard color codes today. When working in a system that you do not have thorough documentation for, be sure to double check the color code.

To be in line with the industry, all new installations should use this standard.

One of the most important product developments in single-zone multi-room audio was the impedance matching volume control. These allow up to 8 rooms of audio to be driven by a single stereo amp. Each room has a specially designed volume control with multiple taps designated as x1, x2, x4 and x8. If there are 3 or 4 rooms being driven, all volume controls are set to x4, if 5-8 rooms are on the zone, all V/Cs are set to x8.

The designation and type of settings vary by manufacturer but the concept remains the same. Some have slide switches, some DIP switches, and some use jumpers similar to those found on computer motherboards or hard drives.

Dedicated Amp for Special Sub-Zone

In some cases the internal power amplifier built into the receiver is adequate for the main room, but the second area may be larger, or outdoors. In this case it is important to make sure there is enough amplifier power.

More power may be required for a number of reasons:

1. Outdoor speakers – no reflections, or "room power" so overall volume can be as much as 6dB lower
2. Outdoor speakers may be considerably less efficient due to the weather resistant materials, meaning lower SPL from the same power
3. Second room may be a handball or basketball court, or a large garage which requires larger (or multiple) speakers.

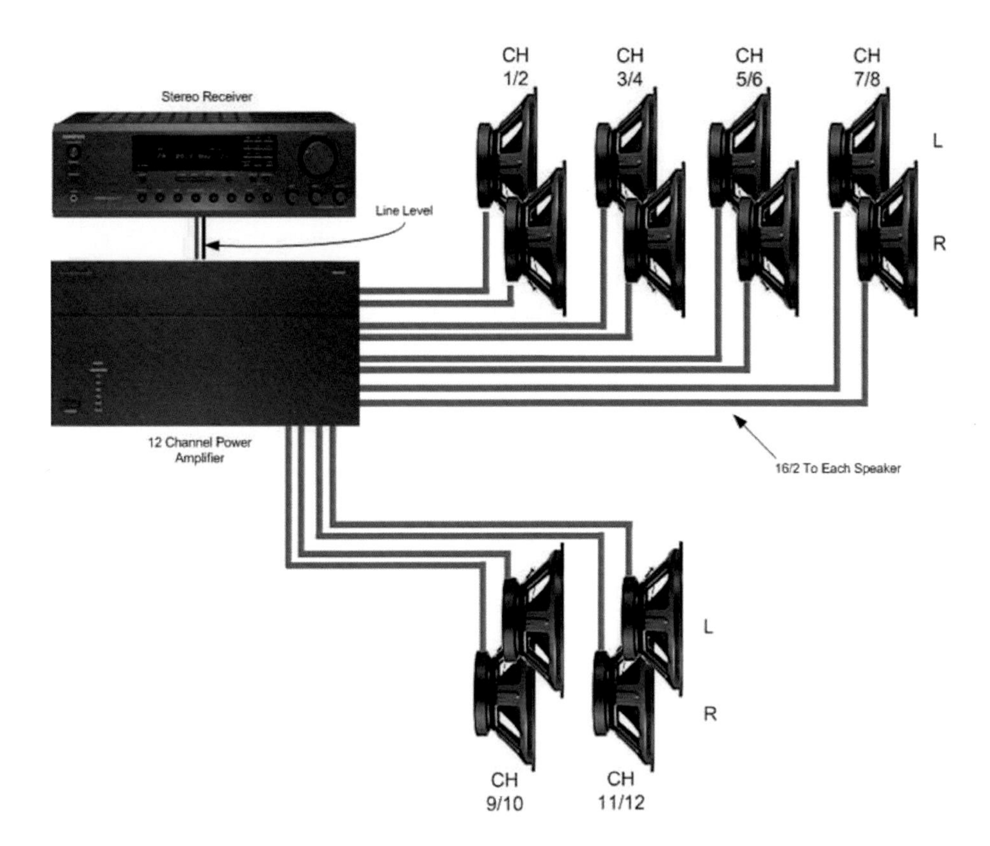

When designing for this type of situation, plan on running signal (Line Level) from the RCA analog stereo outputs of the receiver to the input of a secondary stereo amplifier. Then, from the high-current dedicated amplifier, run 2-conductor speaker cable from the outputs to the speakers. Remember, these speakers still all get the same source material, so this is a Single-Zone system.

Some situations may require many speakers, all on the same zone. This may be a retail location, a doctor's office, or restaurant. In this case a good solution is a multi-channel amp that allows a single monaural or stereo input to be fed to all channels. This same type of amplifier is also useful for many other system designs, as we will see later. Typically, a line level signal is fed from the source (in this case a stereo receiver) to the input of the amplifier, which has the ability to route the signal to all channels as either mono or stereo.

Using "Zone 2" on a Surround Receiver

Many surround receivers have a special feature which provides a second switching matrix of analog sources which delivers the selected audio signal to a "Zone 2" analog stereo pair output – typically a Left/Right RCA pair. This Zone 2 has its own control functionality on the factory remote, which can also be programmed on a custom remote. The user can select Zone 2, then choose an audio source, and adjust the volume level, all independently of the primary listening area. They also have control of the internal sources such as AM/FM/HD radio or in some cases an iPod bridging device that is integrated into the receiver.

Most receivers allow Zone 2 to be turned on and off independently from the surround system, which means the homeowner can use a remote or other user interface to power on just Zone 2, and listen to music or news in some part of the home – without turning on the theater. Some manufacturers also offer a dedicated remote control just for Zone 2.

Some receivers may even have Zones 3 or 4. This means with external amplification they can actually deliver a multi-zone configuration, based on the same sources as the surround system, using just a standard receiver in the primary listening area.

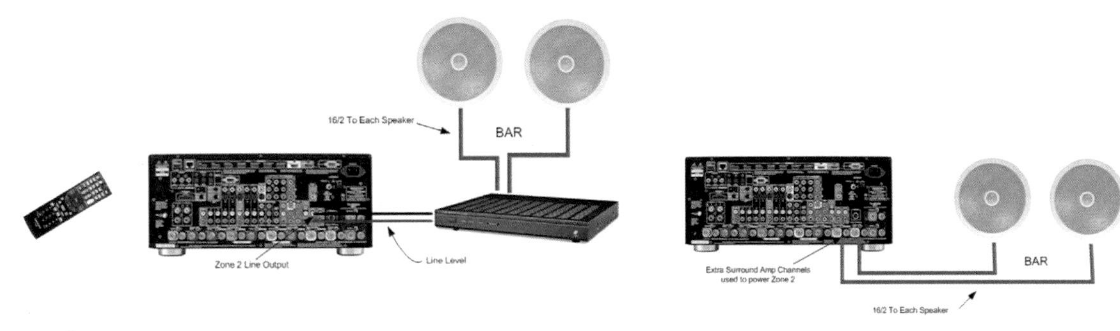

Zone 2 w/External Amp　　　　Zone 2 Using Internal Amp

Zone 2 Line Out to External Amp

In its simplest configuration the Zone 2 stereo line level outputs are fed to the input of a separate amplifier which drives speakers in Zone 2, wherever that might be. It might be ceiling speakers over the bar, or rock speakers outside on the patio. With the right amplifier, multiple rooms may be driven and impedance matching volume controls can attenuate the levels in each room. Remember every speaker pair on the Zone 2 feed receives the same source material (it is a single-zone system).

Extra Amp Channels for Zone 2

Most surround speakers with Zone 2 functionality also add another clever feature which allows the Zone 2 output to be routed internally to two of the surround amplifier channels, if they are not being used at all in the surround system. For instance, a 7.1 receiver is installed in a media room which is configured as 5.1. This means the Rear Surround amplifier channels are dormant (no signal to them) and no speakers attached. This receiver may then be configured via its setup menu to send the Zone 2 signal to those dormant amp channels, which then drive the speakers for Zone 2. This can be a very cost effective way to provide a client with a surround system AND a single zone music system, each with its own control and source selection, without the need for separate sources, switching, or amplification.

There are a couple of important things to keep in mind in this configuration:

1. Most receivers will only recognize analog audio inputs as sources for Zone2. This is because there is no Analog/Digital (A/D) conversion stage available to capture the digital audio from HDMI, Toslink, or SPDIF inputs and convert them to analog. So be sure to specify redundant audio interconnects from devices like cable set top boxes and Blu-ray players. Then the Zone 2 matrix will have an analog signal to work with. Know the functionality of the receivers you sell so there are no surprises.
2. Driving more than one pair of speakers with surround amp channels in this configuration is not recommended. If Zone 2 is going to serve more than one room, or an outdoor space, an auxiliary amplifier should be used.

Constant Voltage (70v) Systems

Constant voltage systems, also known as high voltage, or 70-volt systems (100 volt in Europe), utilize step-up and step-down transformers to drive the network of speakers with very little loss due to cable length. They are widely used in commercial distributed audio systems.

The amplifier output is stepped up to a high voltage, which minimizes line loss (power loss) over the cabling to the speakers. This can also allow for a smaller gauge of speaker cable since the current is reduced. The voltage is stepped back down at each speaker by a transformer. Often it is possible to select how much power each speaker gets, as long as the overall power of the amplifier channel is not exceeded.

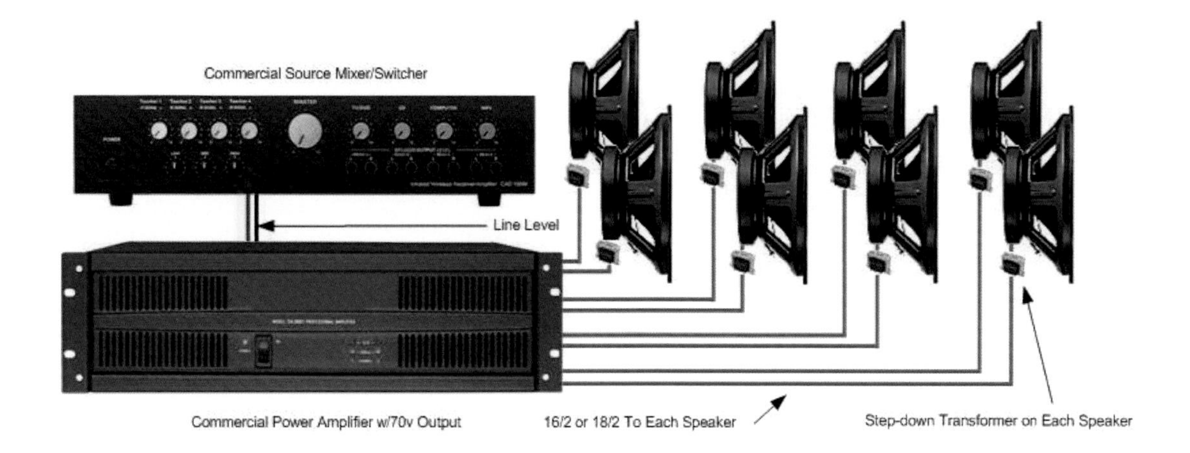

Advantages: Dependable, smaller speaker cable, minimal loss over distance, many speakers can be used without complicated series/parallel calculations, choice of power to each speaker

Disadvantages: Transformers can induce distortion and compromised frequency response, more expensive transformers are needed to reproduce low frequencies, and system is sensitive to line loss due to a short or moisture

Multi-Zone Audio

If different audio sources are desired in different rooms or spaces, a Multi-Zone system is required. This system allows different rooms or zones to select their own choice of audio sources. For example: Mr. Smith is listening to the ball game on the porch, while Mrs. Smith is listening to smooth jazz in the bathroom. Each user was able to select their own source device, and then control it. Mr. Smith can switch channels on the tuner, adjust the volume, or mute the zone altogether. Mrs. Smith can skip tracks, pause, etc. If the source is a CD changer or iPod, it may even be possible that Mrs. Smith receive metadata information, such as artist, track title, etc.

Keep in mind the same source can be selected for two or more zones, which is great if you want the same music everywhere for seamless background music when entertaining, or to allow everyone to hear the Super Bowl at once. However, this can lead to control conflicts.

For instance, if there is one AM/FM tuner in the system, and two users in different zones are listening to it, when one user changes the channel, all users hear the change. In some systems it is appropriate to have two radio tuners (A and B, or His and Hers), for more flexibility.

Now let's discuss the different system types available on the market. This technology has evolved a great deal since the early days of impedance matching and IR routing.

Analog Systems/IR Control

The first Multi-Zone systems utilized a switching matrix, which could be controlled via IR, and typical included rotary volume controls in each room. Before IR keypads, the volume controls included IR receivers. The user simply pointed the handheld IR remote at the volume control (VC), and the commands were carried back to the head end over copper cable, commonly Cat5.

When the IR keypad was introduced, some systems simply inserted them next to the volume control. This keypad essentially took the place of the IR remote. It stored the IR commands locally and sent them to the head end to control source selection and source functions.

Some manufacturers introduced electronic volume controls which still passed speaker signal but were IR controllable.

The Next Step in Analog

The most widely used multi-zone system has a central control processor, which handles all source selection, control routing to sources, and amplifier volume/mute functions. The multiple amplifier channels may also be in this one main unit.

Since the zone volume can be controlled at the amp, passive volume controls are no longer needed in each room. Instead, a system keypad is installed in each zone, which communicates with the central unit and allows control of all functions. In some cases, if two-way communication is available, the keypad may also allow the user to see the status of source devices through metadata.

This configuration allows for a less expensive cabling scheme as well. Speaker runs can go straight from the amplifier to the speakers without passing through a volume control location. Note: If your client is interested in the system being backward compatible, a traditional cabling scheme is still recommended. (Run a 4-conductor to a future volume control location, and barrel connect two 2-conductors out to speaker locations from there.)

The keypads are commonly run with UTP such as Cat5e. This cable is run from the main processor unit out to each keypad location. This allows them to communicate back, and control all functions available to the user. It is highly recommended that, even if you are installing a standard single-zone system with the older, passive-type volume controls, that you include a Cat5e cable along with the 4-conductor. This allows flexibility and scalability in the future.

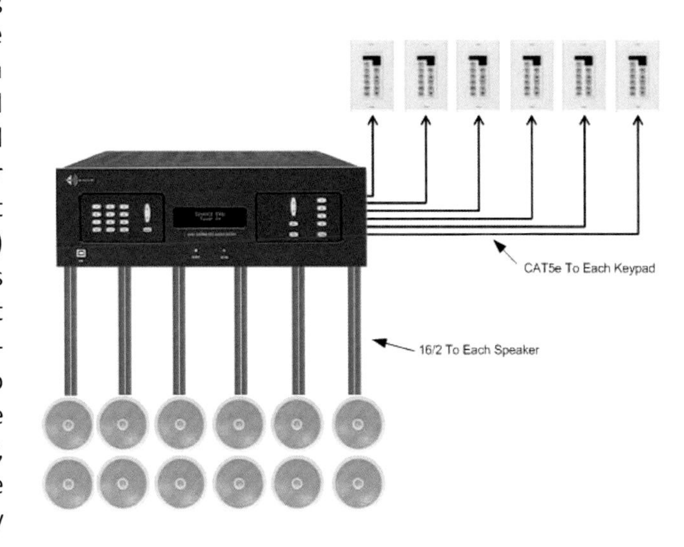

Multi-Source/Multi-Zone Example

The typical 6-Zone system shown has a central processor/amp and a system keypad for each zone. This system utilizes CAT5e connection from each keypad, and 16/2 speaker cable directly from each amp channel to its speaker.

A-BUS

A-BUS systems are yet another option for distributing audio throughout the home. A-BUS technology was originally invented in Australia, but over the years has come into widespread use throughout the world. The patent is held by LiesureTech and the technology is licensed to numerous manufacturers.

The A-BUS system utilizes a local amplifier in each room. The audio signal from the source (or sources) is kept at line level and sent directly to a special keypad, which contains the small stereo amplifier. The signal is then amplified locally and sent on to the speakers. There are many choices when designing and installing A-BUS systems, from the one pictured above (single source, single zone), to systems that are capable of multi source, multi zone output with individual zone control.

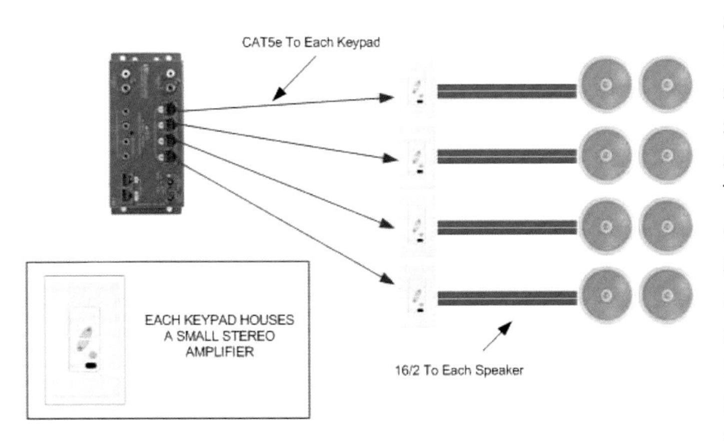

In an A-BUS system, the UTP cable (such as CAT5e) carries three distinct signals:

1. The analog stereo line level signal
2. The 24V DC power supply voltage to power the amplifier built into the keypad
3. The control signal between the keypad and the main hub

Sometimes the term "A-BUS Ready" will appear on other equipment such as surround receivers. This means there is an RJ-45 jack, which contains audio and is compatible with the A-BUS protocol.

Digital "Hub" Style Systems

A digital "hub" type system is similar to A-BUS in that its configuration utilizes a "local amplifier"-type design; however, in this system type, all audio is distributed from the hub to the room modules in digital format. So if an input signal is analog, it is converted to digital immediately. Since the signal to the rooms is digital, it is not susceptible to RF and EM noise.

Alternative Distribution Options

While distributed audio continues to be a staple in the new build market, the demand is also increasing among customers wanting a solution for retrofit applications. Many clients want the experience of a distributed audio system, but do not want to cut holes and run wires all over their completed and furnished homes. With advances in technology, there are now alternatives to traditional distribution, which allow the audio signals to be distributed over existing wires or sent completely wireless in order to minimize intrusion and simplify installation.

One such technology, Power Line Carrier, uses protocols to allow for audio (even data and video) to be carried over existing electrical wiring, even when line voltage is present.

In addition to Power Line Carrier, wireless systems utilize the existing home computer network to obtain content from the Internet, and send it to receivers in various rooms. The handheld controller is also part of the network and provides visual feedback from the system.

The Evolution of Home Technology Architecture

Over the last decade we have witnessed dramatic changes in the manner in which AV content is consumed, distributed, and controlled throughout the home. These changes are largely the result of four factors:

- The widespread deployment of broadband internet access to the home
- The prevalence of wired and wireless home networks
- The number of products that are now sold with network connections embedded in the product (it's not just the computers and the printers connect to the home network – now TVs, receivers, thermostats, cameras, music systems and many other devices connect to the home network)
- The ubiquity of wireless smart mobile platforms in the home and the intuitive AV and home control software applications

It was inevitable that this new "network enabled" environment, along with consumers' comfort level with mobile devices, would lead to new, more versatile, distributed audio options.

The New World of A/V Content and Distribution

Broadband internet connections carry an "always on" stream of entertainment content to the home in the form of movies, music, or TV. Services like Netflix, Pandora, and HuluPlus (to name just a few) allow homeowners to enjoy high quality entertainment content from any device in their home that is connected to their broadband gateway/router. And today, almost every AV device has a wired or wireless network connection to provide access to this broadband stream of entertainment content. These developments are rapidly changing the way music is selected and managed in the home, whether streamed in real time or stored onsite. Using the network as the backbone, and the internet as a source, provides

1. Extensive access to content, more than we dreamed just a few years ago
2. Sharing of sources from room to room, rather than just from a central "head end"
3. Distribution of the content over the network, without proprietary wiring
4. Streaming audio from mobile devices which are not physically connected to the system
5. Control of the system via mobile devices, an interface the client is already familiar with
6. Metadata and other feedback from sources via two-way communication

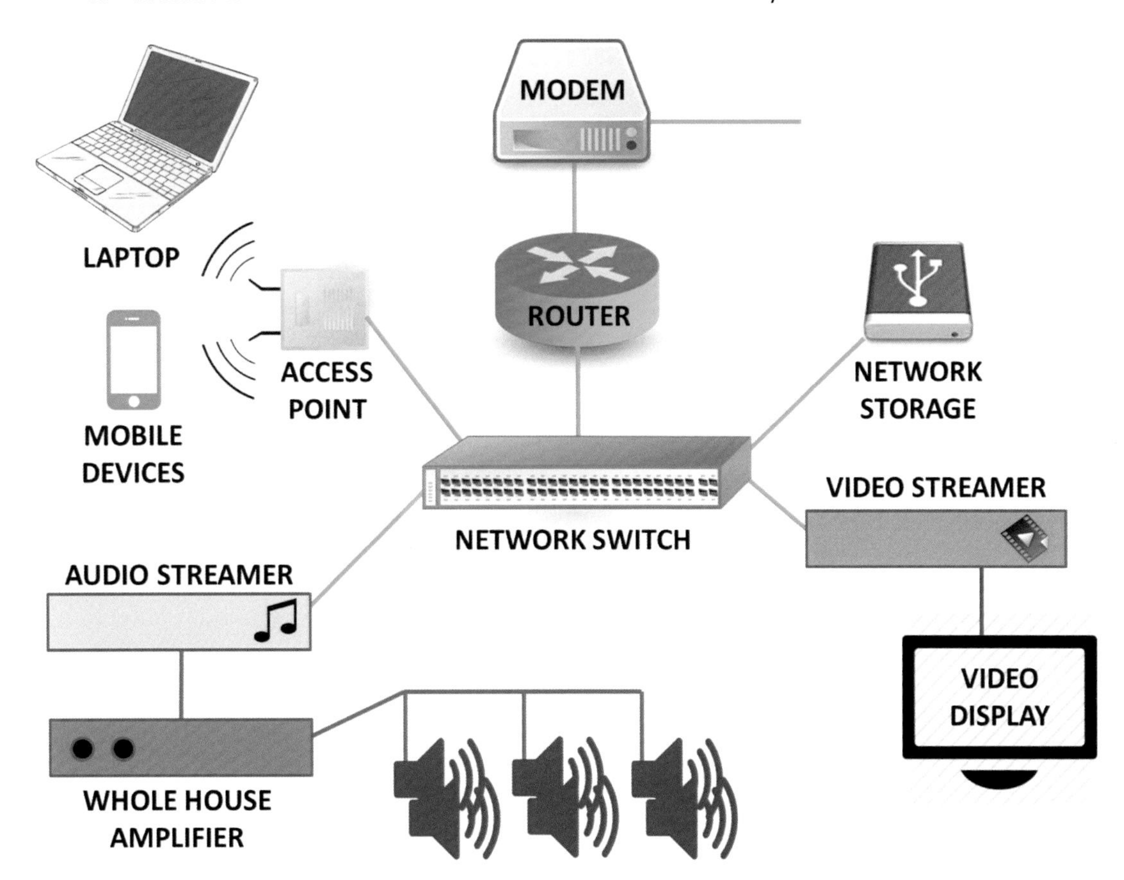

In this new distribution environment, the home entertainment content (not just music but all media) is stored in the Internet cloud, on a computer, on a network drive, or on a mobile device. The key to effectively enjoying this content from every device in the home is the ability to easily locate the content and play it back in various locations throughout the home. The best way to accomplish this task is to make sure the homeowner's content is synchronized to a centralized storage server – either an "always on" network storage drive in the home and/or a cloud based storage service. Not all entertainment content can be easily moved from one hard drive to another due to copyright restrictions. In those cases, it is important to choose a system where the copyright content can still be digitally moved across the home network from one location to another.

Remember, this evolution of A/V streaming and control places new demands on the caliber and quality of the routing, switching, and wireless architecture of the home – demands that must be met to ensure the reliability of these networked distribution solutions. And the pace at which the technologies are changing requires a great deal of diligence and investing the time to stay current with what is available.

In this section we will talk about various system components.

- Audio sources
- Switching and routing components
- System equalization options
- Power amps
- Speaker types
- User interface options

Sources

A source device is where the signal path in a distributed audio system begins. There are many different types of source devices, but we will be discussing the most commonly found devices within a client's home. Much of this is covered in the chapter dealing with sound and audio fundamentals, so this chapter will focus on components which are more specific to distributed audio.

Tuners

Most single room audio systems have a receiver or pre-amplifier with a built in tuner of some kind. However, most distribution switches and amplifiers that are designed for multi room audio systems do not contain a built in tuner. For this reason it is important to consider a stand alone tuner when specifying or installing a distributed audio system. These devices can be standard AM/FM band, satellite radio bands (XM/Sirius), and or digital HD radio bands. Another advantage of stand-alone tuners is the ability to have multiple tuners built into a single device. This will allow the customer's family to listen to multiple radio stations in different rooms at the same time.

Set Top Boxes

Today's TV services such as Cable, Satellite, and IPTV can provide a dependable, convenient source for commercial-free music channels. Many clients prefer the convenience of this type of service, as well as the meta-data functionality which can display the artist, album, song title etc. This also makes it easy to distribute the audio from a television broadcast or on-demand movie to the house system for seamless audio.

CD Players/Changers

CD players are usually very simple, having L/R analog outputs and a digital output. Note that high-end CD players may also have an RS-232 or related connection for remote control.

DVD/DVD-Audio/SACD/Blu-ray

It is a good idea to make the audio from the main DVD or Blu-ray player available on the distributed audio system, so that movie soundtracks can be sent wherever desired. Remember, the digital audio outputs may work on the surround receiver but the analog outputs are needed to be selected on the Zone 2 matrix.

Mobile Devices.... and The Ubiquitous iPod

As mentioned earlier the iPod and other mobile devices such as phones and tablets have become popular as sources. In many cases they are integrated directly through the wireless network, but even in the most basic systems they can be treated as sources.

One common way to integrate these sources into a multi room audio system is with the use of a 3.5mm to RCA cable. You simply plug the left and right signals into the audio system input. REMEMBER, this method of integration will not allow for control of the device except through direct use of the Mp3 player itself.

Docking stations are now available for many popular devices, which provide not only access to the audio they contain, but the ability to control them via various control systems.

Computers/Servers

A digital music server can be any device that contains and organizes music digitally onto a hard drive. Many manufacturers produce chassis style units that may be placed with other traditional audio equipment and hooked up as a permanent source component. These types of servers typically contain hard drives with hundreds of Gigabytes of storage space and allow access and browsing through a GUI displayed on the TV screen.

These options may not be as cost effective as a simple docking station, but often times will provide a wealth of additional multiple audio streams.

Internet Sources

Interfaces for Internet Radio

An increasingly popular source option is to stream music from the internet. Many new receivers or other internet connected devices are adding features to allow customers access to internet radio applications such as Pandora or AOL Radio as well as subscription based music libraries like Rhapsody. Much like the FM and HD radio stations that we are familiar with, many internet radio stations offer access to free music. In some cases the customer may pay a service fee for upgraded features, like no commercials or unlimited listening but they still don't have the ability to search for or choose specific songs that they want to listen to. With subscription based libraries however, the user can search a practically unlimited database, create playlists, choose specific songs, and customize their own listening experience. In either case, the customer does not own the content that they are listening to, and cannot at any time transport the songs to different devices or share them to other systems.

Other Sources

In addition to these popular sources, some customers will request that other sources be integrated for use with their new system. These may include a variety of legacy devices including turntables, tape machines (open-reel, Compact Cassette, even 8-track), musical instruments such as keyboards or electronic drums, video players, and game consoles.

Receivers

In its most simple form the receiver is a combination of radio tuner, auxiliary input selector, and two amplifier channels (stereo). As described earlier, the stereo receiver was the core of the original multi-room systems. Its sources were selected manually (usually push buttons or a rotary switch) and then amplified to power 1 or 2 pairs of speakers. With an impedance matching transformer module, additional rooms could be added, all sharing the available amplifier power, of course.

Today's receivers are much more complex and capable than their ancestors. Receivers are typically able to switch input source audio (analog, digital, decoded surround [6 channels], and audio from HDMI). Surround receivers can also switch video sources in addition to audio sources (DVI, HDMI, S-video, component or composite). Most high-end receivers can also be remote controlled via IR, RS-232 or Ethernet from a separate control system.

In single-zone audio systems the receiver serves as the radio source, the switching matrix and the power amplifier.

Switching

In order to create a mulit-zone/multi-source system, a switching device is needed which can select any source, send that source to any zone, and control that source effectively.

The Multi-Zone Switching Matrix

The multi-zone matrix is the heart of any system with more than one zone. It is simply a switch that allows any one source to be directed to any zone or more than one zone.
Dad can listen to the ball game on the deck, while Mom listens to XM radio in the bathroom. Sources can be any number of different audio devices or the audio signal from an A/V component like a cable box of DVD player. The outputs of the matrix feed amplifier channels. When an output is stereo, a Left and Right signal feeds two amplifier channels (L & R) which then power a pair of speakers in that zone.

Power Amplifiers

Amplifiers make weak audio signals stronger. They increase the "amplitude" of the audio waveform. A good amplifier not only makes the weak audio signals stronger, but does it with as little distortion to the signal as possible. Amplifiers should produce a signal that is as close to the original signal as possible.

Most amplifiers have a minimum output load impedance requirement. This is the lowest impedance the amplifier can comfortably drive, and will be shown in its specifications. If the amplifier is rated only at 8Ω, then any load lower than that should be avoided. If the amplifier shows a power rating at both 4 and 8Ω, then is should be safe to have it drive a 4Ω load. Presenting an amplifier with a load that is too low causes it to run at a higher power output, which means higher current, which may be more than the output devices can handle. This will either damage the amplifier or trigger its protection circuit.

Stand-alone power amplifiers will usually contain from 2 to 8 separate amplifiers, each dedicated to a specific channel. They accept the input from a preamplifier or sound processor such as an equalizer and increase the amplitude of that signal to a level adequate to drive a loudspeaker and create sound.

Each amplifier channel may have a gain adjustment to compensate for differences in audio input

Speakers

Almost all dynamic drivers work the same way. The amplifier drives a coil of wire placed inside a stationary magnetic field. The direction of the current through the wire causes the coil to be displaced in or out, and the amount of current determines the travel distance. The coil is held in place by the driver cone. The cone movement moves air and thus creates sound.

Bookshelf and Tower Speakers

Within most distributed audio systems that have been designed in recent years, bookshelf and Tower (floor standing) speakers have been replaced by less obtrusive architectural style speakers. There are still clients out there however, who are willing to sacrifice the space in order to increase the performance of their system. This includes floor standing models and bookshelf speaker.

Subwoofers

There are a variety of subwoofer designs to choose from, including the popular self-powered, free-standing design, and even in-wall products which require a separate remote amplifier. In all cases, even moderately priced systems, two or more subwoofers are recommended for smooth response around the room.

The designer must be familiar with the specific products being used, and how subwoofers integrate with the regular full-range speakers.

In-Ceiling Speakers

Architectural speakers have become an enormously popular option for multi -room audio systems in the residential market over the last 15-20 years. If fact, the demand for attractive architectural speakers has driven a great deal of innovation in this product category. They are easy to hide, and in most cases, offer more than adequate performance for applications like background music and stereo audio listening.

In-ceiling speakers generally come in two types of configurations:

- **Monopole** – Single voice coil speaker to be used as either a left OR a right audio channel. One PAIR of monopole speakers is REQUIRED to achieve a normal stereo listening environment.
- **Stereo (DVC)** – Stereo speakers, commonly called dual voice coil speakers (DVC) can be used to reproduce both the left and right audio signals from a single speaker enclosure. These speakers offer a very convenient solution for applications like closets, hallways, and bathrooms.

In-Wall Speakers

In-wall speakers are another architecturally attractive option that is available when designing a distributed audio system. These speakers are often a very close representation of their bookshelf cousins and many times can offer extremely good audio performance while remaining hidden.

Like in-ceiling speakers, in-walls are specifically designed to be permanently installed in a framed residential structure. They are made to use the ceiling or wall cavity, and the plasterboard surface, to produce smooth audio response. Like in-ceiling speakers, they can usually be painted to blend in with the wall color.

Back-Boxes

Most manufacturers that produce architectural style speakers also offer back-boxes that can be used with their in wall or in ceiling speakers. Back-boxes provide a host of benefits and should be used whenever possible. In some cases, these back-boxes may even be required for the installation by local codes and regulations. Many commercial speaker designs are already completely enclosed.

Back-box benefits include:

- Increased speaker performance due to calculated/calibrated air space
- Added protection of speaker
- Reduced chance for speaker rattles and vibrations
- Reduced transmission of sound to adjoining spaces
- Increased fire safety

Invisible Speakers

Another option for hiding speakers within the home is to specify "invisible" speakers. These speakers can actually be mounted into the wall, and then textured and painted to match the wall. Once the installation process is complete, the speakers are essentially a part of the wall and fully "invisible".

When using these speakers, be sure to consult with the manufacturer in order to fully understand the installation process and requirements of the system.

This type of installation requires efficient project management and excellent communication with other stakeholders.

Contact Drivers

In addition to the invisible speakers we just discussed, there are options available that will turn any solid, flat surface within the home into a transducer (speaker).

These contact drivers can be mounted to anything from drywall to a window and then connected to the system via speaker wire just like a regular speaker. They then vibrate the surface of whatever they are attached to and created sound waves just like a normal speaker would.

Remember, these drivers are typically mounted to a large piece of flat, solid material that may be very difficult to move at extremely high speeds. This means that they may not be able to perform in the high frequency range very well but typically do quite nicely with background music applications.

It is always recommended that you consult with the manufacturer in order to understand how these products will work in your specific application. Many times it will require extra equalization in a system with contact drivers to make them sound convincing and natural instead of flat and artificial.

Outdoor Speakers

Another consideration when designing distributed audio systems is the client's backyard, deck, or pool area. Outdoor speaker systems can be an excellent added value to your customer, but in order to perform well, they must be properly designed.

Outdoor speakers are specifically designed to withstand the harsh environment of being exposed to extreme temperatures, moisture and the sun. This is helpful in that the speakers are built to outlast in the weather, but it also means that they are likely to be much less efficient than their indoor counterparts. Constructing transducers with tougher materials to withstand the elements also makes them harder to move, which results in a need for more power. As an added challenge, outdoor speakers do not have the benefit of room gain, (where sound waves bounce off of walls and reinforce themselves adding volume) like indoor speakers do. Outdoor speakers come in an extremely wide variety of shapes and sizes. Designers in the custom A/V industry should be familiar with a variety of styles in order to meet the needs of various customers.

Outdoor Subwoofers

As we mentioned earlier, outdoor speaker systems can have problems meeting adequate volume levels for comfortable listening. High frequencies have extremely short wavelengths, making them very directional and not as susceptible to volume loss from the missing boundaries of an indoor system. Low frequencies on the other hand, are very difficult to produce outdoors because bass frequencies get lost in all directions.

In order to solve the problem of inadequate bass in outdoor systems, some manufacturers have begun producing outdoor subwoofers. These subwoofers can drastically improve the overall performance and acoustical balance of an outdoor system and are highly recommended when a customer is looking to entertain in a large outdoor area.

Most outdoor subwoofers are designed to be buried in the ground in order to increase the boundary effect and use the earth around them as a baffle. In these applications, the subwoofer is passive and the amplifier will be remotely located. Remember that the added power requirements will mean larger gauge speaker cable and it must be rated for direct burial.

User Interfaces and Control Options

As with any subsystem, the overall success of any distributed audio system depends greatly on how easy and convenient it is to control. In the most simple systems, there might be a receiver or the Zone 2 of a surround system, and passive volume controls at the room location. In more sophisticated systems the multi-room audio might be just one part of a larger integrated system, controlled on the same touchpanels as all of the other systems.

Passive Volume Controls

Passive volume controls are most often used in distributed audio systems that are designed as a single zone. There is no source or input control with these user interfaces. They simply raise or lower the volume in a specific location of whatever is already playing through the system.

Sometimes these passive volume controls can contain impedance matching circuitry, which eliminates the need for a speaker selector or other impedance altering device at the head end of the distribution system. Passive volume controls come in every possible color one could imagine. They also come in outdoor, slider, or knob style configurations.

In some cases, if the speaker wires are run directly back to the amplifier instead of an outlet location within individual rooms, a speaker selector box with built in volume controls can be used to adjust the rooms from a single location.

System Volume Controls

Some distributed audio systems are designed with electronic volume controls that talk directly to the processor/

amplifier. These volume controls do not need speaker wire, because instead of attenuating the signal being sent to the speaker, the volume control tells the amplifier itself to turn up or down the output level.

Some of these electronic volume controls have the added option of receiving IR commands from a handheld remote.

In some cases, an electronic volume control with LEDs and UP/DOWN buttons may actually be a passive control that works without a knob. Check manufacturer specs as you design the system.

Other Interfaces

The distributed audio system is often just one of many subsystems controlled by a more sophisticated control system. This means the user interfaces with their music via hand held remotes, keypads, touchpanels, and mobile devices. The amount of detailed information available to the user depends on the features of these user interfaces. Remotes and keypads typically allow source selection and control of each source. Touchpanels and mobile devices may allow the user to see metadata and graphics related to their music sources. Designers should, however, be wary of automatically assuming a mobile device is the best, most efficient, method of control. Sometimes a simple keypad or handheld remote is a much more direct way to address a subsystem. Mobile devices utilize "apps" (applications) to perform various functions and sometimes there are a few extra steps involved which make the mobile device less than ideal. A good understanding of the client and how they will use the system is very important. For more about user interfaces, see Chapter 19.

SYSTEM DESIGN

This section will cover the "nuts and bolts" of what makes these systems work.

Basic Electrical Concepts

When designing distributed audio systems, it is extremely important to be well versed in the basics of electrical theory. This knowledge is essential in many different aspects of design and installation, from troubleshooting complicated system problems, to something as simple as proper planning of a speaker cable run. For these reasons, it is critical to understand the basic units of electricity, including voltage, amperage, wattage, and resistance. These fundamentals can be found in the chapter dealing with Basic Math and Electricity.

Impedance vs. Resistance

In alternating current (AC) circuits, the circuit resistance is more complicated than with direct current (DC) circuits. Typical circuit elements such as cables, speakers, and other components resist the change in current direction and add an extra resistance referred to as reactance.

The impedance of a circuit (cable, component, etc.) is made up of the normal DC resistance plus the reactance of the AC part of the signal added together.

The unit of impedance is the same as for resistance = Ohm.

Impedance is referred to in formulas by the letter Z (rather than R).

Why is This Important?

Anytime you are responsible for specifying a multi room audio system it is extremely important to use the Ohm's Law calculations in order to ensure that when the system is installed, both the speakers and amplifier(s) will be operating within the parameters set by the manufacturer. Improper system design in regards to speaker loads on an amplifier will typically result in:

- Very poor performance
- Low volume levels
- Clipping distortion
- Damage to amplifiers

Amplifier Power

Remember that power changes if the load changes. An amplifier channel rated at 100 watts into 8Ω will deliver about 200 watts into a 4Ω load. But if that amplifier is not rated to perform with a 4Ω load, it should not be put in that situation at all. Also, if an amplifier channel is driving two speakers, each of those speakers could potentially receive half of the total power available. All of these considerations are important in system design.

Parallel Circuits

Having already covered basic electrical theory, it is now time to apply it to speaker and amplifier system design. Parallel circuits take one output and send that output to multiple speakers simultaneously. With parallel transmission, current flows independently for each circuit. Parallel circuits work very well with distributed audio because they deliver more of the available power to the load.

Parallel connections with local impedance matching volume controls have historically been the most popular configuration in residential systems. Current flows independently for each circuit, so you can add volume controls to individual speaker pairs. This method is often also preferred because current flowing through one speaker does not go through any other speaker, so the sound quality is usually better in each speaker. Some low-impedance capable amplifiers are current-limited. As their impedance drops, their output increases (as in the example illustrated). As such, each individual speaker gets the same amount of power, whether you have a single pair, or two pairs. Note that you do not want to exceed the impedance limits of your power amplifier.

To calculate the total impedance of speakers wired in parallel use the formula shown below:

$$\text{TOTAL } Z = \frac{1}{\frac{1}{Z_1} + \frac{1}{Z_2} \cdots}$$

Series Circuits

Series connections transmit all currents through the electrical circuits by forcing the current to pass through each load in succession, one after the other. Current stays constant in series circuits, even if voltage is divided. The number of electrons remains constant.

A series circuit has only one path of current flow, and that path is continuous. Therefore, current flows from one speaker to the next in the series.

Series-type circuits allow you to control volume in all of the speakers no matter where you place the volume control. Current flows from one speaker to the next in line. Using series connections is a good way to increase the overall impedance in order to ensure a safe level for the amplifier. Note that if a speaker gets damaged, you will lose audio in all speakers on that line. Also note that the power output of the amplifier is split to each speaker.

Each speaker receives a portion of the total power based on its impedance, so you would calculate the total impedance by adding together each speaker's individual impedance. For example, if you have a series circuit with two speakers whose impedance is 8 Ohms each; the total impedance of the circuit is 16 Ohms

$$\text{TOTAL } Z = Z_1 + Z_2 \cdots$$

Speaker Sensitivity

As discussed in Chapter 9, this specification tells you the SPL that would be produced if a signal level of 2.83 volts (typically a sine wave at 1 Watt at 1Khz) were applied and a measurement were taken at 1 meter from the speaker on-axis.

It is important to understand the sensitivity of a speaker (this can help determine how much amplifier power is needed) as well as its dispersion pattern. For most background music situations, a low efficiency speaker may work just fine, but if higher levels are needed, a higher sensitivity will deliver more sound with the same amplifier power.

Wire Gauge and Length

Here is a quick reference that shows the AWG gauge needed to keep line loss to 20% or less. In most background music situations, only a few watts of power are actually needed, so a loss of 20% is not critical.

Why does the gauge need changed according to the nominal impedance of the speaker? Because a lower impedance (calculated like resistance in the circuit) results in an increase in current, which requires a larger conductor.

	🔈		
	4 Ω	8 Ω	16 Ω
50 ft.	16	16	16
100 ft.	16	16	16
150 ft.	14	16	16
200 ft.	12	14	16

Example:

An amplifier rated at 100W into 8Ω, would output about 200W into 4Ω, and only 50W into 16Ω.

Cable Specifications and Local Codes

Regional and local codes dictate what cable types (especially jacket materials) can be used for different applications. In the US and Canada, electrical codes are written by the NEC and CEC respectively. Other bodies provide similar guidance in other areas. It is important to understand and follow your local codes to provide safe installations as well as avoiding expensive fines.

You will find that certain cable construction techniques are approved for certain uses, and in many places residential and commercial codes may be different. Also be aware that many prefabricated interconnects, such as audio cables and HDMI cables, may not be rated for permanent installation within a wall.

Other Local Codes

Before installing in-wall or in-ceiling speakers, you should be familiar with the fire and building codes for the location. Some jurisdictions may require a fire-rated back can. For example, colder areas may have infiltration requirements that require either a back can or special techniques for sealing the outside wall and ceiling penetrations. Locales that are prone to earthquakes often require safety cables on all installed speakers, to keep them from falling out of the wall or ceiling. Most areas require penetrations of the home's framing to be filled with fire caulk. Sometimes, the insulation contractor does this; sometimes, the home technology professional is responsible.

Frequency Response

Frequency response is a description of how a device, or room, reproduces the audible spectrum of sound frequencies. (20Hz - 20 kHz). In real world applications, true flat response is rarely achieved. But a smooth response curve, with no drastic dips or peaks, is a reasonable goal in most situations. The response curves shown in manufacturer specs is usually the "on-axis" performance, measured in an anechoic chamber where there are no room reflections.

When a loudspeaker is placed in a room, the goal is for the listener to hear smooth reproduction of all audio frequencies. In general, most speakers and room environments have smooth enough response for background music, but more accuracy may be needed for more critical listening or entertaining. Equalization can be used to provide smooth frequency response. In some situations, equalization may be used to correct for the performance of the speaker and/or room.

Level Calibration

The primary goal of calibrating distributed audio systems is to provide the required SPL at the maximum volume setting, but no more. Although the amplifier may have power to spare, we may not want to make this power available to the user. Instead, we want to "dial-down" the amplifier output to the point at which it is only supplying enough power to achieve the target SPL when the user interfaces (UIs) are at their maximum setting. This step is important to ensure long term dependability of the system.

Deciding on a System Type

System design always begins with the client interview, or "discovery" phase. The designer should ask open-ended questions that will help them to understand how the audio system will be integrated into the user's lifestyle.

- What kind of music do the family members listen to?
- How do they entertain?
- What sources will be needed?
- What rooms need to have music?

From this collection of information, the designer can determine what kind of system is needed; how many zones, how many sources, how much power, what type of speakers, and what kind of control.

Placement Guidelines

For many reasons, you will sometimes be forced into less than optimal loudspeaker positions. Try to keep the following in mind as you search for the best possible solution:

Nulls are preferable to hot spots. People notice if music is becoming too loud. It will make them uncomfortable, particularly if they are trying to have a conversation or if they have any hearing disabilities. On the other hand, listeners will happily tolerate the music becoming quieter. If they want to hear it better, they will naturally move to a position where it is louder. If they prefer not to hear as much music, they may actively seek out a null location.

What is Uniform Distribution?

Obviously, every point in a room will measure somewhat differently. How much different can twopoints be and still be considered uniform? Remember that 6 dB is not as dramatic as it might seem, but it is noticeable. A maximum variance in level at different locations in a room of 3 dB would be excellent. A variation of 6 – 9 dB is more typical.

Sound Transmission

Sound transmission is a topic that can take days to cover thoroughly. For the purposes of this book, we will just touch on the fundamentals that must be considered when designing and installing multi-room audio.

When an open back architectural speaker (in-wall, in-ceiling) is installed, it produces sound in the room, and also on the backside of the speaker, in the wall or ceiling. The energy from the woofer is nearly equal between the front and rear. So a lot of sound is projected into the cavity behind the wall surface. In some cases, this is not an issue. Maybe the other side is in an attic or the garage; however, in many cases, such as a room adjacent to a nursery, we do not want sound in the adjacent space. It is imperative to think through all of these scenarios, even before the proposal is submitted. There will likely be some locations which require special attention. Always make sure the client is aware of this challenge, and aware that there may always be a little bit of bleed-through, even when measures are taken to minimize it.

In many cases, there are some steps that can be taken to mitigate sound transmission issues. First, the speaker locations may be modified to redirect the unwanted sound to non-critical spaces. Also, completely sealed speakers (commercial models or optional back-boxes) may be used to attenuate the sound. Joist/stud cavities will resonate less (making the speaker sound better) when they are filled with standard fiberglass insulation or other absorptive materials.

In some situations, it may be necessary to use wall speakers instead of ceiling speakers.

Outdoor Speaker Considerations

When you install speakers outdoors, there is a special set of considerations.

- Listeners may be farther away from speakers in an outdoor setting
- Without room boundaries, SPL levels fall off quickly; 6 dB for each doubling of distance
- Low frequencies are even more difficult to reproduce when non in an enclosed space
- All-weather speakers are constructed to withstand the elements and are sometimes less efficient
- Longer cable runs may introduce substantial signal loss

For these reasons, outdoor systems typically require more power and a close look at speaker sensitivity. The other major consideration is finding the best possible locations for all speakers.

Outdoor Speaker Placement Recommended Practices

- Placing speakers under the eave or against the wall of a house will add boundary reinforcement, delivering more level with the same power
- Using multiple speakers, all located closer to the listeners is generally better than a few larger speakers farther away.
- Proper placement and aiming of speakers can minimize the impact on neighbors.
- Cabling to outdoor speakers should be of a larger gauge and rated for direct burial (even if not buried). Remember, doubling cable length doubles its résistance and larger gauges have less resistance at a given length.
- If more bass response is desired, outdoor subwoofers will deliver the low frequencies and take a load off of the full range speakers.

Digital Rights Management

Another, somewhat misunderstood, aspect to consider when setting up a distributed audio system is content. Digital Rights Management (DRM) must be observed. DRM refers to different technologies used to enforce pre-defined policies that control access to software, music, movies, or other digital data. Copyright laws were originally enacted to protect authors from theft of their words and/or ideas. In recent years, copyright laws have been updated to protect material (words, ideas, etc.) that is electronically transmitted via the Internet and through other electronic means. Thus, you will need to be familiar with any such laws that are related to audio distribution technology.

Be sure you understand and follow the laws in effect in the country you are working.

Since technologies and copyright laws are constantly being updated and changed, you must keep abreast of any new technologies and laws that could affect your work in distributed audio.

RECOMMENDED PRACTICES

- Use impedance matching technology only when control limitations will not be burdensome; provide additional means of control if so
- Expose only a single means of volume control for each zone
- Provide local and remote jacks so that end-users can connect personal devices into the system
- Address the possibility of unacceptable "bleed through" into upper/adjoining rooms during the design of the system
- Use round speakers for ceilings; rectangular for walls
- Use decora-style plates and volume controls; use multi-gang boxes when more than one VC is present at the same location
- Attempt to provide uniform coverage so that all areas within the space remain within a 6 dB SPL range
- Use the quarter points along the longer axis of a room as the default locations for a pair of in-ceiling speakers
- Use multiple smaller speakers located closer to listeners in outdoor applications
- Design systems so they can produce a minimum SPL of 75 dB at the listening height anywhere within the space; some projects may require higher minimums
- Size amplifiers to produce 10x the wattage required to achieve the target SPL
- Calibrate installed systems so that the amplifier is producing the minimum gain necessary to achieve the target SPL
- Ensure that the total resistance of the speaker wire does not exceed 10% of the speaker impedance; preferably it will be less than 5%
- Follow industry standard color-coding conventions for four-conductor speaker wire
- Run UTP cable to the volume control locations even if no other provisions for present or future control have been made

SUMMARY

Distributed audio is considered to be one of the core competencies for any home technology professional. Distributed audio systems are part of almost every job done by an integration company. Remember; this content is not difficult to master, but it must be done right. Also, following the recommended practices outline in the next section will help you to achieve the performance goals desired, while also maintaining high efficiency. Always focus on reliability and ease of use.

Questions

1. What is the desired maximum level variation within one room of a distributed audio system?
   a. 3 dB
   b. 6 dB
   c. 10 dB
   d. 1 dB

2. Outdoor speakers require more power because:
   a. They are usually less efficient than indoor speakers
   b. They often require long speaker cable runs
   c. They don't have the advantage of room reflections
   d. All of the above

3. A system that allows the choice of both radio and internet music, with all rooms receiving the same programming, is:
   a. Multi-zone/Multi-source
   b. Single-zone/Multi-source
   c. Single-zone/Single-source
   d. Constant voltage

4. When the load (impedance) on an amplifier channel output is increased:
   a. Its output (in Watts) goes down
   b. Its output becomes louder
   c. Its output section is likely to overheat
   d. Its output exhibits a significant loss of high frequencies

5. If the "Zone Two" output of a surround receiver is used to provide audio to two additional rooms, what precaution should be taken to ensure stable performance?
   a. Make sure the output on the internal amplifiers is at maximum, to provide needed headroom
   b. Use smaller gauge speaker wire to both rooms, keeping impedance steady
   c. Use the line outputs and a separate power amplifier to drive the extra rooms
   d. Compensate for line loss by using the equalizer to boost high frequencies

Chapter 11
VIDEO IMAGING, SIGNALS AND DISPLAYS

HOW WE SEE

Visual imaging is the science of understanding, capturing and reproducing images for individuals to interpret and appreciate. Understanding how the eye and the brain "see" images provides greater understanding on how our customers will use video products. The eye and the brain work together to provide a visual image. The eye has two different types of receptors, rods and cones, which are distributed across the retina.

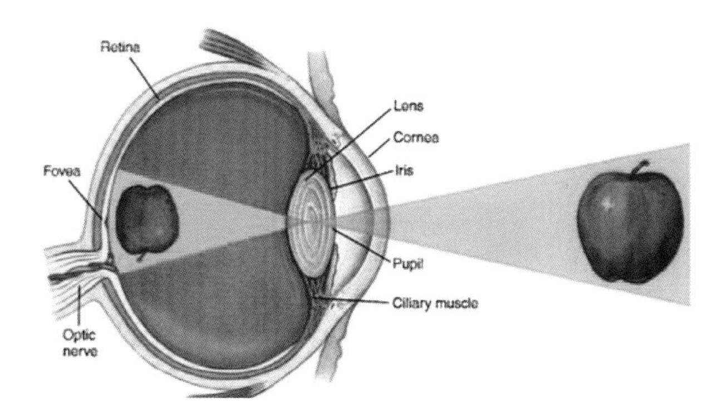

Figure 11-1 The Human Eye

Rods:

More sensitive and are used in dim lighting and at night
Prevail in the periphery of the retina
Have lower visual acuity than cones
See only in monchrome (black and white)

Cones:

Less sensitive than rods and are used in daylight
Predominately in the center of the retina
Three kinds of color sensitive pigments (red, green and blue), which allow each type to respond to a different range of the visible wavelength (If one or two of the pigments are missing, a form of color blindness results.)
Sensitive to color wavelengths that are very similar to the RGB wavelengths of the video signal
Have high visual acuity

How We See Color

A video display reproduces color similar to the way cones in the human eye pick up color. A typical color TV is made up of three colors – red, green and blue – more commonly called RGB. These three colors mixed together yield white, gray, and all other color hues. The absence of them yields black. A good video imaging system has the ability to show nearly all visible colors depending on the ratio of these three colors. By comparison, the eye sees actual colors (wavelengths) for each RGB color but not exactly the same as the color TV. The eye senses the green and blue wavelengths as the color TV does. However, the eye's red receptor is more yellow in wavelength than the red phosphor on a display. Shown here is the visible spectrum. Note that wavelenghts shorter than the visible spectrum are Ultraviolet and those longer wavelenghts are Infrared.

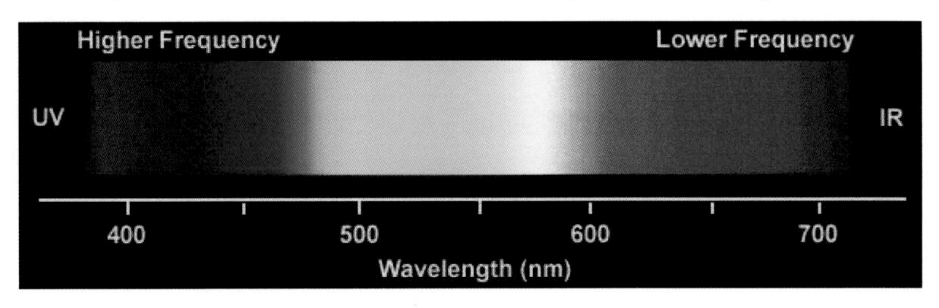

Figure 11-2 Visible Color Spectrum

Image Perception and Persistence of Vision

Image perception is a person's ability to detect a certain level of contrast or the recognition of patterns in an image from its details. This perception is an observer's capacity to respond to low-contrast and fine-detailed stimuli. Image perception is related to the psychophysical skill of detecting light and interpreting it as a perception.

Persistence of vision is the characteristic which allows movies and video to be seen as smooth motions as it happens in nature. The human eye retains an image for a fraction of a second after it views it. This effect is very important in how video displays work. The brighter the image, the shorter the vision persistence is. We know that 24 or more still "frames" per second will appear as natural looking motion.

Scanned Images

Video displays show scanned images. The scanning process breaks an image into distinct pieces of data and keeps them in an orderly sequence. Just as humans visually scan information off the written page (e.g., English readers from left to right and then back to the left to the next line); video scanning occurs at a very high rate depending on bandwidth capacity. Detailed resolution capability is determined by the scanning rate or number of lines scanned over a period. As the scanning rate increases so must the bandwidth capacity increase.

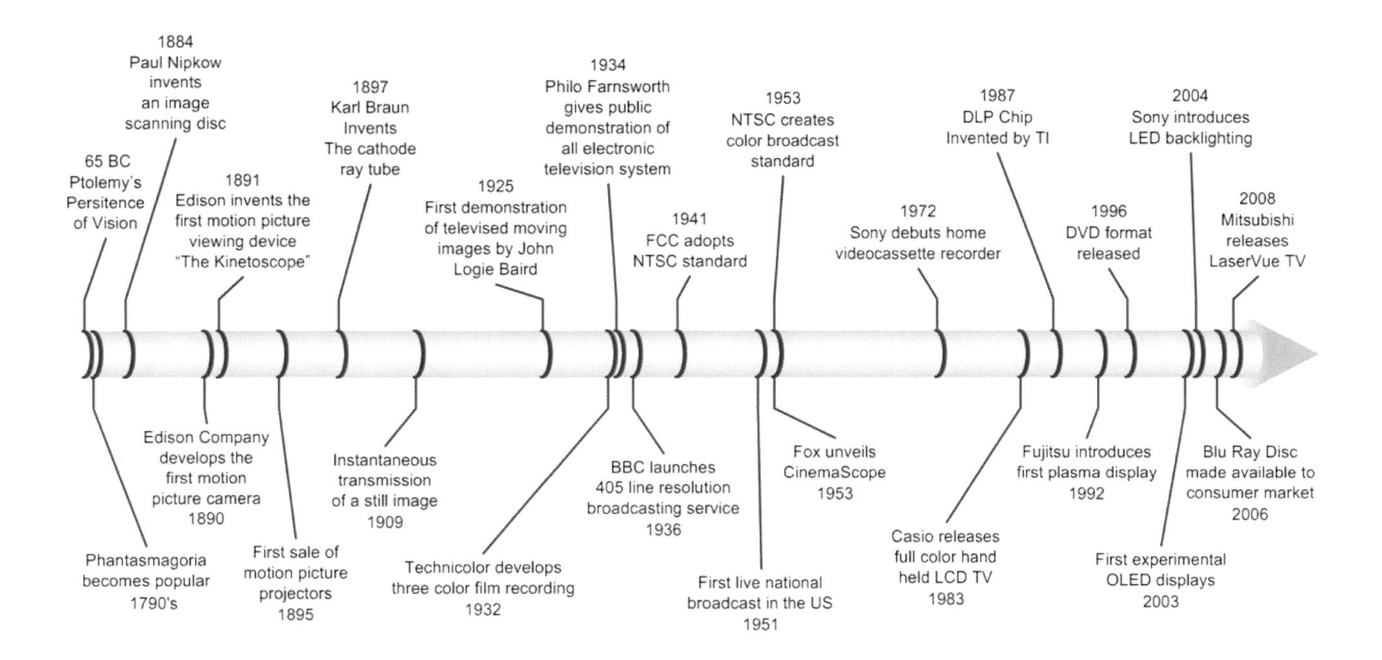

Introduction to Motion Pictures

Unbelievably, the concept of cinema goes all the way back to 65BC. Ptolemy was the first to discover "persistence of vision". That was actually where that whole idea of a motion picture came from. We are talking about something that has been around for quite a while.

In the 18th and 19th centuries, there was a concept known as Phantasmagoria, which was popular in Europe. This strange presentation using shadows, projection, and even smoke, to create the effect of ghosts and other scary images, was so convincing that it often scared people off. It was first used to make séances more convincing and later in other entertaining ways.

There was one famous "ghost showman" named Etienne Robertson who did a stage show in Paris that included sound effects and other clever techniques to terrify his audiences. "I am only satisfied if my spectators, shivering and shuddering, raise their hands or cover their eyes out of fear of ghosts and devils dashing toward them", he once said. It may be hard to believe that the concept of the "scary movie" was really around back in the 1790's.

The "motion picture" is typically considered to have been born in the late 19th century and the real growth of the industry started in the early 20th century. At this time, the first commercial theaters were being built and the silent movie era began to cause the demise of vaudeville. The growth of the industry with sound started coming about in the 20's and 30's. The 1930's through the 1950's was the golden age of cinema in Hollywood. This is where the glamour and excitement of the whole movie or cinema experience started to occur. Then during the 1950's, television began to be seen by Hollywood as a real threat to their revenue. This had a big impact on the movie industry. With television, people had the option of staying home and selecting from a few different programs. To combat the consumer view that they could enjoy programming at home just as easily as at the theater, movie producers and directors started making wider screens and using gimmicks like 3D to attract more customers. It is not just a coincidence that we are now seeing those wide formats in the home theater market. The other technological developments in audio and video have followed the same pattern, with home and public entertainment taking cues off each other.

With High Definition television and 16:9 widescreen formats established as the norm for broadcast TV, the movie industry has answered with 3D formats and even more recently, motion activated seating. Both of these technologies can also be purchased for the home, and so the cycle continues at a faster and faster pace.

Early Film History

Motion pictures as we know them today can generally be traced to 1890, when the Edison Company's head motion picture engineer W.K.L. Dickson invented the first motion picture camera, called a kinetograph. The camera was followed closely by a viewing machine named a kinetoscope, where the viewer would start the machine by inserting a coin and then looking through a "peephole" to watch the film.

In 1894, the Holland brothers opened the first kinetoscope parlor in New York. This storefront business offered patrons the opportunity to walk in and pay per film they wished to view. This also offered the ability to have multiple films for viewing and change them out as the local patrons got bored with the old ones. After the first year of operation, the parlor had seen gross receipts of $16,000, proving the financial viability of the industry.

The true birth of "movies" is considered to have occurred in 1895, when the Lumiere brothers invented and then showcased a new film recorder/projector system (called the Cinematograph) at the Grand Café in Paris to a paying audience of 33 spectators. This date is generally recognized because the Cinematograph was the first film projector to advance beyond the experimental stage and be offered for commercial sale.

Motion pictures took another step as an industry when Harry Davis, a wealthy vaudeville theater owner, opened a storefront theater in Pittsburg where customers could watch an entire program of films for a nickel. Up until this point, films were generally only a few minutes long and customers would pay per viewing, but by charging a single price and offering 10 to 30 minutes worth of programming, this format gained huge popularity with amazing speed. Before the end of 1905, over 9,000 of these "Nickelodeons" had sprung up across the country, and generally signaled the downfall of live entertainment formats like vaudeville.

The Evolution of Television

The first step in the evolution of televised images came in 1884, when Paul Gottlieb Nipkow patented his theory for an image-scanning disc. By using a disc with equal diameter holes drilled in a spiral configuration it became possible to "scan" an image into a sensor where the light and dark variations could be changed into an electrical signal. Then, the signal would be used to control a light source behind a second disc, rotating in sync, at the same speed and direction as the first, and reproduce the original image. This electromechanical concept was used in many design concepts during the infancy of television broadcasting, but with higher resolution and more reliability. Fully electronic systems proved to be a much better solution in the end.Electronic televisions were made possible by the invention of the Cathode Ray Tube by Karl Ferdinand Braun in 1897. Although it was Braun that invented the CRT, it was Alan Archibald Campbell-Swinton (a fellow of the Royal Society in London) that originally theorized the use of cathode ray tubes in the transmitting and receiving of images in 1908. However, it was not until 1934 that the first real all electronic television system was demonstrated to the public by Philo Farnsworth at the Franklin Institute in Philadelphia. With multiple scientists

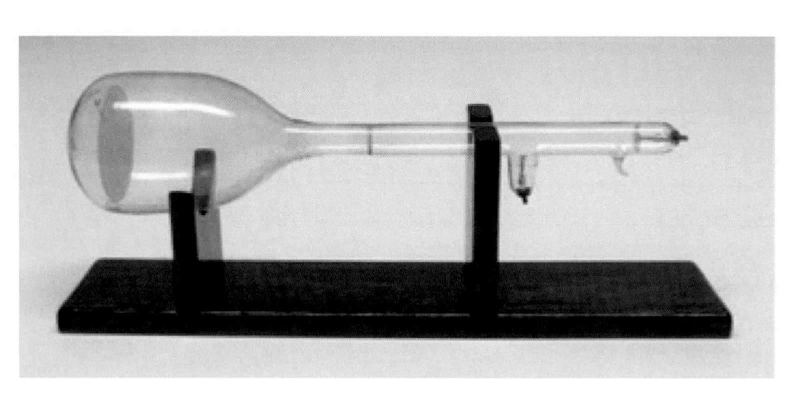

Figure 11-3 Early Cathode Ray Tube

working around the globe, the early television industry struggled to find a format that everyone could agree upon. Even today, there are many different formats still in use around the world, but in order for the industry to develop and grow within the United States something had to be done. That is why, in 1941, the FCC approved the standards set forth by the National Television Systems Committee and mandated that all stations would conform to broadcasting their signal in the newly defined format.

With a standardized format and manufacturing costs coming down, televisions began to enter the consumer market. In 1951, President Harry Truman's speech at the Japanese Peace Treaty Conference in San Francisco, California became the first live national broadcast in the US. By 1954, more than half the American population owned a television set.

As new digital technologies began to emerge in the mid to late 1980's it became clear that the NTSC (National Television Systems Committee) standard would eventually become outdated. During the early 1990's the Advanced Television Systems Committee (ATSC) developed a new set of standards that was officially adopted by the FCC in 1996. With the new format in place, (after a number of years of pushing back the deadline) on June 12th 2009 the FCC finally mandated the termination of all analog broadcasts in the US.

VIDEO TERMS AND SPECIFICATIONS

Digital Signal

A digital signal is based on discontinuous data or events in which audio or video information is represented in binary ones or zeros. The on/off digital video signal is usually a step wave form, whereas an analog signal is a continuous wave form. Digital content can be broadcast, stored magnetically or optically, and shared via the internet.

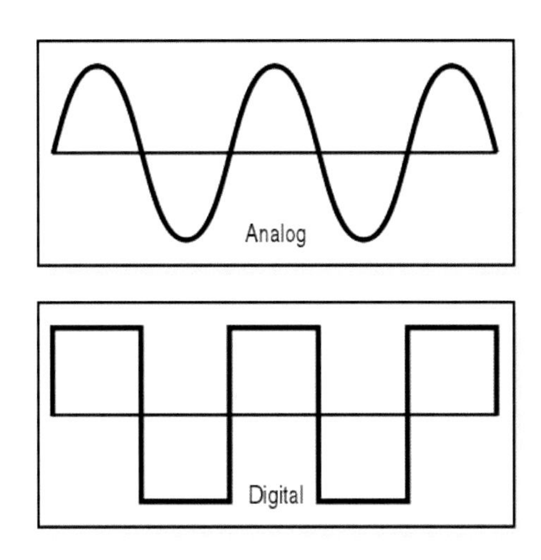

Figure 11-4 Simple Digital and Analog Waves

Analog Signal

An analog signal is a continuously variable signal in which small fluctuations are meaningful: e.g., audio recordings (vinyl records, cassette tapes), video recordings (Beta, VHS) and broadcasting (AM, FM, conventional TV). Like analog audio, each generation of copies loses some quality.

A variety of analog video signals haved been used over the years:

RGBhv -The best quality analog signal is one that keeps all five elements of the image separate: Red, Blue, Green, Horizontal sync and Vertical sync. This RGBhv signal is also known as VGA, the signal often used for computer video signals. The reason this signal is so well transmitted is that when going from an analog source to an analog display, NO processing is needed on either end of the transmission.

Component - The next best signal is called Television Component and in this signal type of signal the five elements are processed in such a way that they can be sent down just three coaxial cables. This type of signal has been widely used for many years to deliver a high quality image.

It will provide an analog image up to 1080i. Currently the component output on many devices is being voluntarily downgraded by manufacturers to carry only lower resolutions, in order to move end users to utilizing digital interconnections.

S-Video is a signal type which processes those same elements in order to be carried over just two coaxial cables. It was developed during the Beta/VHS "wars" back in the '80's. Because there is even more processing needed, its picture quality is not as good as component or RGBhv. S-Video is rapidly disappearing from the marketplace.

Composite Video is widely used as a simple, single cable solution to interconnecting basic video devices, such as VCR's and basic video games. It is a compromise of quality but popular when a high quality picture is not needed. It is usually carried on a yellow RCA cable.

RF - Finally, there is one more type of analog signal which is even more convenient, yet even more compromising. In order to transmit several different video signals, along with their accompanying audio, each one of these "channels" can be modulated to ride "on the back of" a specific carrier frequency. When this signal reaches its destination, the carrier wave is removed, leaving the original signal. This method of transmission is known as RF and includes off-air, cable TV, and even satellite. An analog carrier wave can carry either analog or digital information, so even if the signal is digital, when several channels are carried on one cable it is still referred to as RF.

The primary forms of digital video are HDMI, DVI, and Display Port.

DVI - Digital Visual Interface

DVI was the first High Definition digital video interconnect. There are many different types of DVI interfaces. DVI carries the same high quality signal as HDMI but does not carry any audio signals. DVI is however the only video standard that is capable of carrying either Digital or Analog video using the same connector. DVI is often seen on computer monitors, now that HDMI has largely replaced it for A/V applications.

Figure 11-5 DVI and HDMI Connectors

HDMI - High Definition Multimedia Interface

HDMI has become the de facto interface for digital video. The HDMI 1.0 specification was designed to improve upon DVI by using a smaller connector, adding support for audio, boasting richer colors and utilizing consumer electronic control (CEC) functions. The HDMI specification describes both the physical connection in terms of connector style and pin-outs as well as the signals transmitted through the wire. HDMI was the first uncompressed, all digital video/audio interface. Standard HDMI cables support 720p/1080i (resolutions will be discussed later in this chapter) signal and have been tested to perform at speeds up to 75 MHz or 2.25 Gbps (gigabits per second). High Speed Cables are tested to perform at speeds of 340 MHz bandwidth or up to 10.2Gbps for compliant support of 1080p resolution and also supporting higher end formats such as 4K resolution. This is the highest bandwidth currently available on a HDMI cable. The HDMI Ethernet channel was introduced with version 1.4. It can transmit data up to 100 Mbps (megabits per second) and supports all the networking formats.

In 2013 HDMI 2.0 was introduced, which added several new specifications:

- Bandwidths up to 18Gbps
- Support for 4K (2160p)
- Up to 32 audio channels
- Audio sampling frequency up to 1536Khz
- Dual video streams on one display
- Up to 4 simultaneous audio streams
- Support of 21:6 aspect ratio

All of these features can be carried over a High Speed HDMI cable.

HDMI Terminology

Audio Return Channel - The audio return path in HDMI which allows a TV to send audio data upstream to an A/V receiver, eliminating the need for a separate SPDIF or TosLink audio connection. It supports the same audio formats as a SPDIF cable, which means it does not support high-definition audio or multi-channel PCM, only the compressed surround formats.

HDCP - High-bandwidth Digital Content Protection, an authentication system developed by Intel designed to protect copyrighted audiovisual content. Most HDMI-enabled and DVI-enabled devices employ HDCP.

Source - A device that sends an HDMI signal, such as a DVD player or set-top box.

Sink - A device that receives an HDMI signal, such as an HDTV.

DDC - The Display Data Channel, one of the channels in an HDMI connection. DDC allows devices to assess each others' capabilities and adjust themselves accordingly. For example, a DVD player can discover the maximum resolution of the monitor it is connected to by reading the monitor's EDID chip and optimize its signal output to match that monitor's display capabilities.

EDID - Extended Display Information Data, the data contained (in memory called EPROM) on each DVI display or HDMI sink. The source device checks the display's DVI or HDMI port for the presence of an EDID PROM and uses the information inside to optimize the output video and/or audio format. All sink devices compliant to the HDMI specification must implement EDID.

Cliff Effect - As the cable length increases, the image quality is perceived to be perfect up to a certain length. Beyond that length, the image either degrades in quality or disappears all together. This is known as the "cliff effect". The cliff effect can vary based on the cable's insulation, inter/intra-pair skew, and termination of the cable connectors, among other factors.

Resolution

The term resolution is used in three different ways in the AV world. The first describes how much **horizontal bandwidth** is present (in a signal or a display). This is typically described as horizontal resolution and specifies how many horizontal transitions between black and white can be reproduced.

The second describes the **number of horizontal and vertical pixels** in a signal or on a display screen. Pixels used to create digital displays are measured in a Cartesian X (horizontal) Y (vertical) system.

The third describes the **number or quantity of levels**. Often called quantization levels or color depth, they are measured in bit depth volume (8 bit, 10 bit, etc.).

In describing video signal formats, the vertical number of scan lines vis commonly used to identify signal type. Usual numbers for video signal resolutions are 480, 720 and 1080 lines. Video signal resolutions of 480 and 1080 can be interlaced or progressive (more to follow).

In our electronic world, marketing focuses a lot on resolution. The term resolution is often used to define a video signal or the native resolution of a display. Native resolution is the single fixed resolution of the display. CRT Monitors can usually display multiple resolutions but flat panels need to scale the image if it does not match its fixed native resolution. Resolution is an easy number for manufactures, sales people and end users to use, but it is not extremely important in terms of visual quality. Contrast ratio, color depth, frame rate and bit rate have far more effect on image quality than pixel resolution.

Signal Formats

Video signals and broadcast formats have evolved over the years and vary in different parts of the world. In the US, NTSC has been replaced by ATSC:

- National Television Systems Committee (NTSC) - Standard Definition Analog Television
- Advanced Television Systems Committee (ATSC) – Digital TV and HDTV

Various parts of the world are regulated by different video standards. These standards facilitate the transmission, reproduction and reception of a variety of signals to a variety of video displays. The world's three major broadcast standards are:

- NTSC/ATSC
- PAL
- SECAM

NTSC

In 1940, the FCC established the National Television System Committee (NTSC) to create standards for analog video distribution. The standards were approved in March 1941, and although they have had minor changes over the years, they were in use until 2009. The NTSC standard was used in the United States, Mexico and Canada, as well as many other countries and the transition to digital varies in different regions.

Since video is transmitted over the air, images must be broken up and transmitted or recorded as a series of lines, one after the other. Until recently, television was primarily an analog medium. The NTSC standard defined a composite video signal with a refresh rate of 60 half-frames (interlaced) per second. Each frame contained 525 lines and could display 16 million different colors.

PAL

The Phase Alternating Lines (PAL) standard is predominant in countries with 50 Hz alternating current. (There are some exceptions. Brazil is governed by PAL and operates with a 60-cycle current.) In the PAL system, each frame has 625 lines but the scanning time is different. Sometimes this produces slightly more flicker. Most of Europe, including the United Kingdom, Australia, China, Brazil, Hong Kong and some African countries operate under the PAL standard.

SECAM

In 1956 the French developed the first European standard for analog color television. It is called Séquentiel Couleur Avec Mémoire, translated: sequential color with memory. This standard was also widely adopted in Eastern Block countries for political reasons to encourage incompatibility with Western transmissions.

Designers must be aware of these world video standards in dealing with clients who might desire to bring equipment from Asia or Europe to the U.S. For instance, not all PAL VCRs can play NTSC tapes.

ATSC

Advanced Television Systems Committee is the current digital broadcast standard used in the United States. The ATSC, formed in 1982, has developed standards for digital television broadcasts, which include (but not exclusively) the High Definition standards. As of 2008, virtually all off-air broadcasts in the US have been changed over to the ATSC digital standards. All satellite services are already digital. Cable providers are left to make the transition to the more efficient digital signals at their own discretion and this transition is happening rapidly. Note that all HD signals are digital, but not all digital signals are HD. Standard Definition digital signals provide improved picture and audio transmission but are still in the 4:3 aspect ratio.

VESA

Video Electronic Standards Association (VESA) is an international nonprofit corporation led by a Board of Directors representing a voting membership of more than 165 corporate members worldwide. VESA supports and sets industry-wide interface standards for the PC, workstation and consumer electronics industries. They also establish standards for flat panel mounting configurations.

Interlaced & Progressive Scanning

Video signal systems use two scanning methods:

- Interlaced scanning. In this system, one-half of the lines weave onto the screen in one cycle of the video system and the other half weave onto the screen in the next cycle. The movement is not very visible because it happens 60 times per second (in NTSC). There can be some flickering visible to your eye and the picture may appear jagged if you really concentrate on it.
- Progressive scanning. In this system, the picture is drawn sequentially, and the lines appear at the same time as a whole frame, giving it the same look as in a movie theater.

Progressive or sequential scanning was first used on personal computer displays. Progressive scanning addresses display pixels sequentially from left to right on each horizontal line. As each line is drawn, a picture is completed from the top to the bottom of the picture. The process is accomplished by an electron beam in CRT, or by addressing digital pixels one by one in digital type displays. When each line is completed, a blanking period is applied which shuts the picture. The horizontal blanking period when video is shut off is also known as the horizontal retrace period or flyback time for CRT-type displays.

Similarly, when each picture frame is completed at the bottom of the picture, the video is again blanked and the scanning process returns to the top of the picture. This period of time is known as the vertical blanking interval, vertical retrace or flyback time on the CRT-type displays. This process reduces image flicker.

In general, the signal with the most pixels will look better. But, how the pixels are created, stored and transmitted can have an extremely large impact on image quality. For example, a 720p signal of a fast-moving sports event may have higher image quality than the same event captured in 1080i. However, a film transferred to video in 1080i will most likely be superior to one transferred in 720p.

4K

4K Resolution is double the resolution of High Definition, or 3,840 x 2,160, which means four times the number of pixels (about 8 million vs. 2 million). The 4K standard adopted for television is known as Ultra High Definition Television (UHDTV). Commercial projection uses a 1.85:1 aspect ration rather than 1.78:1 so their pixel count is approximately 8.8 million. While the availability of 4k content is still very limited, the upscaling done by the display still delivers an excellent image. This allows the viewer to be much closer to the image withoug seeing pixels, and allows much larger displays to be used when viewing distance is already established.

3D

The physical world exists in three dimensions: width, depth, and height. 3D imaging is based on the way the human brain and eyes work. The pupils of a person's eye are about 2.5" apart. Each eye views a scene from a different angle and generates a unique image. The brain then merges the images to create a single picture. The slight difference between the image from the right eye and the image from the left eye allows the brain to judge the depth and 3D vision is attained.
In order for the human brain to interpret depth into a two dimensional picture, we must recreate the effect of each eye receiving a different angle of the picture. There are a number of different ways to accomplish this, but the most popular involve the use of either polarization, or active shuttering.
Passive 3D creates a polarized image which can be viewed with polarized glasses to get a different image to each eye. This is the modern version of the old red and green glasses of early 3D movies. Active 3D uses glasses which have LCD shutters which open and close each lens alternately. 3D technologies which require no glasses have been demonstrated but have not proven themselves in the marketplace. 3D is widely enjoyed in commercial theaters, and has seen less acceptance in the home.

Aspect Ratios

To begin, we need to explain what an aspect ratio is, and re-define the terms when it comes to home theater. For example, most of us know that the current HDTV aspect ratio is 16:9. This means that the screen is 16 units wide by 9 units tall. Divide the width by the height to calculate the ratio. If we were in the film industry, we would refer to this aspect ratio as 1.78:1, which is actually easier to understand: the image is 1.78X wider than it is tall. Both numbers express the same ratio – 1.78:1 is the same as 16:9. For purposes of this discussion, it is easier to think of 16:9 as 1.78:1, where the image is 1.78 times wider than it is tall.

Figure 11-6 Motion Picture Aspect Ratios

Today, there are three common aspect ratios in motion pictures:

- **1.33:1 STANDARD** – Rarely used today for theatrical motion pictures, this format essentially died out for film presentation during the 1950s. It is still used today for some television production; however, it is quickly being phased out in favor of 16:9 (1.78:1) HDTV. Notable films shot in this format: CASABLANCA, WIZARD OF OZ, and GONE WITH THE WIND
- **1.85:1 ACADEMY FLAT**- This is the closest cinema aspect ratio to the HDTV 16:9 format. While this is a widescreen format, this aspect ratio does not give the full immersive effect of 2.35:1. Instead, this format is generally used for dramas and comedies where the director wants a more "intimate" effect. With this format, the people are generally the largest items in the frame. Notable films shot in this format: E.T., ROCKY, THE GODFATHER, and SAVING PRIVATE RYAN.
- **2.35:1 / 2.40:1 SCOPE** – This format is the most popular for big budget film productions and has been since the 1950s up through today. In fact, over 75% of the top grossing movies of all time have been shot this way. The director chooses this ratio when he or she wants a huge, immersive image that envelops the viewer and makes them feel part of the action.

The "original" aspect ratio developed for film is 1.33:1, where the image is 1.33 times wider than it is tall. The 1.33:1 image is almost square with the image only slightly wider than its height. This aspect ratio actually represents the physical frame size of traditional film stock. The exposed area of 35mm film has an aspect ratio of 1.33:1, which is true even of film and still cameras today.

When television came along during the late '40s and early '50s, it adopted this ratio, except it was re-named 4:3. Again, these are different representations of the same ratios.

Figure 11-7 Film 1.33:1 = Television 4:3

During the 1950s, movie studios were feeling the effects of television on the film industry. Instead of going out to the movies, people were staying home and watching TV. To counter this, movie studios came up with new ideas to get people into the theater.

One of the first strategies tried by the movie studios was widescreen, panoramic film production and projection. The idea was to give theatergoers an immersive, giant screen experience they simply could not reproduce in the home.

Figure 11-8 TV vs. Cinemascope

The thinking behind widescreen photography is based upon solid science. As human beings, our whole sense of being immersed in the world around us is derived from engaging our "peripheral vision." Widescreen photography engages our peripheral vision and allows us to be fully immersed in an image. This has to do with the fact that our eyes are spaced apart horizontally rather than vertically.

Anamorphic

Motion picture engineers developed a system to create widescreen images that utilized the full 35mm film frame by placing an anamorphic lens in front of the camera when shooting the film. An anamorphic lens horizontally "squeezes" the light coming into the camera so that a panoramic, widescreen image could be stored on what was essentially a square film frame. The resulting image is stretched vertically, with objects and people being too tall and skinny. This "anamorphically squeezed" image is what is stored on the 35mm film frame itself. When projected in the theater, another lens is used to stretch the compresses image back to its original wide format.

COLOR SPACE, COLOR TEMPERATURE & VIDEO ADJUSTMENTS

Color Space and the Color of White

Color space refers to a color model that describes the way colors can be represented as numbers. This way of describing colors is used in calibration. Keep in mind that video displays are capable of a large color space, but not equal to that of the human eye.

Black is the absence of light. White (and all shades of gray) are created by a combination of red, blue and green. It is important that we establish a "color of white" in order for all of the other colors to look right. We all know we can compare several objects, all of which are considered white, and they will not be exactly the same color. So standards were established by the International Commission on Illumination (CIE) so that producers of content, manufacturers, and installers can all refer to the same color of white. The color of white known as D65 or D6500 is the color of sunlight at mid-day in Western Europe. We use this as our reference. A lower color "temperature" has a red hue, a higher temperature has a blue hue. Hence the terms "cool" or "warm" light.

Brightness (Black Level)

The brightness control on a display actually establishes the level of black, on which all other colors is based.

Contrast (White Level)

Contrast is the relationship between the lightest and darkest areas on a display device or picture. A small difference means low contrast; a great difference means high contrast. Based on the black level, the contrast determines the white level.

Color

The color level control adjusts the gain, or color saturation, setting for the color decoding system. There is no accurate method for adjustment of color level other than using a test pattern designed for this purpose.

Hue

The Hue or Tint control adjusts the balance between red and green and is important to achieving accurate color, especially for flesh tones. This adjustment is not used in the PAL broadcast standard.

Sharpness

The sharpness control, as with many controls on a video display, is not calibrated to a reference standard. Most sharpness controls add a highlight around images to give the impression of greater sharpness. Too much sharpness results in visible artifacts on sharp edges, too little makes the image seem out of focus.

Grayscale

Grayscale calibration can only be done with sophisticated equipment. It ensures that all levels of gray, from black to white, contain an equal balance of red, blue, and green.

DISPLAY TECHNOLOGIES

Display Types

Direct View

Front direct review refers to a displayed image from a flat panel, such as LCD or PDP or a Cathode Ray Tube (CRT). These displays are getting thinner and lighter, as well as larger, as the prices trend downward. The retail price of a 60" display has dropped by as much as 90% in just over a decade. The technologies are maturing and performance is excellent.

Front Projection

Front projection refers to an image projected on a screen by a video projector. The traditional front projection system is still the most widely used for serious home theater applications mostly because the image can be made big enough to provide a true cinema experience for several people in a large room. Disadvantages include the presence of the projector (and its heat and noise) in the viewing room, and the fact that ambient light washes out the image.

Rear Projection

Rear projection refers to an image projected on a screen surface from the rear of the screen. For many years, High Definition rear projection televisions were a cost-effective option to get very good performance in a package that was lighter and more shallow than a tube television. These displays have nearly disappeared as flat panels have come into wide use. However, the rear projection concept is still very viable using a projector and screen. This configuration requires a darkened space behind a special screen, and has the advantage of a bright clear image, even if the viewing room is not darkened.

Making the Right Choice

Selecting the appropriate display style for use within an environment is a critical choice. The designer must take into consideration image size, viewing angles, ambient light, and the type of content to be delivered. Within each display type there are various technologies that also must be considered.

Display Technologies

CRT (Cathode Ray Tube)

CRT is a specialized vacuum tube in which images are produced when an electron beam strikes a phosphorescent surface. Karl Braun invented the CRT in 1897, but it was not until 1931 that Allen Du Mont made the first commercially practical and durable CRT. Up until recently, the CRT was the basis for virtually all video displays. CRT's are emissive (emits light to create an image). The CRT picture is excellent because the blacks are produced literally by the absence of light. This means the contrast between dark and light is very good. The CRT, however, is nearly gone from the consumer marketplace, having been replaced by flat panel displays and other projector technologies.

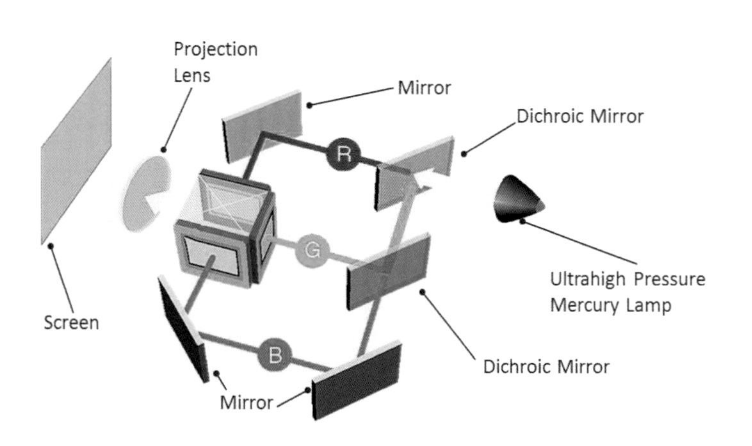

Figure 11-9 LCD Technology

LCD (Liquid Crystal Display)

LCD displays first gained popularity in laptop computer displays and their sizes vary from quite small (portable DVD players) to over 100-inch units available today.

LCDs provide high definition, are long lasting and offer high performance in bright environments. They are lightweight and energy-efficient.

LCD direct view displays are transmissive (light is transmitted through the panel). The actual light originates in a florescent, electroluminescent panel, or LED array mounted behind the liquid crystal passive panel. Each pixel is made from three liquid crystal elements (one for each color) that are either on, off, or in between. This allows a variable amount of light through each of the Red, Green and Blue pixel elements.

LED TVs are LCD panels that use "light emitting diodes" in place of CCFLs allow for localized dimming which results in improved black levels and contrast ratios. LEDs also produce a wider color spectrum and allow for more vibrant and accurate coloring than traditional LCDs. Finally, the newer LED technology has allowed for even slimmer panels through the use of "edge" lighting, where the LEDs are placed around the perimeter of the screen as opposed to directly behind it.

A rear projection LCD display uses the same technology as a direct view system. However, the light source is concentrated in a high intensity "point" source. Light passes through the liquid crystal pixels and is projected onto the back of a screen material. There are few of these still on the market at this time.

PDP (Plasma Display Panel)

Plasma flat screen displays consist of two layers of glass making a sandwich of pixel cells containing gases, electrodes and phosphor. Passing a high voltage through these low-pressure gases causes them to become plasma and emit Ultraviolet (UV) light energy. This UV light energy stimulates the phosphor to produce visible light and create a picture.

Donald Bitzer and H. Gene Slottow invented the plasma display panel at the University of Illinois in 1964. Plasma sets actually work in a way similar to that of CRT sets. However, instead of using an electron beam to excite the phosphor-coated screen surface, plasma sets use a highly ionized xenon and neon gas in thousands of micro-chambers. Because, like a CRT, a plasma display creates light when needed and leaves the pixel unlit when not needed, the blacks are accurate and the contrast ratio is high.

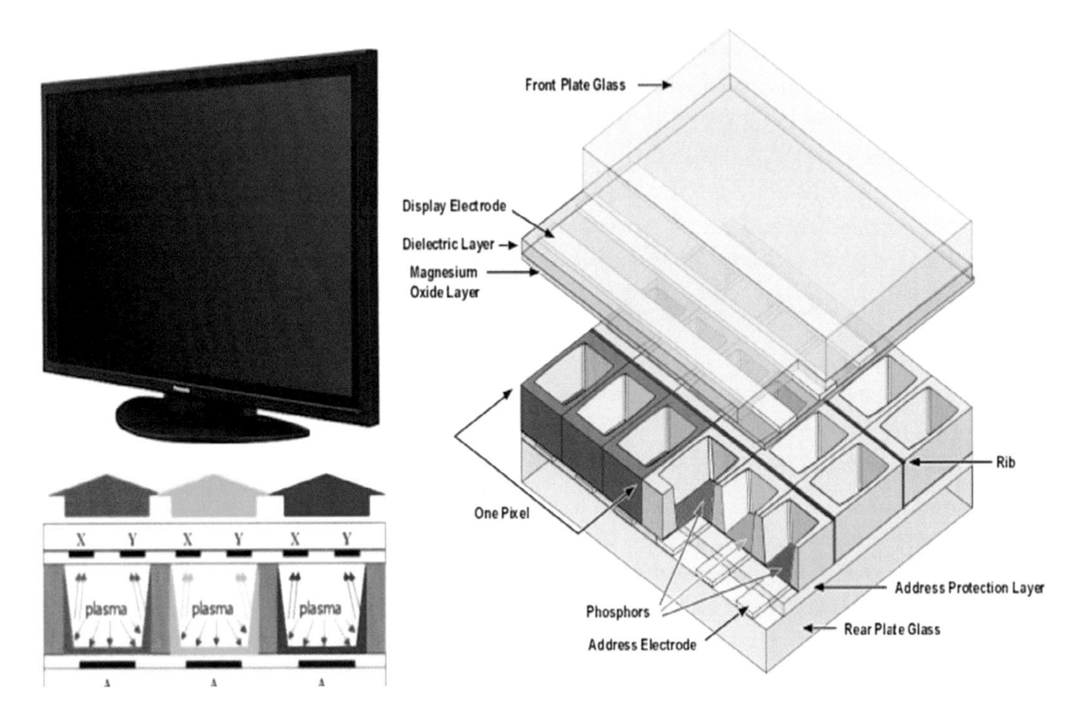

Figure 11-10 Plasma Technology

An addressable electric grid controls the firing action of each chamber (sub-pixel). Each time a chamber fires, it releases a burst of UV light that in turn excites the phosphor to release red, green or blue visible light. Multiply this action by one million and an image is formed.

OLED (Organic Light-Emitting Diode)

Formerly known as organic electro-luminescent (OEL), organic light-emitting diode technology applies an electrical field to an organic electroluminescent compound, producing a self-luminous diode that does not require backlighting or diffusers.

OLED displays are currently used in a number of colorful, small-format displays, such as car radio faceplates and digital camera screens. The technology is emissive (emits light to create an image) and, therefore, does not require a backlight to function, unlike flat panel LCDs.

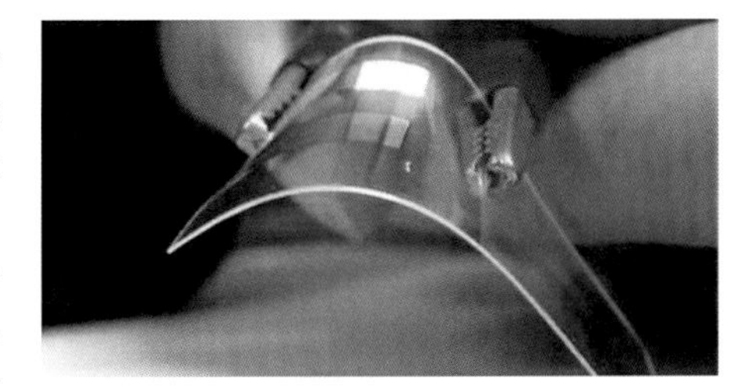

Figure 11-11 OLED Panel

DLP (Digital Light Processing)

In 1987, Texas Instruments (TI) developed (under the inventive creativity of Dr. Larry Hornbeck) the unique chip design that makes the digital light-processing display a popular video performer. DLP technology is an all-digital display technology that is used in projectors and TVs. DLP is a reflective (light is reflected to create an image) technology. At the heart of DLP technology is the "digital micro-mirror device" (DMD), a semiconductor-based light switch array of thousands of individually addressable, tilt-able, mirror pixels. The DLP chip is perhaps the world's most sophisticated light switch. It contains a rectangular array of up to two million hinge-mounted microscopic mirrors. Each micro mirror measures 16 x 16 microns (less than one-fifth the width of a human hair).The DMD is the only mass-produced spatial light modulator and TI has shipped more than eight million systems

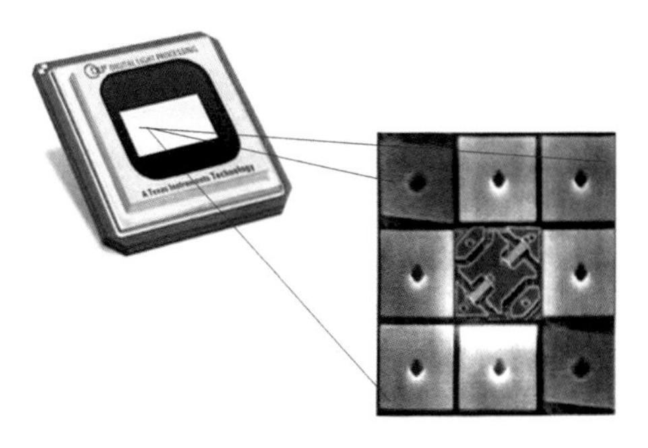

Figure 11-12 DLP Chip

since 1996. The DMD has proven its resilience and dependability, and brings many advantages to light-steering applications. DLP has been successfully implemented in rear projection TVs, but remains primarily today in projectors.

DLP is available in single and three chip versions. The single chip system uses a color wheel. The image changes for each segment of the wheel, and since this happens so fast the eye sees it as one continuous video image. Unlike the single chip system, a 3-chip DLP system does not require a color wheel. Instead, using a prism, the light source is first broken down into the three primary colors (red, green, and blue) and then each color stream is sent to an individual DLP chip. These chips then process the three independent light sources back into a single image to be output through the lens and onto the screen. By eliminating the need for moving parts, the three-chip system increases reliability as well as improving picture quality.

LCoS (Liquid Crystal on Silicon)

LCoS technology is like a cross between DLP and LCD. It is also a reflective technology. Light reflects from microscopic elements on a chip, but the elements either absorb or reflect the light. There are usually three separate LCoS chips in a typical display device, one each for Red, Green and Blue.

LCoS consists of a liquid crystal layer that sits on top of a highly reflective material. Below this material is another layer that contains the electronics used to activate the pixels. LCoS uses a reflective LCD display panel with a high open area ratio. By placing the wiring area and switching elements under the reflective layer, it is possible to achieve an almost seamless image. LCoS has been used in rear projection televisions, but is now primarily used in projector/screen applications.

PROJECTION SCREENS

When a projector and screen are used to produce an image, they both play a critical role in the process. They must be specified, set up, and calibrated perfectly to ensure the best performance. It is important to fully understand both components, as well as the viewing area, in order to properly design the system.

Screen Types

The formulation used in the screen can have a dramatic effect on the way it reflects the projected image to the viewer. Here are some examples:

- **Matte White screens** reflect the light evenly across a wide area. They are good for situations when viewers are located both on axis and off to the sides. This type of screen surface is said to have a gain of 1.0. It neither absorbs light, nor focuses it back on the center axis.
- **High Gain screens** are formulated to minimize the reflection off axis, and send more light directly back toward the projector. Since most home theaters have dedicated seating areas directly in front of the screen, this can be advantageous in a couple of ways; more brightness for most viewers and less light reflecting off of the walls. A high gain screen not only prevents off axis reflection of the projected image, it also minimizes how much off-axis ambient light affect the image. These screens have a gain of over 1.0 and have a narrower viewing angle.
- **Gray screens** were developed to enhance the contrast ratios of projectors which did not deliver deep blacks. Their gain is low, so some brightness is sacrificed. But the gray surface helps produce blacker blacks.

- **Rear Projection screens** are also available in different formulations, but are specially designed for the projector to be behind the screen. In this configuration the space behind the screen must be completely dark, but the ambient light in the viewing room has little effect on the image; ideal for spaces that are not darkened. If adequate throw distance is not available, a mirror system can be implemented.
- **Acoustically Transparent Front Projection screens** have tiny perforations which allow loudspeakers to be placed behind them (the ideal location for a center channel speaker is in the center of the image). Although the screens pass sound well, some equalization is usually required to allow the speakers to deliver the full spectrum of sound when firing through a screen.
- **Curved screens** are often used when a 2.35 or 2.40:1 image is projected. This wide image can distort both in geometry and color at the edges of a flat screen. The curved screen minimizes this distortion and provides the viewer with an image with more consistent color and brightness from end to end.
- **Screen Masking** allows the unused portions of a screen to be covered with a non-reflective black surface, making the viewing area look better. Today's masking systems usually change only the image width, as image height is constant at all aspect ratios.

Projection screens are available in all of the typical aspect ratios. Most are 16:9 (1.78:1) but extra wide 2.35:1 and 2.40:1 screens are gaining in popularity for viewing motion pictures.

Every projector/screen system must be designed with viewing distance, throw distance, and projection angle in mind. For more on these, see Chapter 12.

SUMMARY

Video will always be a key component in any entertainment/integration system. As the delivery of content evolves the video display becomes the center of attention for a wide variety of viewing options; from movies and television to telepresence, internet, and more. A solid understanding of the history and fundamentals, as well as current technologies, will serve you well as you progress in your career.

Questions

1. HDMI V 2.0, introduced in 2013, added what feature?
   a. 4K support
   b. Up to 32 channels of audio
   c. 4 simultaneous audio streams
   d. All of the above

2. Which HDMI communication allows the source device to understand the capabilities of the display device?
   a. Hot Plug
   b. DDC
   c. EDID
   d. ARC

3. The standard aspect ratio for High Definition content is:
   a. 4:3
   b. 16:9
   c. 16:10
   d. 2.35:1

4. Which display technology utilizes millions of tiny mirrors to create an image?
   a. DLP
   b. OLED
   c. CRT
   d. LCD

5. When a rear projector system is installed is it important to remember:
   a. The area behind the screen must be kept dark for the picture to look good
   b. The throw distance will be exactly half that needed for front projection
   c. A matte white screen must be used for best contrast
   d. All of the above

Chapter 12
HOME THEATER DESIGN & INSTALLATION

THE HOME THEATER EXPERIENCE

For much of this industry's history, home theater has been a major part of most integrators' business model. From the most elaborate dedicated home cinema to the popular multi-use media room space, this combination of subsystems has been central to our mission of enhancing the lifestyle of our clients, and adding value to their homes. Of all the possible improvements or options a homeowner might select, few are so certain to add enjoyment (and value) to a home. The key to this value is the fact that, unlike a swimming pool for instance, a theater or media room is capable of such a wide variety of functions; movies, sporting events, TV, games, and more. And it can be used year-round, by all family members. In this chapter we will look at the key elements that make up a well-designed theater space, and how proper design can ensure a great experience for your customer, regardless of price point. The technical details about audio and video equipment, control systems, etc. are covered elsewhere in the book, so in this chapter we will focus on how these various systems work together to deliver an experience that is greater than the sum of its parts.

The History of Home Theater

To understand how today's home theater evolved, we need to look back in history. Various forms of public performance have always been adapted for presentation in the home. Before technology, all performances were live. In some cases, the performers were in a public setting, and in some cases music, drama, etc. were performed in the home. For instance, many wealthy families in the 18th and 19th centuries had pianos in the home. And in many parts of the world, families have entertained themselves over the years, by singing and playing stringed instruments in the home. Early motion pictures were only available in a public setting. Despite the popularity of radio, nothing could match the "big screen" experience. Hollywood added sound and color and the golden age of movies was under way. When television was introduced, advancements were made to get people back out to the movies. Wider screens, surround sound, even early attempts at 3D. With the advent of video tape and eventually DVD, it became possible to have a great cinema experience at home. Eventually digital technologies and great audio made home theaters a highly appealing concept, and today the wide selection of quality content, along with great audio and video technologies, makes the home theater a popular addition to any home.

Definition of Home Theater

What is a home theater? You may get many different answers to this question, depending on who you ask. In fact there are many very different scenarios which all qualify as home theaters. For our purposes we will define the minimum requirements for a home theater as:

- A high-definition display of the appropriate size, calibrated for best perfomance
- A minimum of 5.1 digital surround sound, properly configured and calibrated
- High quality sources such as DVD, Blu-ray, digital cable or satellite, and streaming HD
- A control system that is easy to use for the whole family
- An acoustical environment which enhances the audio system and has a low noise floor
- Controllable ambient light

Custom vs. Out of the Box

Many consumers have been led to believe that a few components from a retailer can be easily be set up at home to create a home theater. By the end of this chapter you will clearly see why this is an oversimplified and misleading message. A properly designed theater not only takes into consideration many elements beyond just audio and video; it is also designed with the actual client and their lifestyle in mind.

A custom home theater is tailored specifically to the client's needs and budget. If properly designed and installed it can deliver a very convincing cinema experience at a modest cost. The key is in the designer's ability to take into consideration all of the necessary elements, not just audio and video.

The Cinema Experience

Many might suspect that our goal in a home theater is to duplicate the experience of a large commercial theater. This is not the case. Our goal is to deliver to the client an experience which closely resembles what the engineers and producers were seeing and hearing when the actual content was created. They may have spent days fine tuning the sound design on a segment that only lasts a few minutes. And when they got every detail just right, that's when they declared "That's it. - that's exactly the way we want it!" If we can provide our customer an experience similar to this, then they will be seeing and hearing the content as it was intended to be seen and heard. Which means a well-designed home theater can actually deliver a better experience than some commercial settings, if done correctly.

Home Theater Types

Every designer should have one goal in mind when designing and installing a home theater; to create a room that meets the homeowner's objectives AND the functional goals of a home theater. Throughout your career, you may be asked to design home theaters that can be categorized as either dedicated or multipurpose. Either type can deliver a great cinema experience.

A dedicated home theater is a room purpose-built for the enjoyment of movies, television, music, video games or other types of audio/video entertainment. Such rooms are usually very specialized and isolated from the rest of the home, physically and acoustically. This isolation allows the viewer to become completely immersed in the presentation.

Multipurpose home theaters or media rooms are commonly found in homes today. An open floor plan, which includes other entertainment areas such as a bar, billiards table, or additional seating makes the space much more suitable for entertaining and family time. These rooms are usually designed to serve many purposes, so you will be faced with a variety of challenges not usually found in a dedicated theater. The room may have an unusual shape or windows for instance, which can adversely affect acoustics and image quality. A good designer will meet these challenges and manage compromise well.

Home Theater Done Right

Installing a home theater in someone's home has advantages for both the homeowner and the designer.

Advantages for the homeowner include providing a versatile space that is appealing to everyone in the family, regardless of age or gender. A theater or media room enhances the family's lifestyle in a variety of ways and never goes out of style. Also, for the homeowner, there is the potential for added value to the home, and appeal to potential buyers at resale.

As the designer, creating extraordinary home theaters that provide that emotional involvement discussed above differentiates you in your marketplace, and will undoubtedly lead to more referrals and jobs when your clients show off your work to their friends.

There are, however, some challenges which must be dealt with:

- A complex system like a theater can only be enjoyed when it is extremely easy to operate. The control system must be fool proof and "idiot proof".
- Multipurpose rooms pose a challenge because they are used for several different activities. And these spaces are not easily isolated, so background noise can be a factor.
- There is always the risk of cost increases during installation due to unexpected construction and décor issues. Careful planning is necessary to control costs.
- There may be a conflict between the desired "look and feel" of a theater room and certain design requirements needed to provide the desired performance. Communication and cooperation between stakeholders is key to resolving these issues.

KEY ELEMENTS

In this section we will discuss:

- **Location**
- **Layout**
- **Seating**
- **Equipment**
- **Acoustical Considerations**
- **Lighting**
- **HVAC**
- **The User Interface**

A custom home theater is much, much more than just audio and video. To create the best design possible, you must look at every element of the room as part of an integrated system. It is likely that your client may not have even considered some of these as part of the theater design. It is up to you, the designer, to make them aware of how each of them impacts the overall experience.

There is a universal design guideline which states "form follows function." Homeowners usually have a specific objective in mind before contacting you to request your services. Meeting with the homeowner to discuss and explore that objective will be essential to your success. Always remember to listen carefully to the client during the initial interview process. Once the objectives have been established, the preliminary design process can begin. If the project is in an existing home, the thought process will be very different than in new construction. But the way the elements interact is the same. As we discuss these various elements, remember that they all affect each other, and making a compromise in one area may require changing something elsewhere.

Location

The home theater experience can be affected by the location of the room within the home. Just like in real estate, location is one of the key elements to consider when designing a home theater. Some challenges with regards to the location of a home theater include noisy neighboring spaces, difficult-to-isolate spaces, light control, and ventilation systems.

Home theaters located adjacent to garages and laundry rooms, theaters with "outside" walls, and theaters located above or below bedrooms are generally undesirable. Sounds from adjacent areas such as running water, opening garage doors, traffic etc. are prone to distract the viewer, and sounds from the home theater can be disturbing to bedrooms and other parts of the home. And everyday sounds can be distracting to those in the theater. Remember the mantra; if some elements of the theater reduce the possibility that the viewers are going to get lost in the experience, those elements should be changed, if at all possible.

When considering a theater in an existing home, you will have to look at many factors. Locations with lots of glass, open archways or open doorways can be difficult to acoustically isolate. You should also look closely at floors, walls, ceilings, pipes, air ducts, doors and electrical conduits which might let sound into and out of the theater room.

Lighting control is made difficult by windows that allow ambient light to leak into the room, seriously degrading the image quality, particularly if a front projection video system is used.

Shared ventilation systems can allow sound to move easily from one room to another, which causes distractions for the individuals using the home theater. It is equally important to control airflow and fan noises created by the HVAC system to keep them from impinging on the home theater experience.

If the theater is being designed into a home as it's being built, many of these issues become much easier to manage. As the designer you will have the opportunity to have a lot of influence on the location and construction of the room.

Layout

Dedicated home theaters built as part of a home will often have the desired layout and shape since most are purpose-built. On the other hand, multipurpose rooms can pose a challenge for the designer and are often irregularly shaped, with many open spaces and/or windows.

Most dedicated theaters will be in a rectangular room without open archways to other rooms. A rectangular room provides predictable acoustical characteristics, and should always have a good traffic pattern designed to avoid disruption. The shape not only helps determine the type of furniture and equipment to use, but also defines where the furniture and equipment will be placed in the room, in relation to each other. Placing the furniture and equipment in the correct position is important to achieving high quality sound for multiple locations within the room. Remember, home theater tends to be a group activity and you want every seat to be a good seat. In fact, in new construction the size and shape of the room may actually be driven by first establishing the seating configuration, and then designing around it.

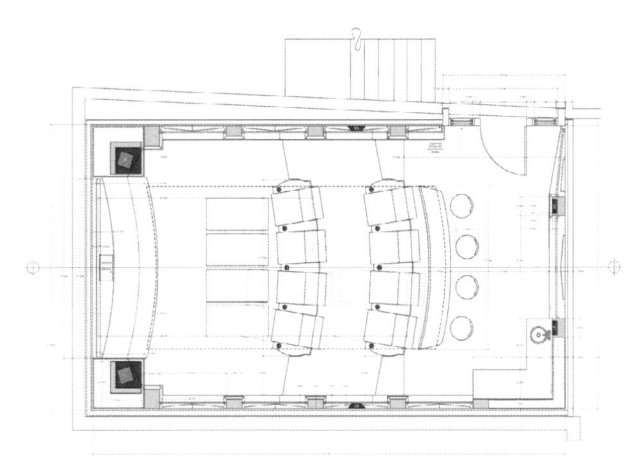

Figure 12-1 Dedicated Theater Plan

In a multi-use space there should always be a specific area which is optimized for the theater experience. This area will follow the recommended practice guidelines for viewing distance, image size, and surround sound configuration. The other areas of the larger space should also have a clear view of the display and may need additional speakers to deliver primary audio to areas like the bar or game room. In the plan shown, a flat panel display is used for everyday viewing and a large motorized screen drops down for movies and special events.

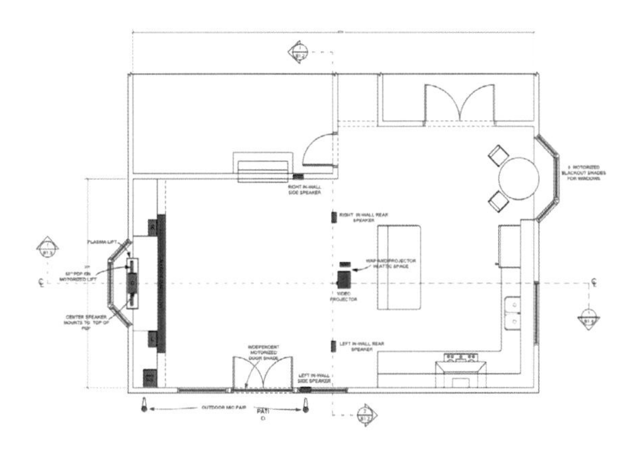

Figure 12-2 Multi-Purpose Space

Seating

Every seat in a home theater should be a good seat in terms of sound and video. Seating is a vital consideration in home theater design. Ideally, seating should be designed and positioned so every viewer has a direct sight line to the screen without turning his or her head.

In a dedicated home theater, designers have the option of installing specialized theater seating which is extremely comfortable and often reclines. Seat dimensions will determine row spacing, the capacity of the theater and riser height and depth. Manufacturers proved detailed drawings which must be referred to in the design process.

The Dreyfuss Human Factors Charts are sources used by industrial designers to design everything from cars to wheelchairs, bathtubs and furnishings. These charts also include corresponding dimensions for typical theater seating, however different theater seating manufacturers have different specifications. So always make sure to have accurate information in the design phase.

In multipurpose home theaters, there are many choices when it comes to seating. Seating in this type of room can consist of couches and chairs in a variety of styles. In this type of theater, be sure to remember that the most important seat is where the homeowner prefers to sit. This seat should be placed in the best acoustic location or the "sweet spot." This same location should also be the ideal viewing location. As in a dedicated theater, try to avoid placing any seating closer than 3 feet from the wall. A popular option in this type of room is stool-height seating just outside of the main viewing area as well as in adjacent spaces.

As part of the layout and seating conversation, we will discuss screen size. Image size and viewing distance are really two ways to look at the same thing. If the viewing distance is already known, then the image size can be calculated. If a particular image size is desired, then we can calculate where the ideal viewing location should be.

In recent years, the same technology used in applications such as training simulators has become popular in home theaters in the form of motion activated seating. Special programming and actuators move the seats in a manner that is accurately synchronized with the motion picture itself. This adds an impressive tactile dimension to the cinema experience, giving the client an enhanced experience which audio and video alone cannot deliver.

Viewing Distance Formula

Viewing distance is discussed here because this concept involves image size and seating location. This chart shows the current recommendations, per CEDIA/CEA CEB-23 Home Theater Recommended Practices – Video Design. Assuming constant screen height, the viewing distance can then be calculated at 3 times the HEIGHT of the screen, regardless of its width.

This standard assumes the screen system will maintain "constant image height" regardless of the aspect ratio. In this scenario, vertical masking will never be needed, just horizontal masking.

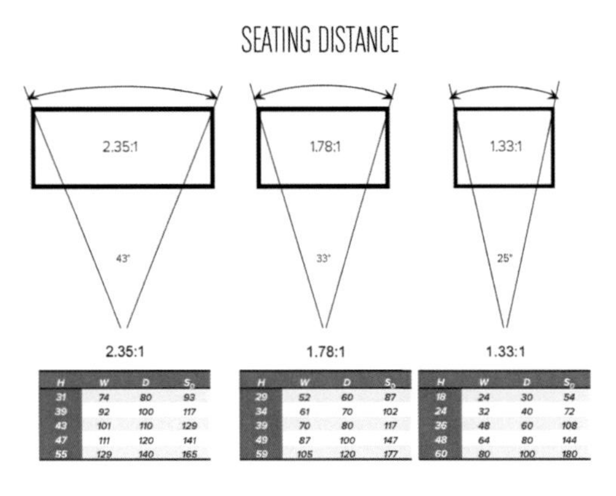

2.35:1

H	W	D	S_D
31	74	80	93
39	92	100	117
43	101	110	129
47	111	120	141
55	129	140	165

1.78:1

H	W	D	S_D
29	52	60	87
34	61	70	102
39	70	80	117
49	87	100	147
59	105	120	177

1.33:1

H	W	D	S_D
18	24	30	54
24	32	40	72
36	48	60	108
48	64	80	144
60	80	100	180

Figure 12-3 CEDIA/CEA CEB-23 Viewing Distance

Sightlines

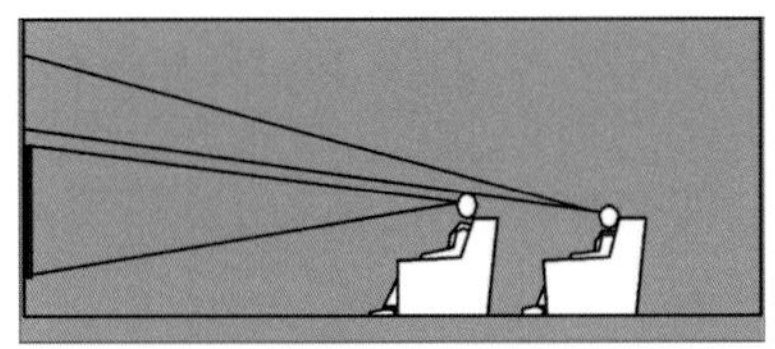

SECOND ROW BLOCKED

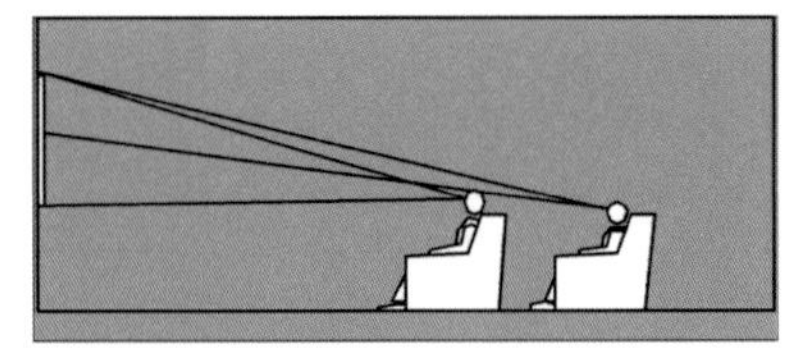

SECOND ROW STILL BLOCKED

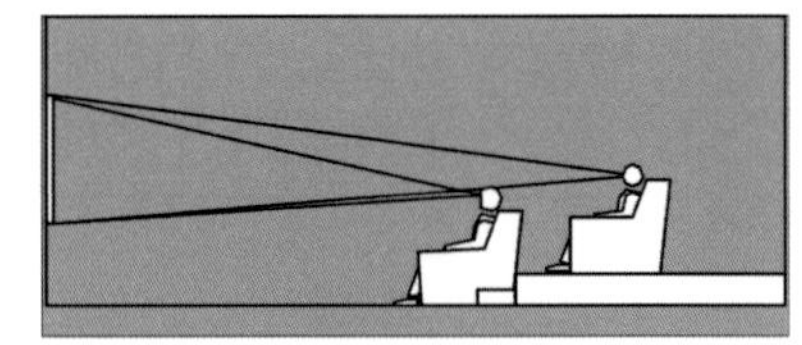

GOOD SIGHTLINES FOR ALL

Once the screen size and viewing distance have been established, a sight line study must be conducted to determine the proper viewing angle and a clear sight line for all viewers. It is up to the system designer to make this determination and explain it to the client. The primary viewer's eyes should be on a vertical viewing angle within 5° of the center of the image. All viewers should be within 15° of this center point. The only way to ensure clear sightlines and proper vertical viewing angle for all viewers is by doing detailed elevation drawings.

In the examples shown it is clear that in the first one the image is too low, blocking the sightline from row two.

The second example moves the image higher but this makes it a bit too high for row one and still does not give row two a clear view

Example three solves the problem by raising row two, providing a good viewing angle and sightlines for all. In this case the designer would probably choose to make the second row the primary viewing/listening location, since their angle of view is perfect. But of course if the image size is designed for row two, the front row will have a larger than ideal image. This may be acceptable if the family has members who prefer an extra-large image and are willing to sacrifice ideal surround sound. This is the kind of compromise that can only be done successfully with a full understanding of the client's needs.

Equipment

The audio and video equipment do the important work of delivering the sound and picture, although you should now be realizing that they are just a part of a much larger ecosystem which affects the overall theater experience. Much of the audio and video technology has been covered in previous chapters so we will provide a bit of a refresher, and focus on the specific role they play in the theater system.

Audio

Whether designing a dedicated or multipurpose home theater, the essential elements of the audio system are:

- Left and Right speakers
- Center channel speaker
- Surround and Rear speakers
- Subwoofers
- Speaker configuration
- A/V sources
- Processing and amplification

Home theater speaker systems are available in a wide variety of styles and price points. For the best performance the system should be able to deliver 105dB of full range sound with minimal distortion and adequate bass. Low cost "home theater in a box" systems may be appropriate for some consumers, but are not capable of delivering acceptable performance for a true cinema experience. Loudspeaker design and types have been discussed elsewhere in this book, so in this section we will focus on how they are implemented and configured in a theater application.

Left and Right Speakers - Before there was home theater there was two-channel stereo listening. In this setup the listener had to be in exactly the right spot in order to hear the correct stereo imaging and properly locate sounds that were intended to be in the center of the soundstage. In fact, early surround systems utilized those Left and Right speakers and added two surround speakers directly to the left and right of the listener. These Left and Right speakers still deliver music and soundtracks in stereo, creating a left to right soundstage. Their location in relation to the primary listener will be much like they would be in a stereo listening configuration. These speakers may be free-standing, on-wall or in-wall. They should be aimed in at the listener so the primary listening position is on-axis to their tweeters.

Center Channel Speaker - Today's surround sound systems use a center channel speaker which does a great job of anchoring dialog and other primary sounds to the center. The center channel speaker is the most important speaker for movies, because it produces most of the dialogue. When all three front speakers are identical they are called LCRs. Ideally the Left, Right, and Center speakers should all be at ear level, but unless a perforated screen is being used, the center channel cannot be placed in the center of the image, so it must be placed just underneath the screen, or in some cases, just above the screen. If placed beneath the screen, ensure that it is not placed too low to provide adequate coverage of the listening area. This will probably be the most common compromise made when designing a home theater. Because the Center channel speaker is rarely at ear level behind the screen, special speaker designs have emerged which optimize performance while maintaining a timbre, or tonality, similar to the Left and Right speakers.

When the Center speaker is combined with the Left and Right speakers, a smooth front soundstage should be created and dialogue should sound centered, even if a listener is a little off the center line.

Surround Speakers – These are often mistakenly called rear speakers. In reality a 5.1 system has two surround speakers which are located to the sides, just behind the listener – per the recommended practices and ITU standard (to be covered later in this chapter). In a 7.1 system there are two additional rear speakers which are positioned at the back of the room as shown in the illustration. These speakers all work together to provide the ambient sound effects that create a sense of space and envelopment. Ambient sound effects include the wind, rustling leaves, rain falling on pavement, crowd noise, etc. Occasionally the surround channels will contain directional action sounds like a plane flying overhead, explosions, bullet ricochets, etc.

In the production studio, the surround speakers are identical to the Left, Center and Right speakers. In a home theater, these speakers may be designed differently, to enhance the envelopment and make the space feel larger. However, all full range speakers in a system should be from the same manufacturer and series, in order to ensure similar sounding performance, especially in the high frequencies. Look for sets of speakers designed to be used together. Unique surround speaker designs called di-pole and bi-pole have dispersion patterns intended to "scatter" sound over a very wide area, making it harder for the listener to "localize" sounds. When properly placed, these types of speakers can help create a more diffuse and involving surround field, particularly in a narrow room where the side walls are close to the listeners. You must thoroughly understand how these speaker perform before specifying them.

Subwoofers are generally the largest of the speakers in the system and are designed to reproduce the deep bass sounds. Subwoofers are also available in flush-mount, in-wall and in-ceiling models. Incorporating subwoofers in a system allows the other speakers to be smaller, because the full range speakers will not be required to generate the deep and powerful bass information provided by the subwoofer. A properly designed system directs frequencies below 80Hz to the subwoofers along with content which is assigned to the LFE (Low Frequency Effects) channel in the encoded audio.

Most free-standing subs are self-powered or "active". They require an analog audio signal from the LFE output of the surround processor, and power for the internal amp. In-wall subs are "passive" and require a separate amplifier, usually located with the other equipment. In this case heavy gauge speaker cable will be run to the subwoofers. Free standing subwoofers will often deliver smoother performance and less sound transmission to other parts of the home when they are "decoupled' from the floor using a decoupling foam material.

The latest research shows that the most predictable bass coverage can be achieved by using multiple subwoofers, for example one in the center of each wall. This will result in smoother bass response throughout the room.

As a general rule, multiple subwoofers should always be specified. This technique is fully explained in CEA/CEDIA CEB-22 and in the CEDIA Technical Reference Manual. Observing these guidelines will ensure predictable low frequency response in any rectangular room.

Speaker Configuration – Now that we have covered the speaker types and their role in the system, let's move on to their configuration in the room. As discussed earlier, the goal of the home theater is to recreate, as closely as possible, the environment in which the content was produced. So we will look closely at how the surround system is set up in the production studio, and then see how the recommended home theater configuration relates to this standard.

The control room in a recording studio is where the engineer and producer create the final product, whether it is a stereo music recording or a full-blown motion picture sound track. This space is carefully designed to allow for critical listening and accurate reproduction of sound.

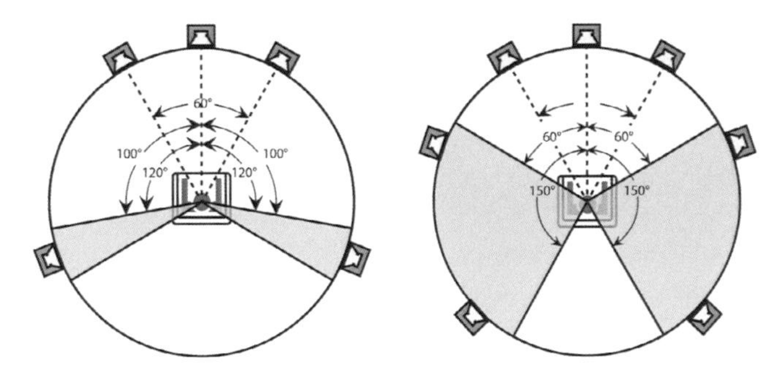

Figure 12-4 ITU Control Room Configuration

It is unlikely that a home theater will be designed to these standards however we can come pretty close by following some well established guidelines.

The speakers in a control room are set up according to standards established by the International Telecommunications Unions (ITU). Virtually every motion picture soundtrack is mixed in a room configured to this standard. The key specifications in this standard are:

- All full range speakers are identical and at ear level
- All speakers are equidistant from the primary listening position
- The Left and Right speakers form an equilateral triangle with the primary listener

The accepted guidelines for home theater surround sound configuration are adapted from the ITU standard. CEA/CEDIA CEB-22 Home Theater Recommended Practices for Audio Design outlines in detail the goals of the audio system and room acoustics, and provides specific guidance as to speaker placement. Application Notes for all three of these documents can be found in the Appendix.

First, the Center channel speaker – we know that in the studio, this speaker is on the center line and at ear lever. But the only scenario in a theater which would allow this is one which utilizes an acoustically transparent screen. In all other situations the video display will be located so that the center of the image is approximately at eye level. So the Center channel speaker in most cases will be relocated to either just below or just above the image. This will be a common compromise in your theater designs. Remember in all cases the center channel should still be aimed at the primary listener, and to ensure clear dialog for all, every listener in the theater should have a clear sight line to the Center channel speaker. This is the only way to guarantee they are hearing direct sound and not reflected sound – very important for understanding dialog.

The Left and Right speakers have traditionally formed an equilateral triangle with the listener. This has been the case since the beginning of stereo recording, and continues to be the standard in the control room. However in a larger home theater when the listener is farther back from the screen, these angles may make the front soundstage too wide, and not "connected" to the image. So some flexibility is permitted in the location of the Left/Right speakers. They may be brought in closer to the image, as close as 22.5° off the center line. The goal is a perceivable stereo image which is still anchored to the image. Left and Right speakers should be at seated ear level for the primary listener. Free standing speakers should be aimed inward so that the listener is on axis to both speakers. If in-walls are used they should be amiable so that the listener is on axis for at least the tweeters.

Surround Speakers – In a commercial theater, there are several Surround speakers arranged down the side and rear walls. Everyone feels immersed in the soundtrack. In order to enhance the envelopment of a surround system in a smaller room, and deliver more balanced sound to all listeners, surround speakers should be located 2 feet above the listeners seated ear level. For 5.1, the surround speakers are located just behind the listener. When rear speakers are added (7.1), there is more flexibility in the location of the side surrounds. The goal is smooth envelopment. In a larger home theater it is not unusual to use two Surround speakers on each side wall, plus two in the rear of the room, to provide good envelopment for all listeners. If the amplifier channel driving an 8Ω Surround speaker is also rated to perform into a 4Ω load, adding a second speaker to that channel works well, and each speaker still receives adequate power.

Surround speakers may be bookshelf type (sometimes mounted behind fabric in columns), on-wall, or in-wall. Some situations may require that all Surround speakers be in-ceiling. In this case, the proper angles should still be observed, and the speakers should be aimable to provide on-axis performance for the listeners.

Surround speakers will often be standard front firing monopole speakers. In many situations, speakers optimized for surround applications will deliver a more diffuse sound and more envelopment. These bi-pole and di-pole designs are available as on-wall or in-wall models.

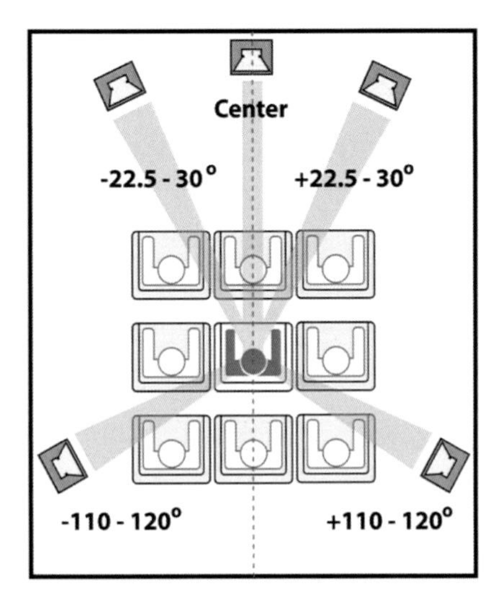

Figure 12-5 5.1 Configuration

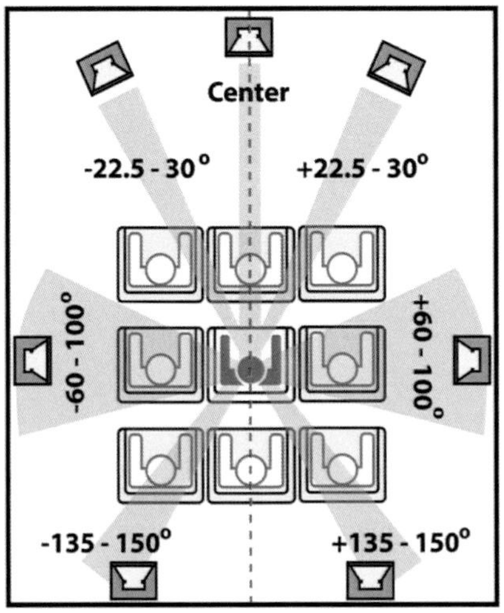

Figure 12-6 7.1 Configuration

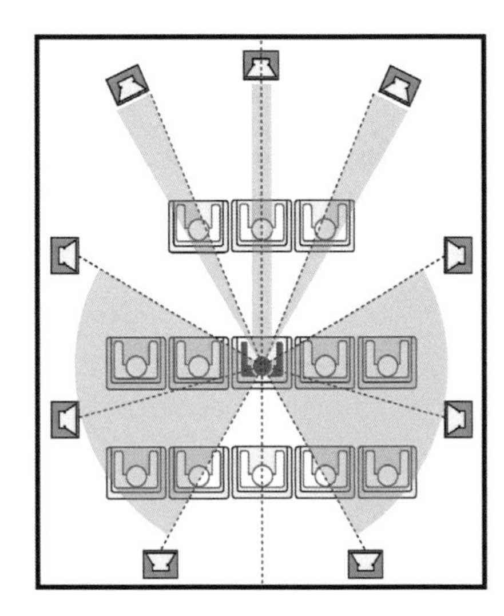

Figure 12-7 Large Room Configuration

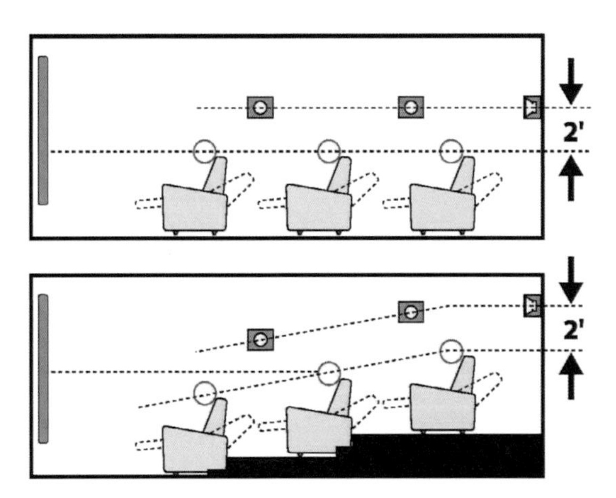

Figure 12-8 Surround Speaker Elevation

Installation Notes:

- To ensure proper performance, speaker cable must be of adequate gauge to deliver power to the speakers without compromise.
- The speaker configuration in the studio shows all speakers equidistant from the listener. This is so that sounds reproduced in multiple speakers all arrives at the listener at the same time. In a home theater it is unlikely that this layout will be possible. This is why the calibration process includes setting delays. By delaying the signal to speakers which are closer, all speakers can be "virtually" equidistant to the listener.
- When using in-wall or in-ceiling speakers for Surround locations, make sure they have frequency response characteristics similar to the front speakers. Usually a manufacturer has flush mount models which utilize the same internal components as their free-standing models. It is the responsibility of the designer to do this research and make sure speakers match as closely as possible. When these speaker types are used, the same recommended angles should still be observed.

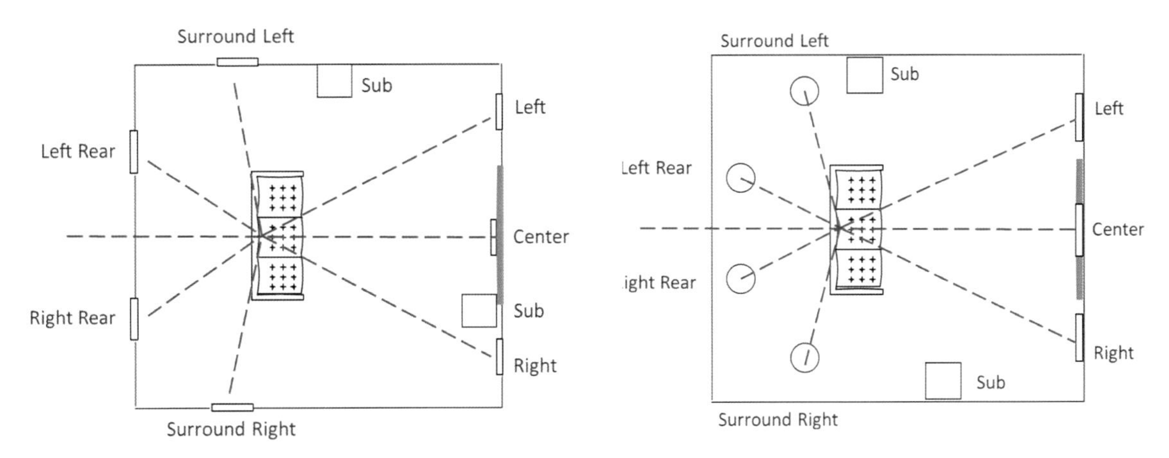

Figure 12-9 In-Wall Compromise

Figure 12-10 In-Ceiling Compromise

Sources - There are a variety of sources that may come into play in a theater or media room system. The most fundamental are some kind of HD television sources (Cable, Satellite, or IP) and a Blu-ray player for movies. Always use the interconnect which delivers the highest quality audio. In most cases this will be HDMI. It offers the highest bandwidth and is compatible with all surround formats, including uncompressed signals like Dolby TrueHD and DTS-HD Master Audio.

Other popular sources include a CD player, game console, Mp3 player or IPod, media server or computer. There may also be legacy devices such as a cassette deck, open-reel tape recorder, Laserdisc, or turntable. These devices will perform just fine in a new system, however compared to HD content the quality may be noticeably lower, and they may not be easily controlled by the programmed user interface.

Processing/Amplification - In many theater systems, a receiver will serve as the "hub" for all video and audio routing and processing. A surround receiver is a sort of "Swiss Army Knife" in that it includes a variety of separate functions, all in one chassis.

- Tuner – AM/FM plus perhaps XM, HD Radio, etc.
- Network access
- Audio and video switching, analog or digital as needed
- A/D and D/A converters
- Processing such as equalization, delay, level setups, video up scaling
- Audio amplifiers – up to 7 channels plus LFE output to powered Sub.

Figure 12-11 Full Featured Surround Receiver

In larger, more sophisticated, systems, the design may incorporate separate components rather than an integrated surround receiver. This would typically include the main processor, an outboard equalizer, video switching, separate radio tuners, a high quality multi-channel amplifier, and perhaps video processing. Advantages of separate components include the option to use different manufacturers, as well as each component having its own features and power supply.

Video

The video display is the focal point of the theater. As technologies evolve and prices come down, more and more options are available. The actual functional details of the various display types are covered in a previous chapter so we will focus here on how the displays are used in a theater, and how the signals are managed.

Video Displays

Depending on the situation, the video display may be provided in a variety of different ways. There are many considerations that must be in play when deciding which type of system to use. Above all, the display system must be fully controllable via the programmed user interface.

Plasma - The plasma display is now considered a mature technology and offers very "film-like' performance with deep blacks and excellent motion reproduction.

Each fixed pixel on a plasma display is a gas filled chamber that, when stimulated by electricity, illuminates to create the display. Because of this, no backlighting is required. In this way, the plasma display is self-illuminating. Another benefit of this gas technology is the ability to display highly accurate blacks, as a pixel turned off is "off." There is no backlighting seeping through that could affect the black level.

They can be very heavy in larger sizes, and do run fairly hot. Plasma displays are available in excess of 100", but keep in mind what is required to mount a heavy display!

LCD - The flat panel LCD display is very popular, due to its high resolution performance, and the fact that it is lighter and less costly to operate. Some LCD displays utilize LED (Light Emitting Diode) technology to produce a more consistent back light. LCD technology currently supports delivering 4k, and even 8k, images and is also used in HD projectors.

Organic LED (OLED) - Organic Light Emitting Diode (OLED) displays are self-illuminating like a plasma display, but can be made in very thin, flexible formats. This technology shows a lot of promise, but faces some very real challenges related to durability and cost of production. It is a technology to watch, but will not likely be a major player in the short term.

The most popular display option for dedicate theaters is still the traditional combination of a front projector and screen. This setup allows for large images and also makes it possible to use a perforated fabric and locate the L, C, R speakers (or just Center channel) behind the screen, as they are in a commercial theater.

This type of installation also makes it possible to conceal a front projection screen and projector by using motorization when not in use. Since the ambient light in the room becomes part of this image, it is imperative that the room be made dark when in use in order to achieve adequate contrast. The other negative aspect of a front projection system is that the noise and heat of the projector is in the room and this may be distracting or objectionable form an aesthetic standpoint. These issues can be addressed by enclosing the projector or locating it at the rear of the room, if these plans are brought into the design process early. The standard aspect ratio for theaters is 1.78:1 (16:9), the High Definition standard. The most widely used projector technologies are LCD and DLP (Digital Light Processing), which is a reflective system with millions of tiny mirrors on a chip about the size of a computer processor. Today's projectors offer such options as 3D, UltraHD (4K) resolution, and native 2.35:1 images.

HDMI should always be used to get the optimum video signal to a projector. Some 4K or 3D applications may require more than one HDMI connection. CAT6A cable with HDMI Tx/Rx units may be used in place of prefabricated HDMI cables. Always make sure any cable installed permanently in a wall or ceiling meets local code requirements.

In certain scenarios it may possible to have many of the advantages of front projection, with none of the disadvantages. If the space behind the screen wall can be made totally dark, and the projector installed in that room, a rear projection screen can be used. In this situation, the ambient light in the seating area has little effect on the image, so the system can be enjoyed during the day or when other lighting is on in the area. The projector must still have the same throw distance and offset as if it were in front of the screen.

If adequate throw distance is not available, a mirror assembly may be used to achieve the same results with as little as one meter of depth in the adjacent space.

When designing a video projection system, it is imperative that the designer fully understand all of the specifications, including brightness, resolution, throw range, lens shift, screen surface type, and viewing distance. All of these factors work together for a successful installation. If the design is not fully thought out, including accurate drawings, difficult and expensive problems may arise in the final phases of the project.

Throw Distance (Throw Range) – the distance the projector lens needs to be from the screen in order to fill the screen properly. Modern projectors have a zoom lens which allows for a "range" of possible distances. The best location for the projector is near the rear of this range, because zooming in sends the light through more of the edge of the lens, where the lens may not be as accurate.

Keystone Correction – a feature used to correct for the keystone effect, a distortion of the image caused when the projector is placed beyond its proper location in relation to the screen. While this adjustment will return the image to a rectangle, it should NEVER be used in a theater installation, as it compromises resolution.

Lens Shift is another way to compensate for keystoning of the image, using optics rather than electronics. This method is acceptable and necessary in some situations. If the projector is designed to be located even with the top of the screen, and must be mounted above that point, then optical lens shift is the ONLY way to return the image to a rectangle without adversely affecting the image quality.

Screens

In any projection system the screen is the link between the video projector and the viewer's eyes. Several options are available:

- Aspect Ratio; may be 1.78:1 (16x9) for HD applications, but black bars will be seen when viewing DVD content. An anamorphic lens system or native 2.35:1 projector may be used to fill a 2.35:1 screen with no loss of resolution. In either of these cases, motorized black masking may be used to cover the unused portion of the screen, either top and bottom or left and right when other aspects ratios are viewed.
- Motorized screens can be used in order to be concealed when not in use. They may come down out of the ceiling or up, out of the floor or a cabinet. Also available for use in widescreen applications is the curved screen, which means every part of the screen surface is an equal distance from the lens, and the viewer.

Different screen materials are used for different applications. The right screen material can help the projector deliver its best possible performance. High gain screens direct more light to the center of the room and reject ambient light from the sides. Certain formulations help projectors deliver deeper blacks. Consult the manufacturer for the best choice in each project.

System Calibration

As mentioned earlier, the home theater system should deliver a performance environment similar to what is found in the post-production studio. This cannot be achieved without proper calibration of the audio and video systems. Professional calibration should always be specified in every theater installation project proposal. This is the only way to get the best performance from the equipment.

Audio calibration ensures that all speakers are balanced in level and tonality, and optimizes the performance of the subwoofer(s). Video calibration takes full advantage of the High Definition signal and delivers the most natural image the system is capable of, while reducing energy use and extending the life of the display or projector.

Video: At least use a DVD and blue filter to do 5 basic steps. More advanced calibration done with colorimeter and software (gray scale)

Audio: Do basic setup, channel delays, level calibration (SPL meter) and EQ (RTA)

Your eyes and ears are very important tools for calibration – use them ALONG WITH your test equipment for best results. Calibration should always be specified and professionally done to ensure the best possible performance.

Acoustical Considerations

Since much of the sound heard in a home theater is reflected, the room must be recognized as a critical part of the audio system. In this section we will discuss the goals of the audio system and some very basic concepts related to the acoustical characteristics of the room. More advanced training in these areas can be found in CEDIA courseware, the Technical Reference Manual, and CEA/CEDIA CEB-22.

The goal of a home theater audio system is to provide the listener with the best sound experience possible, depending on the budget. The system should allow the listener to be transported from his or her home into the movie or other program being watched. To accomplish this, the audio system should be able to deliver:

Clarity - Clarity is the prime acoustical goal because its perfection depends on the successful attainment of all other goals. Of paramount importance is dialogue intelligibility in movies, but one must be able to understand musical lyrics, detect quiet background details, and sense realism for acoustical sounds. Elements that affect this goal are varied including equipment quality, room reverberation levels, ambient noise levels, and listener position among others.

Focus - The ability to precisely locate each reproduced sonic cue or image in a three-dimensional space is defined as acoustical focus. Recordings contain many such images superimposed side to side and front to back in every direction for 360 degrees around the listener. A system is said to have pin-point focus if, from the perspective of the listener, each of these images is properly sized, precisely located, and not wandering. Good focus also provides that individual images be easily distinguishable from amongst others within the limits of the recordings quality.

Envelopment - An audio system should reproduce virtual images of each recorded sound presenting the listener with its apparent source location in a three-dimensional space. Each sonic image relates a part of the recorded event and together these sounds compose a wrap- around soundstage that envelopes the listener. Proper envelopment requires that the soundstage be seamless for 360 degrees without interruption by holes or hot spots caused by speaker level imbalance or poor placement. While envelopment requires three-dimensional imaging of all sonic cues, of pivotal importance is the realistic recreation of the ambient sound field of the recorded venue. Focused sounds become more realistic as they move side to side and front to back with the backdrop of the ambient sounds of the intended venue.

Smooth Response - The frequency response of the speakers, as well as the room itself, contribute to the smooth response. Smooth response is achieved when there are no peaks or dips across the entire audio spectrum.

Dynamics – Dynamic range is the difference between the softest and loudest sounds heard at the listening position. Lowering the noise floor increases dynamic range, just as increasing maximum volume does – and is a much more effective approach.

Seat-to-Seat Consistency

A home theater is rarely a single person environment. There is always a "sweet spot," but the goal is to deliver a great experience to everyone in the room. This means that every listener hears dialog directly from the Center channel, and experiences a smooth front soundstage, and good envelopment. A good quality audio system, properly installed and calibrated, will provide the foundation to achieve these goals, but if the room has serious problems, the potential of the audio system will never be realized.

How Can the Room Affect the Audio Performance?

There are many ways that the room can negatively impact the audio system's performance. We will discuss the most common of these and some basic solutions that can be implemented, however room acoustics is a highly specialized field which requires advanced training beyond the scope of this book. Here are the most common challenges presented by a room:

- Reflections (flutter echo, ambience); Hard surfaces are reflective, especially at higher frequencies. Depending on several things, these reflections may be heard as a ringing, flutter, or "slap echo". All of these can have a negative impact on clarity, especially dialog.
- Unbalanced tonality; dimensions and surfaces can alter the frequency response heard by the listener. A good speaker has smooth response, and the room must also have smooth response if the listener is going to hear the sound as intended.
- Uneven bass response; low frequencies have long wavelengths. These long waves

react within the room to create "standing waves" which cause different locations in the room to have very uneven bass response.

- Noise and Sound Transmission; a room which is not acoustically isolated from adjacent spaces may allow distracting sounds to enter, and unwanted sound to escape. This problem is different from internal room acoustics and requires a different set of solutions.

Acoustical Treatment

Acoustical treatment is not always needed. In fact, the average multipurpose media room with furniture, bookcases, plants, carpet, etc. usually has enough variation in the materials within the space to take care of most sound reflection issues. But when designing a brand new dedicated theater, it is important to address the acoustical issues with some level of acoustical treatment. We will take a quick look at the challenges and some common solutions.

Reflections – It is important to point out that reflections are not inherently a bad thing. Our brains have become very good at using reflections to help us better understand our surroundings. We use reflections to determine where a sound is coming from, if it is moving, and in what direction. A room with no reflections (anechoic chamber) is great for engineering audio equipment but is a terrible environment for humans. So the last thing we would want to do to a listening space is to absorb too much of the natural reflections.

Figure 12-12
Absorptive Panels

Rectangular rooms are not ideal from a reflection standpoint. Because their surfaces are parallel, reflections tend to be linear and repeating. This is what is heard as ringing or fluttering. But rectangular rooms are very predictable. Reflection problems can be easily treated, and low frequency response can be managed in the design phase.

Figure 12-13
Diffusive Panel

Absorptive treatment includes glass fiber panels and acoustical foam. These materials allow the sound to enter but much of it does not return to room but is converted to heat energy instead. Acoustical treatment is not the only thing that absorbs sound. Everyday items like furniture, carpet, and curtains are all absorbers. Remember, however, that all absorbers must follow the laws of physics, and this means the frequencies they absorb best are those whose wavelength does not exceed twice the thickness of the absorber. So a 1" absorber is most effective at high frequencies, while an absorber that is several inches thick will be effective across a much wider bandwidth.

Another way to combat harsh reflections is by scattering the sound waves. This is called diffusion, and in a typical residential space there is lots of diffusion; cabinets, furniture, etc. all randomize the way sound is reflected. In a theater, purpose-built diffusors can scatter sound waves and minimize harsh reflections without attenuating the high frequency response of the space. In general, theaters are treated with more absorption at the front of the room (to tighten up dialog and overall clarity) and more diffusion at the rear of the room (to enhance the envelopment of the Surround speakers.

Ultimately we want the overall frequency response of the room to be smooth and not overly reduced or increased in any specific range. So when we specify absorption it is best to use thicker absorbers, so that a wide range of frequencies are absorbed more equally. A room with too much 1" absorption, for instance, will lose its high frequencies leaving mids and lows and creating a dead, but boomy environment.

Low Frequency Problems – just like mid and high frequencies, low frequencies reflect off of hard boundaries. But in this case the wavelengths are much longer. In fact in a room the size of a typical home theater the wavelengths can be longer than a room dimension. In this case the reflected sound combines with the original wave in a unique manner. When the two waves are in phase (both positive) the energy combines to create very powerful bass at that location. When they combine out of phase (one positive and the other negative) the result is a cancellation and a location in the room with little or no bass at that frequency. This phenomenon is known as a "standing wave" or "room mode". Understanding and dealing with this problem requires advanced training and is necessary for someone to be a successful theater designer.

Noise and Sound Transmission – any background noise in a theater affects the experience by distracting the listener and reducing the dynamic range of the audio system (the difference between the softest and loudest sound). This includes projector noise, ventilation, and sound which is transmitted into the theater through boundaries (walls, floor, ceiling). The theater can also present a problem for family members in other areas of the home if too much sound is allowed to escape.

Sound transmission is considered an acoustical issue but its solutions are completely different than those used to control reflections within the room. Preventing sound transmission in and out of a theater room requires extensive engineering and a combination of mass and decoupling. This is a highly specialized field which may require partnering with an acoustical engineer or home theater design consultant.

The important point is this: If a client inquires about "soundproofing" a room, it is important to explain that a high level of sound transmission control is a complex and expensive undertaking. Some basic steps can be taken in the construction of the room shell which will deliver a noticeable amount of isolation, but total elimination of sound transmission is very rarely going to be needed (or possible) in a typical residential setting.

Aesthetics - Every client wants an attractive room. Good design should take this into consideration. Acoustical treatment and equipment can be concealed and in some systems the entire wall surface is covered with stretched fabric for a smooth, finished appearance. Some projects may require that you work with an interior designer to provide a look that is appropriate for the home while still providing good performance related to audio and video.

General Guidelines for Acoustical Issues

- Acoustical treatment for room reflections should generally cover no more than 50% of the wall surface (25% absorption, 25% diffusion).
- Absorption at the front of the room will help with clarity, especially dialog
- Diffusion at the rear of the room will help by randomizing reflection and enhancing envelopment
- Thicker treatment (both absorption and diffusion) is more effective over a wider frequency range, for more balanced room response
- When concealing speakers or acoustical treatment behind fabric, the material must be acoustically transparent. Test the fabric by holding it in front of a speaker and verifying that it does not attenuate high frequencies.
- Standing waves are found mainly in dedicated theaters where boundaries are tight and solid. Media rooms which are open to larger areas are less likely to have room mode issues.
- Acoustical isolation is a complex and specialized discipline and should be left to experts
- Some simple wall systems can deliver enhanced isolation. Doors should also provide a tight seal, and all wall penetrations such as outlets, light fixtures, and architectural speakers should be fully enclosed to minimize sound transmission.

Lighting

Lighting design is often ignored or misunderstood by theater designers and audio visual (AV) integrators. However, lighting is one of the key elements of theater design due to the impact it will have on the viewer's experience.

Improper lighting design can:

- Negatively impact the performance of the video display
- Improperly emphasize design elements
- Add unnecessary heat to the space
- Contribute to unwanted sound transmission through lighting fixtures
- Add background noise due to lighting fixtures or transformers producing noise

A well thought out lighting design will enhance the aesthetics of the room and become an integral part of the theater experience. Pre-programmed scenes can deliver the perfect amount and type of light for viewing and intermission and include very slow fades which will not shock the viewers' eyes during transitions. Every theater system should include an integral lighting system which is controlled by the theater's control system.

Lighting Guidelines

- At a bare minimum, lighting at the front of the room should be on a separate load from other lights, so that light on the screen or display can be minimized regardless of how the rest of the room is lit
- Scenes should be programmed for viewing and intermission with slow fades
- Indirect light is best and architectural features should be highlighted
- Step lighting or down lighting should be included to enable safe passage when the room is darkened
- Avoid fixtures or transformers which can produce noise
- Fixtures mounted in walls or ceiling should be fully enclosed to reduce sound transmission
- Avoid fixtures that might rattle when sound levels are high
- Lighting system should be controllable via the theater control system

HVAC

HVAC systems are used to control heating, cooling, air circulation (supply and return), and moisture control (humidity) in the home.

A dedicated theater room presents some special challenges the HVAC contractor may not be familiar with. The tightly sealed room retains the heat of occupants and equipment and may need fresh air to be added to the system. If possible this room should be on its own controllable zone. In all cases some basic design principles will help to minimize any negative impact of this system.

HVAC Guidelines

- If possible the theater should be on its own zone
- Supply and return for theater should be dedicated and not shared with any other room
- Ductwork to and from the air handler should be oversized to reduce whistling
- Supply and return runs should be either serpentine or have multiple right angle turns
- All ductwork should be lined with insulation to absorb sound
- Special baffles can be used at registers to act as "mufflers" of air handler noise
- An approximate calculation of heat in BTUs (British Thermal Units) can be made by assuming 300 BTU for each occupant and 3.4 BTU for each watt of electrical draw in the equipment. This is only an approximate calculation but will serve to start the conversation with the HVAC contractor.

User Interface

The control system is perhaps as important as any other system in the theater. A theater that is difficult to use will not be fully enjoyed by the family. The control system should control every subsystem, including lighting and HVAC as well as audio and video.

The user interface is part of the control system that is visible (and touchable) to the user. It should be simple and intuitive, allowing the user to issue commands and receive information by way of a visual display (screen), keyboard, mouse, menus, graphics, touch controls, etc. It should be user friendly for not only the main user but children, babysitters, and others who visit the home.

A wireless hand-held remote or touchpanel is best. Even though system control via phones and tablets is possible, that doesn't mean this is the best approach. The best controller for a theater is one which is dedicated to the task and can be operated easily in a darkened room.

THE ROLE OF THE DESIGNER

Theater Design is a Specialized Field

You now understand that a good theater or media room system is much more than just audio and video. Always take into consideration every aspect of the project when designing the system. Remember that these various elements all work together (and influence each other) so if one changes another may need to be revisited. CEA/CEDIA Recommended Practices should be fully understood and used as a benchmark to strive for. There will always be compromise and learning where to compromise (and where not to) will define you as a good designer.

A home theater is a very complex and interactive system. A good designer has a full understanding of each of the key elements, not just a few of them. This skill set makes a theater designer a specialist who should be able to establish themselves as the expert and charge accordingly.

An important prerequisite for charging for design and engineering is making sure the client understands the level of expertise needed to the job. You should be able to start establishing this during the client discovery phase, when you are asking open ended questions about their lifestyle, entertaining, family profile etc. They will realize that you are really working to meet their needs. When a proposal is created and submitted, it should further underscore the idea that you are designing specific to their needs. The proposal should also include sample drawings and engineering documents which show the client the scope of the design and engineering work and help them understand what you will be doing on their project.

Working with Other Stakeholders

Every project will be different. In some cases you may have complete control of every aspect of the theater. In others you may find yourself working with an architect or interior designer and sharing responsibilities. In all cases, communication is the key. The most important thing is making sure that interior design elements do not interfere with the performance of the theater.

Make no mistake; working with other stakeholders can be a challenge. But if you can establish a good rapport with these professionals, and work together to make the whole project successful, odds are they will want to work with you again. And that means more business, doing what you are good at, and working with people you know you can cooperate with.

SUMMARY

- The goal of a home theater is to provide a true "cinema experience" to the client. This is accomplished by emulating the environment of the production studio.
- A theater or media room serves many purposes, and adds real value to a home..
- A good integrator adds value through design, engineering, installation, and calibration.
- A complete theater project takes into consideration all of the key elements, not just the audio and video
- Much of what we hear in a theater is reflected sound, so the room must be considered part of the audio system.
- Acoustic treatment must be attractive and fit with the look of the room.
- Lighting is important from both a technical and aesthetic perspective.
- HVAC systems affect not only client comfort but system performance.
- The theater's control system and user interface should be simple and intuitive.

For more information on Home Theater Design; CEDIA Advanced Residential Electronic Systems and online training, CEA/CEDIA CEB22, 23 and 24.

Questions

1. The ideal primary viewing distance from the image is:
   a. 2 X the image width
   b. 3 X the image diagonal
   c. 3 X the image height
   d. 1.75 X the image width

2. When designing a dedicated home theater, the best way to ensure good sightlines for viewers in two rows is to:
   a. Make the screen larger
   b. Raise the position of the screen
   c. Offset the seating
   d. Elevate the second row on a riser

3. When a wavelength is longer than a room dimension it can cause:
   a. Flutter echo
   b. Long decay times
   c. A Standing wave
   d. Comb filtering

4. The international standard which is the basis for surround sound speaker configuration was originally established by _____.
   a. The ITU
   b. TIA 570-A
   c. Dolby Laboratories
   d. VESA

5. The ideal way to compensate for image keystoning is with:
   a. Digital keystone correction
   b. Optical vertical lens shift
   c. Video processing
   d. A curved screen

Chapter 13
THE HOME NETWORK

INTRODUCTION

The topic of network design, installation, and configuration is broad and complex. As the network has become more and more important in the residential systems market, home technology professionals have adapted by learning new skills and changing the make-up of their staff. Since networking is essentially the same everywhere in the world, and there is so much training available, this chapter will only touch on some very basic concepts; and introduction to higher level training which is readily available from CEDIA and elsewhere.

TODAY'S HOME NETWORK

What is a Home Network?

Millions of people now have an IP or computer network in their homes. They may not realise they have one, but the simple little box provided by their ISP (Internet Service Provider) to deliver their Internet connection is the foundation of a small computer network. Many home owners look at the box and think of it as just 'the thing that connects us to the Internet.' They would be right......but it also does a vast array of other tasks.

In the early days of the Internet a connection at home was only possible for one device at a time, usually the main PC. The box that sat between the PC and the Internet was called a 'Modem' and access to the internet was restricted to that single PC. As connected devices became more widely available customers demanded multiple, simultaneous Internet connections. The answer was to form a small, 'local' network and the new device that delivered this facility became known as a 'router.'

Adding a router to the home delivered two important changes:

1. More than one device could access the Internet at the same time
2. Communication between 'local' devices became possible

At one time, home owners were only interested in connection to the Internet, but for more tech-savvy people communication between local devices was a huge advantage too. Music, movies and other files can be moved, shared or streamed between devices. Music on a fixed PC in the home could be played on other devices around the house. A single shared network printer could be used to print letters and pictures from any device, just like in an office environment.

Today, most ISPs (Internet Service Providers) provide a proper 'router' when you buy Internet access for the home. Their specification varies widely from the 'cheap and cheerful' to the fairly complex. Routers can be upgraded, if more features or greater reliability are needed. Knowing how and when to install a more advanced router is a fundamental requirement of the modern residential technology professional.

What Functionality Does the Network Provide?

For residential custom installers a network can perform a vast array of tasks. We've already discussed Internet access, but there are many more types of communication that the network can also deliver:

- Media delivery
- File delivery (Documents, spreadsheets, etc)
- Internal system control messages (between system components)
- Integration control messages (between different systems)

In- house systems that operate over a computer network now include multi-room audio, TV film and other video distribution, lighting system control, CCTV and other security devices, telephone and other communication methods (Skype, FaceTime, VoIP, etc) - not to mention automation of mechanical devices, heating systems, etc.

While computer networks appeared in homes quite quickly their role in systems integration took longer to evolve. Few manufacturers adopted the network as their preferred communication or control method, since established protocols like Infrared (IR) and RS-232 were more widely available, and already well understood. However, as the ability of the computer network to carry more than one service (ie; streaming media and control signals simultaneously) advanced, the advantages of networks became clearer, and their adoption began do advance rapidly.

Why is it Important to the Customer?

In recent years Internet connectivity has changed. Where once it was a luxury it can now be considered a utility. Every area of communication, from personal to 'broadcast', has now moved to an 'online' model meaning the delivery of Internet access to a huge array of devices in the home is essential.

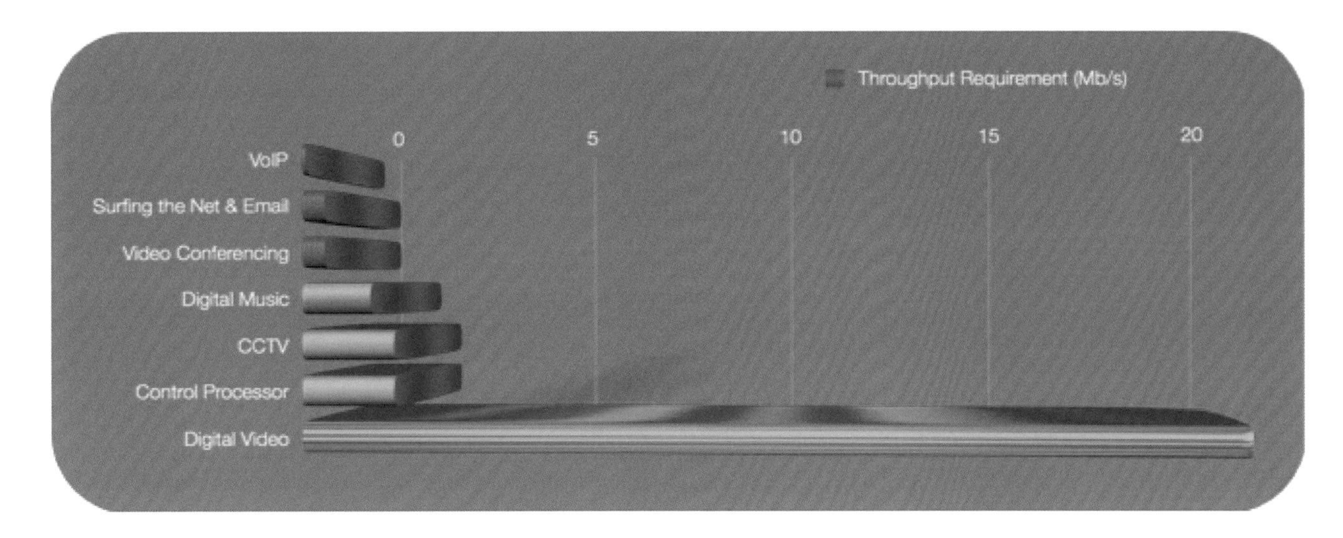

Figure 13-1 Relative Throughput Needed

Media (especially music) led the move from cable and terrestrial RF to "online" content delivery. TV and movies soon followed as the speed of Internet connections grew (video requires much more data to be delivered per second than music). Now, just about any sort of data can be stored in, and delivered by, the Internet, without the need to know exactly where it's coming from. This principle of Internet storage, data processing and delivery is known as 'Cloud' computing, so named because the Internet is like a "cloud" of information.

Why is it Important to the Integrator?

The modern computer network connects so many devices, and controls so much information within a home, that Home Technology Professionals now face an almost complete dependence on its availability. With this in mind, many integrators have chosen to "Own the Network" in the client's home, thus ensuring complete control of the service. While this requires a proper understanding of network design, setup, and maintenance, the dividends of the extra effort are huge in terms of reliability of service and ongoing business.

Many clients may be skeptical about the need for a more advanced network than that delivered by the 'free' router they received from their ISP. In this situation it is imperative that the integrator is able to explain the advantages of a better network infrastructure. These include:

- Equipment reliability - Basic routers are built for the least money possible for basic Internet connection only. They are not designed for the constant, high bandwidth traffic of the typical whole-house installation.
- Advanced features - A basic router will only offer the features required by basic users. Advanced setups requiring remote access, bandwidth controls and additional security require more advanced equipment.
- Service dependence - Simple network hardware is fine for simple use. If basic web surfing and email are the only concern they may perform perfectly well. However, custom installations also deliver a multitude of additional services such as on-demand TV and control of lighting and heating. To guarantee availability of these services we must ensure a reliable network infrastructure.

In short, investing in a better network is an investment in all of the connected services and forms the foundation of all other systems in the home.

Binary Data

All information carried by a computer network is, by nature, digital. Information carried has only two states at any given time - a 'One' (some voltage in the cable) or a 'Zero' (no voltage on the cable). Each digital digit is called a 'bit' of information and is abbreviated with a lowercase 'b'. For example:

"A sample on a CD disc is 16bits"' (16b)

Groups of bits are often arranged in longer groups of which there are two common types; the 'Nibble" which contains four 'bits' and a 'Byte' which contains eight 'bits'. A Byte is abbreviated with an uppercase 'B'. For example:

"'An IP address is made up of 4Bytes"' (4B)

Standard International Unit multipliers also apply to computer data. We often see the terms 'Kilo' (1 thousand), 'Mega' (1 million) and 'Giga' (1 trillion) used in this context. Again, here are some examples:

"This hard drive has a capacity of 100GB » - (100 Giga Bytes)
"My Internet speed at home is 20Mb/s"' - (20 Mega bits per second)
"I ripped the album at 320kb/s"' - (320 Kilo bits per second)

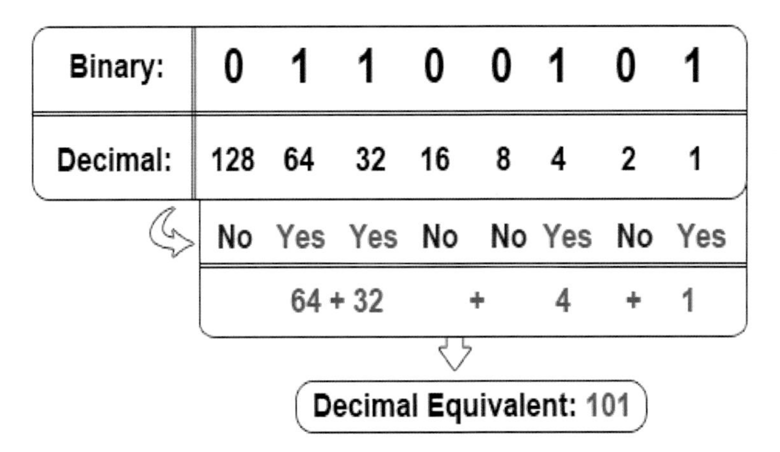

Figure 13-2 Binary vs. Decimal

Packets

Unlike serial data (audio and video for example) a network moves large files by splitting them up into much smaller 'packets' of data. An IP packet on a computer network consists of two types of data; the packet "header" (a minimum of 160 bits) and the packet "payload". The header contains information about the message, where it came from, where it is going and if it is part of a larger message. The payload is the data that needs to be moved (the music, movie, control message, etc.).

To deliver packets from a sending device to a receiving device the computer network uses a collection of protocols. The two most common, Transmission Control Protocol (TCP) and Internet

Protocol (IP), are abbreviated to the collective term 'TCP/IP' and this pairing can deliver a vast array of content. Using the TCP/IP protocols, network devices can deliver information between devices with reliability as TCP has a system in place for recovering any data that is lost en route. This ensures every piece of data that is sent should always be received.

As an alternative, the Unified Data Protocol (UDP) removes some of the traceability of each packet in exchange for a much smaller header. This reduces the network bandwidth (capacity) required and is very useful for data such as audio and video where individual packets can be lost without the intent of the content being disrupted.

Addressing

Every network device must have an identity. This is formed in two parts; the fixed identity of the actual hardware (one laptop vs another otherwise identical laptop) and the current location of that laptop within its current network.

The fixed identity of a device can be likened to a human being's name. It doesn't change no matter where in the world it is. The current location of a device can change in the same way that an individual human might be at home some of the time and at work at other times.

Hardware devices with a network connection are given an identity during manufacture which will not change during its life. This is called a 'Media Access Code' or 'MAC' address. A MAC address is made up of 48 bits and expressed as 12 hexadecimal characters (each Hexadecimal character has a decimal value between 0 and 15 and represents 4 binary bits). An example MAC address might look like this:

1D:23:4F:11:0D:3A

Note that neither Microsoft, or Apple, use the term 'MAC' in their operating systems. Microsoft uses the term 'Physical Address' while Apple uses 'Hardware Address'. In reality they all mean the same thing...

Each unique hardware device on the network is called a 'host' and is given a unique network address when it joins a new network. This is akin to a human guest being assigned a room when arriving at a hotel.

The current system of IP addressing consists of 32 bits of data in 4 groups of eight bits (i.e. 4 Bytes). This combination offers 4.3 Billion possible addresses which, at first seems like a very big number! However, with modern homes often having a large number of host devices (every family member has a smartphone, plus tablets, PCs, smart TVs, etc.) it's clear that this number isn't big enough to cope with the demand of a modern world populated by over 7 billion people and counting. In fact we need to split up 'local' network devices from devices on the internet and address them as two separate groups:

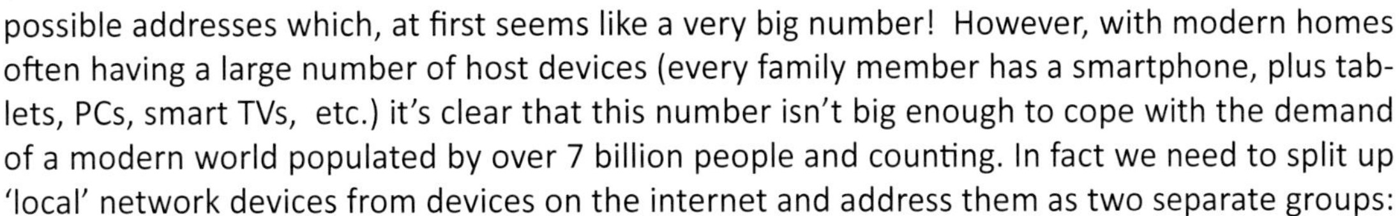

1. The Wide Area Network (or 'WAN')
2. The Local Area Network (or 'LAN')

To "share" IP addresses, each LAN is effectively closed off from other LANs, and the WAN. It has its own 'scope' of host devices which is independent of other LANs. Because each LAN is separate from other LANs (and the WAN), all LANs can all use the same range of IP addresses.

NETWORK HARDWARE

Planning and Design

When planning a network for a home, we must first consider the infrastructure that will be required to achieve the goals of the systems within the home. These include:

- The subsystems within the home that require network access
- The individual subsystem components and their planned locations
- The location of the Internet service where it enters the home
- The location or locations available for network hardware
- The total bandwidth requirements of all subsystems

Once these factors are understood, a design (both physical and electrical) can be drawn up.

Network Hardware

A home network may require a number of hardware devices to be present to deliver on the design requirements outlined earlier. Here are the most common devices that may be required:

Device	Likely Location	Usage
Router	Equipment rack or service entry point to home	Connects the LAN to the WAN, issues IP address to LAN hosts devices, offers basic switching between LAN devices, sometimes offers basic wireless access to LAN
Switch	Equipment rack or remote rooms	Directs traffic between LAN devices
Wireless Access Point (WAP)	Various locations around the home depending on coverage requirements	Offers a wireless connection to the LAN for wireless devices like smartphones, tablets, control handsets, etc.
Network Attached Storage (NAS) Drive	AV Rack or other convenient location	Offers storage capacity for media, documents and other files, also can contain 'server' devices to deliver content to network devices (eg. an iTunes server to deliver music to music streamers)

NETWORK SECURITY

Achieving Adequate Network Security

To ensure privacy of data and security of any service connected to the home network, the network itself must be made as secure as is reasonably possible. Three key factors affect the security of a network:

1. Hardware - This much be chosen for function and reliability under attack (router setup, remote access, wireless security)
2. Software - Needs to be configured and maintained correctly (user accounts, password

changes and software updates)

3. Wetware - 'The Client' - Client education as to likely threats and attacks (and what to do about them) is the most difficult and yet most important aspect of securing a network

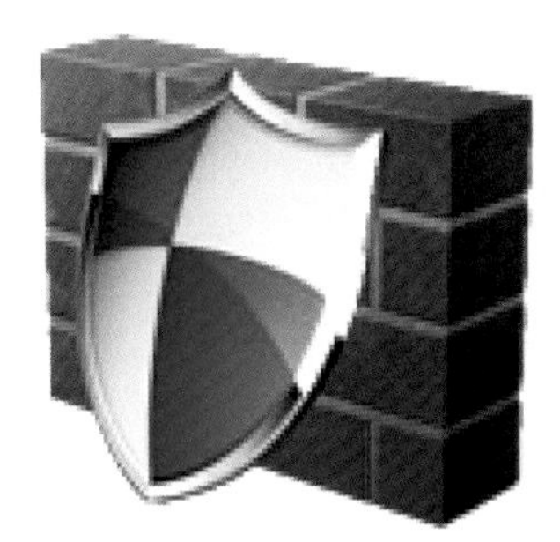

Network security is absolutely critical for high-end clients. Your ability to design and configure a secure network is a skillset that will be key to your company's success.

THE FUTURE

Network Dependency

While the future is never completely clear, a growing dependency on the computer network for technology communication is certain. So many aspects of our lives are now dependent on the availability of data and information that it can be safely assumed that the network will feature as the key component of the modern home's infrastructure for years to come.

IPv6

The IP addressing scheme discussed earlier in this chapter has, to all intents and purposes, "run out" of space. To counter this, a new scheme (IPv6) is already in place and is starting to be rolled out at various levels. At the time of writing it is only being used within the confines of the internet infrastructure itself but, in time, it will reach the home network.

The new scheme for IPv6 uses 128 bits rather than the current 32 of IPv4. In fact, it delivers a staggering number of possible addresses. The number is so big it's difficult to express it, even in mathematical terms. However, here are a few ways to write down that huge number:

- 2128
- 3.4 x 1038
- 1 address for every atom on the surface of the earth, for 100 earths
- 340,282,366,920,938,463,463,374,607,431,768,211,456

Experts consider this to be enough for the foreseeable future...

SUMMARY

It should be clear after reading this chapter that the network can be considered as the most important electronic subsystem within the home. One by one every system of home technology has moved to using the network for communication, at least in some part. Our reliance on data and digital communications as an instantly available commodity is now set and our dependency on it for everyday tasks is only likely to grow.

A systems integrator operating today, and in the future, is likely to be working with networks at a deeper and deeper level as time passes. With this in mind, further education and updating of existing knowledge is a paramount undertaking for all involved in this fascinating topic.

Questions

1. How many bits are in a byte?
   a. 2
   b. 4
   c. 8
   d. 12

2. Which service cannot run via IP
   a. Video Distribution
   b. Telephony
   c. Control
   d. All can use IP

3. TCP/IP is:
   a. the basic communication language or protocol of the Internet
   b. the part of a system that converts analog to digital
   c. the only protocol that enables computers to communicate with each other
   d. None of the above

4. Which of these is needed for two or more devices to access the internet at the same time?
   a. Switch
   b. Hub
   c. Modem
   d. Router

5. A packet contains two things, a header and :
   a. Contents
   b. A footer
   c. A payload
   d. ASCII text

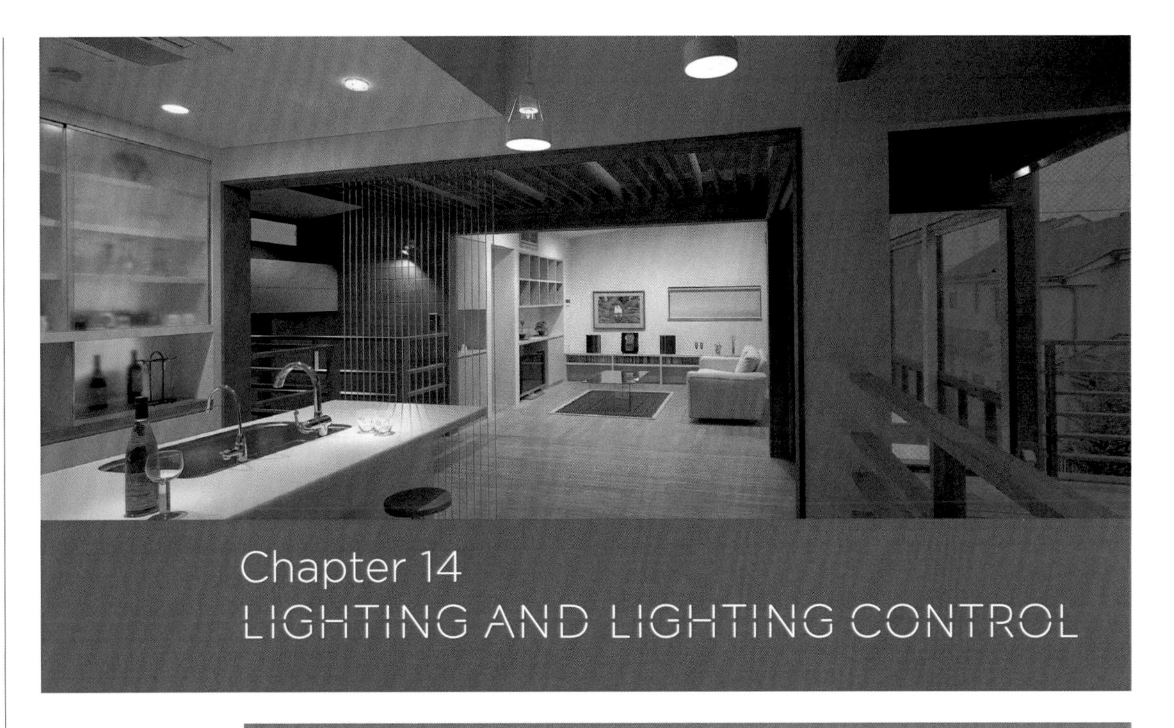

Chapter 14
LIGHTING AND LIGHTING CONTROL

THE BENEFITS OF LIGHTING CONTROL

Lighting in today's homes has evolved from an overhead fixture or two that provide general illumination in a space to multiple lights in a room, installed for specific functions. And while lighting designed to meet certain needs makes a space more usable, adding lighting control to that room—or an entire home—takes aesthetics and functionality to a new level.

Lighting control was made possible (and affordable) by the development of the solid-state dimmer, which was smaller and more efficient than previous technologies. This made dimming (controlling the brightness of a light) practical for use in the home. Taking lighting control beyond "on" and "off" offers many benefits, including enhancing ambience, increasing safety and security, adding convenience, and even saving energy. Here are just a few examples of how lighting control accomplishes those things.

Enhancing Ambience

Lighting control can personalize a space and help create just the right mood or ambience for any activity. A homeowner might want dim lighting for hosting a party or watching a movie, and brighter light for making dinner or reading the newspaper. Lighting control also allows a homeowner to highlight artwork, architectural details, collections, or specific pieces of furniture.

Increasing Safety and Security

Being able to conveniently turn on hall lights from a tabletop keypad in the middle of the night, rather than fumble for a switch on a wall, is an easy way to create a path of light from bedroom to bathroom.

And turning lights on from the car via a wireless remote means a homeowner doesn't have to walk into a dark home. Lighting control also provides a convenient way to automatically turn on exterior lights at night, even when no one is home.

Adding Convenience

With lighting control, a homeowner can adjust lights to the desired level at the touch of a button, in one room or throughout the entire home, as well as turn all lights on or off (depending on the type of control). The homeowner always has the right amount of light where it is needed. When several different lighting loads in one room are integrated together different "scenes" can be created and programmed for one-touch access. A scene is a preset combination of lights at specified levels. This means rather than adjusting each load separately, a single button stroke takes all of the lights to the proper level. Examples of preset scenes would include:

- Home
- Away
- Morning
- Good Night
- Entertain
- Intermission (in a home theater)
- Vacation (scenes that actually change to give the impression the home is occupied)
- Reading
- All On
- All Off

Saving Energy

Dimming lights is one of the easiest ways to save energy without sacrificing comfort. Every time a homeowner dims the lights they save energy—and the more they dim, the more they save. Dimming lights also extends the life of light bulbs. When integrated with a comprehensive energy management system, lighting can play an even more important role in energy savings.

LAMPS AND LUMINAIRES

The foundation of lighting control begins with the basics—an understanding of lamps and luminaires.

Lamp Types

In industry terms, a light bulb is called a lamp. The following are the main types of lamps used in residential lighting.

Incandescent Lamps

Incandescent lamps come in a variety of wattages, shapes, sizes, and colors, and historically have been the most common type of lamp used in homes. They use basically the same technology as the original electric light bulb— a tungsten-wire filament. These lamps are relatively inexpensive, emit an amber, warm light, and work well with dimmers. But they aren't very energy efficient.

Incandescent lamps are slowly giving way to newer, more efficient technologies. In fact, in the United States under the Energy Independence and Security Act of 2007, 40, 60, 75, and 100 watt incandescent will no longer be manufactured in the U.S., and are being phased out in favor of new lamps that will use fewer watts and less energy (such as LEDs).

Figure 14-1
Incandescent Lamp

Halogen Lamps

Halogen lamps are an improved version of incandescent technology. These lamps produce a bright, warm light, and last longer and are more energy-efficient than incandescent lamps. They also work well with dimmers.

Fluorescent Lamps

Fluorescent lamps produce light by passing an electric arc through a mixture of an inert gas (argon or argon/krypton) and mercury (a minute amount). The mercury then radiates ultraviolet energy that is transformed to visible light by the phosphor coating on the bulb.

Figure 14-2 Halogen Lamp

These lamps use almost a third less electricity than incandescent lamps and last up to 20 times longer. Fluorescent lamps are available as linear (tube) and screw-in Compact Fluorescent Light (CFL) types. Because fluorescent lamps contain a small amount of mercury, they should not be discarded with common household trash.

Linear, or tube fluorescent lamps, are ideal for utilitarian spaces in a home, such as garages.

CFLs are often suggested as more energy-efficient replacements for traditional incandescent lamps. Originally, these lamps emitted a blue-ish light, but many CFLs currently on the market produce a warm-toned light, similar to incandescents. Dimmable CFLs are also available, but need to be used with a compatible dimmer for the best dimming results.

Figure 14-3 CFL

Light Emitting Diode (LED)

LEDs differ quite a bit from traditional incandescent lamps in that they lack a glass bulb or tungsten filament. Instead, an LED is comprised of a chemical chip embedded in a clear plastic capsule. LEDs are the most energy efficient light sources available, and they last a long time (they're generally rated for over 50,000 hours of lamp life).

Dimmable LEDs are available, and as with CFLs, need to be used with a compatible dimmer for the best dimming results.

Types of Luminaires

The term luminaire refers to the light fixture, as well as the parts around it that help position, protect, and connect the lamps. Luminaires can range in price from very inexpensive to extremely expensive, and are available in a wide variety of types designed for specific applications. Some of the most common types of luminaires found in homes include:

Figure 14-4 LED Lamp

- Portable fixtures (such as table, floor, and desk lamps)
- Recessed downlights (installed in ceilings for general or ambient light)
- Track lighting (for direct light)
- Surface-mounted lighting and sconces (for general or ambient light)
- Cove lighting (for indirect lighting)
- Pathway/landscape lighting
- Chandelier fixtures

Figure 14-5 Portable Fixture

Figure 14-6 Wall Sconce

Figure 14-7 Recessed Fixture

Figure 14-8 Chandelier

Figure 14-9 Outdoor Fixture

Figure 14-10 Track Light

TYPES OF LIGHTING

Most types of residential lighting fall into one of three categories: ambient, task, or accent lighting. Often the type of lighting needed in a space determines the type of luminaire used for that lighting.

Ambient Lighting

Ambient lighting, also called general lighting, is the overall illumination in the room and can come from a variety of sources, including overhead lights, table lamps, recessed lighting, or track lighting. Ambient lighting provides quantity and is the starting point and base of a lighting plan.

Task Lighting

Task lighting uses light fixtures that direct light to an area for a specific use, such as reading. Task lights usually include desk lamps, ceiling and pendant fixtures, and appliance lights. Task lighting can be used to supplement ambient lighting or used alone. Adding task lighting to ambient lighting allows for less overall lighting in the room, creating a warmer feel.

Accent Lighting

Accent lighting adds interest to a room by highlighting certain objects, such as artwork, or drawing attention to decorative or architectural detail. Accent lights include wall-mounted fixtures, track lighting, and recessed lighting.

Lighting control can be as simple as solutions for one room in a home to complex systems designed for an entire home.

Basic Solutions

Basic solutions are ideal for a single room or a few rooms in a home. These solutions include stand-alone in-wall dimmers, lamp dimmers, and even automated shades that operate independent of one another. Basic solutions can be controlled manually via the component itself (such as an in-wall dimmer), a wireless remote, or via another control device, such as an occupancy or vacancy sensor. (An occupancy sensor works in conjunction with a dimmer to automatically turn lights on when a person walks into a room and off when he leaves. A vacancy sensor also works in conjunction with a dimmer to automatically turn lights off, but a person needs to manually turn the lights on.)

Some basic solutions designed for larger rooms allow a homeowner to create "scenes," which are predetermined light and shade settings. These settings are executed by touching a button on a keypad. Preset scene buttons allow a homeowner to easily and conveniently use a room in a number of different ways.

For example, one touch of the "Movie" button in a great room can lower multiple lights and close shades for the perfect viewing environment. The "Relax" button in that same room might lower shades and dims lights to just the right levels for unwinding after a long day.

Whole Home Systems

Whole home systems provide the most advanced light control, delivering customized control for a homeowner's lifestyle.

A whole home system can comprise various components, including switches, dimmers, sensors, keypads, time clocks, and processors and/or power panels. You can also integrate other components and systems, such as automated shades, temperature control, security, and audio/visual systems, with a whole home system.

Just as with basic solutions for larger rooms, a homeowner can create lighting scenes with whole home systems, but on a more extensive scale. He can control every interior and exterior light from master keypads, which work in conjunction with dimmers, switches, and sensors. Other systems which are integrated into the whole home system (such as shades and temperature control) can also be included in scenes.

Additional benefits of a whole home system include:

- All On/All Off buttons—turn on and off all of the lights in a home with a single button press on a keypad
- Time clock—lights can be set to turn on and off at predetermined times during the day, or according to any set schedule

Types of Whole Home Lighting Control Systems

There are three basic types of whole home lighting control systems:

- Localized
- Centralized
- Hybrid

In a **localized system**, lights are controlled by dimmers and switches in every room. (A dimmer or switch is needed for each fixture or group of fixtures in a room—which leads to the potential for many dimmers/switches on a wall.) Keypads are located conveniently throughout the home for one-touch control of lighting scenes. A localized system can also include wireless components such as tabletop keypads and occupancy or vacancy sensors.

Centralized control uses panels located out of sight to power all home lighting, rather than dimmers and switches in each room. Keypads minimize wall clutter.

A **hybrid system** is a combination of both types of control. It utilizes remote dimming panels to power lighting in rooms with multiple lighting zones. (Keypads minimize wall clutter in these rooms.) Local dimmers are used in areas with fewer lights, such as kids' bedrooms and guest rooms, for simple, intuitive control. A hybrid system, like a localized system, can also incorporate wireless components.

Wired or Wireless?

A wired lighting control system requires special wiring. Wired lighting control is used in localized and centralized systems, and is ideal for new construction or renovation that involves opening walls.

A wireless lighting control system utilizes the existing wiring in a home. It's used in both localized and hybrid systems and is ideal for retrofit applications.

Most residential integrators work in a region where electrical wiring and devices must be installed by a licensed electrician (even wireless dimmers). In order to do lighting control projects it may be necessary to have a licensed electrician on staff or partner with an electrical contractor. Check your local codes to make sure you are in compliance.

LIGHTING CONTROL TRENDS

The lighting control industry is constantly evolving, with new products and solutions regularly becoming available to meet customer needs. So where is lighting going right now?

The phase-out of standard incandescent lamps in many parts of the world, coupled with a focus on saving energy, is driving the popularity of LEDs for the home. Though LEDs still cost more than halogen lamps or CFLs, their cost is coming down. LEDs are versatile, last a long time (making them ideal for hard-to-reach places, since a homeowner won't have to change them very often), and use little energy. Plus, many brands of LEDs are also dimmable.

Lighting has always been a subsystem which lends itself to integration. Lighting control systems can easily be interfaced with whole house control and automation, creating even more ease of use for the homeowner. Today's trend toward network enabled devices and network based control make integration easier than ever.

Lighting control apps for mobile devices are another emerging trend. Most apps are available for free, and when downloaded to a mobile device provide a homeowner with yet another level of convenience for controlling lights (and other devices integrated with his lighting control system). Being able to control lights while away from home also gives a homeowner an added security feature.

SUMMARY

Automated lighting control systems provide benefits homeowners might not have known were available. A simple remote controlled dimmer in a master bedroom or media room can add convenience, while enhancing the ambience in the room. At the other end of the spectrum, a whole home lighting control solution creates unmatched convenience tailored for the homeowner's lifestyle, while redefining the way they live. Lighting control in a home theater or media room can be an extremely attractive option which adds to the cinema experience. Integrating lighting with other systems can provide added security and help guide police or fire personnel to a home in case of a burglar or fire alarm. Every residential integrator should become familiar with the fundamentals of lighting control.

It should also be pointed out that shades and drapes are a form of lighting control. They not only regulate the light in the room, they protect furnishings and art from Ultraviolet light, and can provide a great deal of energy savings. For more on motorized shades, see Chapter 15.

Questions

1. Lighting control provides:
   a. Convenience
   b. Security
   c. Energy savings
   d. All of the above

2. LED lamps can have a lifespan up to:
   a. 1,000 hours
   b. 5,000 hours
   c. 10,000 hours
   d. 50,000 hours

3. The development which marked the beginning of widespread use of lighting control in the home was the:
   a. Solid-state dimmer
   b. MOV (Metal Oxide Varistor)
   c. Incandescent lamp
   d. LED lamp

4. CFL lamps should not be carelessly disposed of, because they contain:
   a. Lead
   b. Mercury
   c. High voltage
   d. Phosphorus

5. A free-standing lamp used specifically for reading would be considered:
   a. Ambient lighting
   b. Accent lighting
   c. Task lighting
   d. None of the above

Chapter 15
MOTORIZATION IN THE HOME

WHAT IS MOTORIZATION AND WHY IS IT IMPORTANT?

By definition, motorization means moving something with a motor. In a broader sense, motorization can be important for many reasons:

- Moving objects which are hard to reach or manipulate
- Moving items that are heavier than we can move by hand
- Providing freedom of motion for those with disabilities
- Making life easier, more convenient

It is the last point which comes into play most often in residential systems. Motorization makes many of the technologies we deal with easier to use or provides concealment for aesthetic reasons. This allows a room to serve two completely different roles in the homeowner's lifestyle. It can allow technology to be invisible when not in use. And it can mean the difference between using and not using something like blinds in a location which is hard to reach.

When integrated with other systems, one button push can initiate multiple functions, so that the motorization becomes just one part of a series of related functions. When truly automated, these functions can take place with no user input, such as when blinds open and close depending on time of day and time of year.

MOST COMMON APPLICATIONS

The most widely seen examples of motorization in the home are blinds and devices associated with video displays and projectors. But a designer with the right skillset can go well beyond these applications to meet some very unique motorization challenges.

Manufactured Motorized Systems

Most motorized systems are pre-fabricated and relatively easy to install and control, as long as detailed drawings are done and structural considerations are addressed. These include:

Blinds and Drapes

Motorized blinds and drapes are a huge technology all their own and have become more popular in recent years. These systems can be considered a form of lighting control but they are much more. They provide not only convenience but some very practical functions:

- **Energy savings** – by controlling sunlight they can save on both heating and cooling. Some blinds have a honeycomb construction that actually increases the R Value of a window opening
- **Protection from UV light** - Ultraviolet light can damage furnishings and artwork. Blinds block sunlight when it is not needed.
- **Room transformation** – a family room or game room with several windows can be darkened and instantly become a home theater with vastly reduced ambient light

Figure 15-1 Motorized Blinds

Blinds can be part of a larger integrated system or simply a single room solution. Models are available which require no hard wiring at all and operate on batteries; ideal for retrofit situations.

TV Lifts

These motorized lifts have become more widely used as flat panel displays have replace heavy, bulky CRT televisions. A lift can bring a TV out of a credenza or a cabinet at the foot of a bed. Larger ones can be implemented to lift a large display up out of the floor. Some have the ability to both lift and rotate.

Figure 15-2 TV Lift

Flat Panel / Picture Frame Devices

A variety of motorized devices are made for concealing wall mounted displays. Some surround the television and cover the screen with art work. Some move an entire framed picture away to reveal the television. This type of option is very appealing to those who seek a traditional décor and only want to see the technology when it is in use.

Figure 15-3 Motorized Projector Mechanism

Projector and TV Drop-downs and Pivots

When a room is going to serve as a media room / theater only part of the time, the projector can be concealed in the ceiling by using a motorized lift. There are several different styles of lifts which can be chosen, depending on many factors. Another ceiling mounted device is a pivot-style mechanism which simple tilts up and down as needed. This is often used to hide and reveal a flat panel display.

Projection Screens and Masking

Large screens are often motorized to disappear into the ceiling or even the floor when not in use, returning the room to its traditional appearance. The combination of a motorized screen and projector can provide a dramatic transformation.

Fixed screens need to accommodate images in several different aspect ratios, so motorize masking is very popular to cover the unused screen surface.

Figure 15-4 Motorized Screen

DESIGN AND INSTALLATION

We have discussed several types of motorization mechanisms which are manufactured either in stock configurations or custom fabricated on a special order basis. But perhaps the most impressive motorization projects are those which are completely custom, and require a great deal of design and engineering to accomplish. These might involve false walls, doors, tables, shutters, or just about anything! The following are some considerations which apply to all motorization projects to some degree (especially the unusual one-of-a-kind projects) but should always be part of the design, specification, and installation process.

Solution Oriented Design

This is a concept which all integrators are familiar with; designing solutions specifically to meet the needs, desires, and lifestyle of the client. In the case of motorization, it is critical to start at the desired outcome, and work your way back to the beginning, paying attention to all of the important aspects shown here.

Cable Management

Cabling in a situation where there is movement requires some special considerations.

- **Bend Radius** – make sure all cables are installed with the minimum bend radius in mind
- **Flex Life** – these cables may move hundreds, even thousands of times. Solid core coax

and solid conductor twisted pair cables are designed for permanent, fixed installation. For moving applications, stranded conductors should be used.

- **Proximity to power** – just as in any cabinet or rack, low voltage signal and data cables should be separated form line voltage electrical wiring. If they must cross they should be at a right angle.
- **Strain Relief** – cabling should be attached in such a way that the cable bends in an acceptable, and predictable, manner. Attachments should be solid but not so tight as to damage the cable and impact performance.

Safety

The safety of the home's occupants must be a primary concern in every design and engineering decision made. Motorization presents a wide variety of challenges in this regard.

- **Mounting and installation** – as with any dynamic load the entire mounting system must be extremely robust, including the wall or ceiling structure, fasteners, mounts, etc.
- **Pinch / Pivot / Shear Points** – places where moving pieces can pinch cables or fingers, or catch clothing or other objects
- **Collision Path** – a place where the moving device could pose a risk to someone walking or standing
- **Obstructions** - location where something could be placed prior to the motor moving and be in the way
- **Open spaces** – any area or space where children or pets might go

Sensors

Custom motorization often requires custom engineered safety measures which include sensors to stop the motion if there is a situation which might present a safety concern. An example would be to stop the motion if someone was in its path, or if there were a physical resistance to the motion, such as unexpected weight. Sensors might also be used to provide status feedback to the control system, allowing it to "know" what position the mechanism is in. Sensors include:

- **Photoelectric** – "through-beam" or reflective sensors which recognize the presence of a solid object in a specific location
- **Motion Sensors** – detect unwanted movement to stop the motorization
- **Torque Sensors** – measure the pressure being presented to the motor

Control

Most motorized mechanism can be controlled with a simple switch (up/down, in/out, etc.) but for the most convenience and efficiency they should be integrated into a comprehensive control system. Like any component in any subsystem, there are a number of possible control methods. The designer needs to be familiar with all control protocols and the capabilities of both the motorized device and the control system processor. These include contact closure, IR, RF, relays, RS232/485, and IP. Control must be absolutely foolproof and reliable.

Maintenance

Because these systems are complex, have moving parts, and involve safety issues, it is recommended that a routine maintenance schedule be established. The first visit should be within a month of installation and subsequent visits planned to ensure dependable, safe operation. These visits also provide an opportunity to make sure all other installed systems are operating properly and the client is satisfied. Routine maintenance should include the following:

- Examine all moving parts for proper operation
- Lubricate all moving parts with high-pressure lithium grease
- Make sure all fasteners are tight
- Inspect cabling for proper attachment, location during motion, and any pinch points
- Inspect millwork and surrounding materials for any rubbing, scraping, etc.
- Check operation from all possible control interfaces

SUMMARY

Motorization mechanisms, whether pre-fabricated or custom, can add a lot of value to an integrated system. Lifts and drop-downs allow a room to have two distinct identities; it can be a family room AND a theater. This allows the integrator to sell and install more technology. Motorized blinds provide actual energy savings and protection of valuables in the home. As part of a larger integrated system, motorization can greatly enhance the client's enjoyment of the home and drive referrals. Many homeowners many not realize the possibilities that exist to conceal technology, which enhances the décor and flexibility of the home. A good understanding of these systems will allow you to bring more options to your customer and ultimately more value.

Questions

1. Motorized blinds can:
   a. Provide protection from UV light
   b. Transform a room for a different use
   c. Save energy
   d. All of the above

2. Cables which will be flexed repeatedly should;
   a. Be plenum rated
   b. Have stranded conductors
   c. Be attached tightly to the mechanism in many places
   d. Bundled together

3. Which type of motorized blinds would be the best choice in a retrofit situation?
   a. Wireless
   b. IP
   c. Venetian
   d. Hard-wired

4. Projector screen masking:
   a. Covers unused portions of the screen, depending on the aspect ratio
   b. Covers the entire screen when not in use
   c. Is used to protect the screen when the surrounding walls are being painted
   d. Is one of the few subsystems which cannot easily be motorized

5. Power cables and signal cables should:
   a. Be different colors
   b. Cross only at a right angle
   c. Be tied together for convenience
   d. Have the same jacket material

Chapter 16
ENERGY MONITORING AND MANAGEMENT

INTRODUCTION

Why is Energy Management Important?

There are many theories and scientific studies about greenhouse gases, global warming, global cooling and climate change. This chapter will not go into any of these topics. It will, however, look at the undisputed fact that energy is becoming more and more expensive, and people everywhere are looking for ways to reduce their energy costs and carbon footprint. By controlling our use of energy (how we use it and when we use it), we can lower our cost of living and also reduce the amount of greenhouse gases in the atmosphere.

Increasing Energy Consumption

The world's appetite for energy is growing every year and is projected to increase by more than 40% by the year 2035. This increased demand is due to population growth that is expected to increase by 25% in the next 20 years, as well as the growing economies in emerging markets like China and India. The decisions we make today about our own personal energy consumption may have far-reaching consequences in the years to come.

Outdated Power Grids

We have been seeing more power outages, and the time for restoration has grown longer and longer. This is due in part to the way we build and maintain our power grids. Energy sources like solar and wind are not yet abundant, and present challenges for regional and national electrical grids to incorporate. Smart grids are developing in some areas like California and Texas, but our aging and overstretched power grid still needs a multibillion-dollar upgrade, making it increasingly important to manage our personal energy consumption.

Governments Driving Change through Regulation and incentives

Many national and regional governments are driving change through regulation designed to curb our increasing energy consumption. Steps are being taken in many countries to reduce the use of coal, however coal is still used to produce over 40% of the world's electricity, and despite the growth in natural gas production, energy costs will continue to increase. Government incentives are also being used to entice the population to take advantage of renewable energy sources like solar water heaters and photovoltaic solar panels, which can reduce the upfront costs of these devices and lower energy consumption, and in turn, energy bills.

The Resultant Market Opportunity and Business Investment

Where there is a need, there is an opportunity. Electronic systems integrators have the skill-set and experience with control systems to be a driving force in this new industry sector. This chapter will help to introduce the fundamentals needed on which to build a larger base of knowledge and eventually be able to sell, design, and install systems that monitor and manage energy usage, and deliver added value to the client.

LEED V4

In the U.S. the U.S. Green Building Council plans to incorporate Life Cycle Assessment (LCA) requirements into the next version of Leadership in Energy and Environmental Design (LEED). LEED is the most widely adopted green building rating system. With the release of the new version of LEED (LEED v4), when designing and building sustainable units using the new LEED format, emphasis will be placed on reducing building operating energy use and becoming net-zero. LEED v4 uses a Life cycle assessment (LCA), a system that evaluates the environmental impact associated with all stages of a product's life. In order to achieve a LEED status, many things can contribute credits to this end. Understanding these credits and how to obtain them can increase the integrators involvement in the design phase and lead to a larger overall project

The Consumer's Perspective

Some consumers simply want to do the right thing philosophically, but on a larger scale, the motivation for energy savings is generally financial – to save on energy costs. Many are already aware of the instant savings that can be had by making minor changes to a lighting system. But the demand is growing for more knowledge, more control, and more aggressive management of energy consumption.

VALUE OF AN ENERGY MANAGEMENT SYSTEM

To the Homeowner

Energy savings can drive the interests of energy management because the client is "doing the right thing" in regards to the environment. Cost savings will drive the sales of energy management because of the long-term savings. When you get a client interested in energy management, you will always need to have the cost savings clearly demonstrated. Once they see the vast array of ways that they can save money and also contribute to the well-being of the planet, they will be more likely to enter into the world of energy management

To the Integrator

As we have shown, becoming knowledgeable in these areas will increase your standing with the client, therefore it can and will drive interest into other integrated systems. You should always show the Return on Investment (ROI) to the client as they can readily understand this type of value.

TERMS AND DEFINITIONS

Power – Electrical power is measured in Watts, which is calculated:
Volts x Amperes = Watts.
So when a 120 Volt circuit is powering a dishwasher that uses 10 Amps, the power usage is 1,200 Watts.

Energy – Energy is power used over time. For instance, Kilowatt Hours.
Your electric utility may charge you about $.15/Kwh.
In the dishwasher example, the appliance uses 1,200 Watts over a 3-hour cycle.
That is 3,600 Watt/Hrs, or 3.6 Kilowatt/Hrs. (kWh)
At $.15/kWh, the dishwasher costs $.54 to run one 3 hour cycle.

Rate Structures

- **Flat Rate** – A utility rate that stays constant. This is the most common rate structure for residential applications
- **Tiered Rate** – A rate program whereby the more electricity you use, the more you are charged per unit of electricity. The baseline rate is the lowest rate charged; as you use more electricity the rate charged increases (typically over the course of the month).
- **Time-of-use (TOU) Rate** –Utility rates assessed based on when the electricity is used (i.e., day/night and seasonal rates). Typically the production cost of electricity is highest during the daytime peak usage period, and low during the night, when usage is low

Seasons

Winter (11/1-5/1)	Summer (5/1-11/1)
Tier 1 (0-11.4 kWh per Day)	Tier 1 (0-12.9 kWh per Day)
$0.12233 per kWh	$0.12233 per kWh
Tier 2 (11.4-14.8 kWh per Day)	Tier 2 (12.9-16.8 kWh per Day)
$0.13907 per kWh	$0.13907 per kWh
Tier 3 (14.8-22.8 kWh per Day)	Tier 3 (16.8-25.8 kWh per Day)
$0.30180 per kWh	$0.30180 per kWh
Tier 4 (22.8-34.2 kWh per Day)	Tier 4 (25.8-38.7 kWh per Day)
$0.34180 per kWh	$0.34180 per kWh
Tier 5 (Above 34.2 kWh per Day)	Tier 5 (Above 38.7 kWh per Day)
$0.34180 per kWh	$0.34180 per kWh

Figure 16-1 Tiered Rate Example

Seasons

Spring (4/1-6/1)	Summer (6/1-10/1)	Winter (10/1-4/1)
Off-Peak (All kWh)	Off-Peak (All kWh)	Off-Peak (All kWh)
Mon-Fri: 12 AM-10 AM, 9 PM-12 AM Sat-Sun and Holidays: All Hours	Mon-Fri: 12 AM-10 AM, 9 PM-12 AM Sat-Sun and Holidays: All Hours	Mon-Fri: 12 AM-6 AM, 1 PM-4 PM, 9 PM-12 AM Sat-Sun and Holidays: All Hours
$0.05003 per kWh	$0.05003 per kWh	$0.05003 per kWh
Peak (All kWh)	Peak (All kWh)	Peak (All kWh)
Mon-Fri: 10 AM-9 PM	Mon-Fri: 10 AM-9 PM	Mon-Fri: 6 AM-1 PM, 4 PM-9 PM
$0.06377 per kWh	$0.06377 per kWh	$0.06377 per kWh
Demand Cost: $3.73 per kW	Demand Cost: $5.02 per kW	Demand Cost: $3.73 per kW
Demand Interval: 15 minutes	Demand Interval: 15 minutes	Demand Interval: 15 minutes

Figure 16-2 Time of Use Example

• **Demand Rate** – Utility rates assessed based on the highest peak demand during the (monthly) billing period. Typically based on a 15- or 30-minute average peak demand.

Energy Efficiency vs. Energy Conservation - Energy conservation and energy efficiency both mean using less energy, but energy conservation refers to any behavior that results in not using energy at all, such as turning off the lights when leaving a room. Energy efficiency is a technological approach to using less energy—requiring less energy to perform the same function. In this case, less energy is used and it lasts longer. An example is an LED light bulb that will use less energy to illuminate a room.

Energy Monitoring vs. Energy Management – As mentioned above, the act of monitoring energy use can really raise awareness and result in substantial savings, but that is not the same as managing the energy used.

"Energy monitoring enables energy management...
You can't manage what you don't measure"

"Energy management" is a term that has a number of meanings, but we're mainly concerned with the one that relates to saving energy in homes. When it comes to energy savings, energy management is the process of first monitoring, and then controlling, and conserving energy in a building. Typically this involves the following steps:

1. Metering your energy consumption and collecting the data.
2. Finding opportunities to save energy, and estimating how much energy each opportunity could save. You would typically analyze your meter data to find and quantify routine energy waste, and you might also investigate the energy savings that you could make by replacing equipment (e.g. lighting) or by upgrading your building's insulation.
3. Taking action to target the opportunities to save energy (i.e. tackling the routine waste and replacing or upgrading the inefficient equipment). Typically you'd start with the best opportunities first.
4. Tracking your progress by analyzing your meter data to see how well your energy-saving efforts have worked.
5. Then back to step 2, and continue the cycle.
6. Kilowatts and Kilowatt-hours – kWh is a measure of energy, while kW is a measure of power. This can be further explained by defining energy. Energy is a measure of how much fuel is contained within something, or used by something over a specific period of time. The kWh is just a unit of energy.

Peak Demand - A period in which electrical power is expected to be provided for a sustained period at a significantly higher than average supply level.

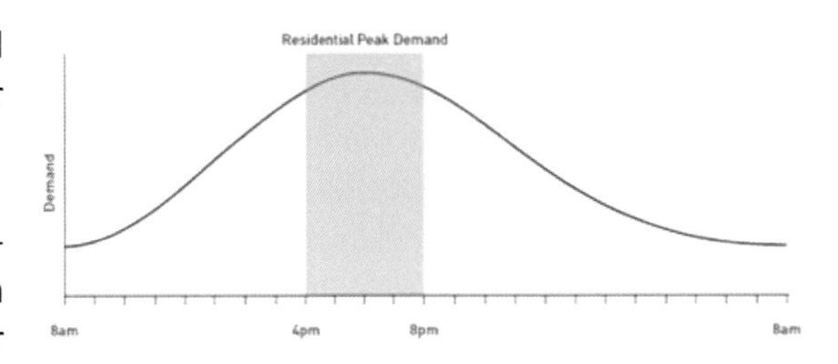

Figure 16-3 Tiered Rate Example

Demand Response – Changes in electric usage by end-use customers from their normal consumption patterns in response to changes in the price of electricity over time, or to incentive payments designed to induce lower electricity use at times of high wholesale market prices or when system reliability is jeopardized. (Voluntary DR participants allow the utility to turn down or turn off the air air conditioning during peak usage hours in the summer, for instance.)

Smart Grid – A smart grid is a modernized electrical grid that uses analogue or digital information and communications technology to gather and act on information, such as information about the behaviors of suppliers and consumers, in an automated fashion to improve the efficiency, reliability, economics, and sustainability of the production and distribution of electricity.

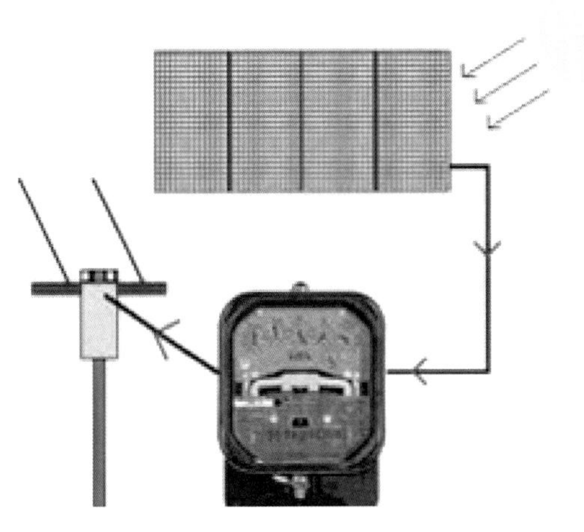

Figure 16-4 Net Metering

Net Metering – A system in which solar panels or other renewable energy generators are connected to a utility power grid, and surplus power is transferred onto the grid, allowing customers to offset the cost of power drawn from the utility.

Phantom/Vampire Loads – Electric power consumed by electronic and electrical appliances while they are switched "off" or in a "standby" mode. They may appear to be turned off but continue to draw power. Some such devices offer remote controls and digital clock features to the user, while other devices, such as power adapters for disconnected electronic devices, consume power even when they aren't actually doing anything (sometimes called no-load power).

STATE OF THE INDUSTRY

The Many Players in Energy Management

There are many different groups focused on energy applications from several different angles:

- Utilities – They have a lot to gain if they can help customers limit their energy consumption. Their potential scope in this area is huge.
- Video/Internet service providers – Also see this market as a big opportunity.
- Security companies – Layering on new automation and energy management services.
- Many companies are introducing new energy related products and solutions which will integrate seamlessly with the rest of the subsystems being installed by systems integrators.

The Many Standards in Energy Management

Several standards are being talked about for energy applications. Existing homes and retrofit applications both need to be well-supported. ZigBee appeared to take the early lead with a lot of support from utilities and home automation companies in the U.S. and elsewhere. Many smart meters are expected to use wireless ZigBee protocols. However, many products using rival Z-Wave wireless mesh networking are available at retail and installation levels and work with numerous control and security systems. Other wireless protocols that may be used include Green PHY (HomePlug Powerline Networking), ZigBee Green Power (low-power wireless networking) and Smart Energy Protocol 2 (SEP2), which allows for home networking ZigBee signals from utilities. This is an area where things will evolve over the next few years, and the most important thing may not be what protocol is being used, but how disparate systems can communicate among the various communications protocols.

THE TWO MOST IMPORTANT SYSTEMS TO ADDRESS

HVAC

HVAC generally consumes the greatest amount of energy in the home, so is a primary focus of any effort to control energy consumption. Several solutions can be implemented by integrators to this end:

- Programmable thermostats.
- Automated settings by zone, by occupancy.
  - "Home" modes within control systems.
- Tie into security system to set when armed.
- Remote access.
- Outdoor temperature or other weather data.

The most effective solutions are the ones which address many aspects via one central control system and energy management application.

Lighting

Lighting can account for up to 20% of residential energy usage. Like HVAC, there are some very good solutions available today:

- Dimming – Energy savings is approximately proportional to the amount of dimming. Dimming also extends lamp life. Remember, not all lamp types can be easily dimmed.
- Lighting Scenes & Schedules – Controlling groups of lights and individual fixtures in a logical manner. Starting with the most heavily used areas of the home;
  - Program home/away/vacation modes
  - Allow for remote access via internet/mobile devices
  - Automate lighting and dimming based on time of day and ambient light

A variety of other benefits come into play with lighting control, including added security and convenience for the homeowner. The energy savings can be the main motivator, or one of many selling points for an automated, integrated, lighting system.

ENERGY MONITORING

The first step toward controlling energy consumption is knowing how it is being consumed. Many systems can be monitored in the home, including electric, gas, water, and oil. No "smart meter" is needed to set up a monitoring system. The users then take advantage of the "Prius Effect";

changing behaviour based on having better knowledge of energy use. In the case of an automoblie, seeing the current fuel economy is sometimes all it takes for the driver to alter their driving habits and save 10-15% on fuel. There are a number of ways to see this information, including web browsers, mobile devices, and integrated control systems. A good energy monitoring system can easily monitor electricity usage on systems such as: main power to the home, HVAC power, A/V racks or theaters, other subsystems. The customer can get answers to questions such as:

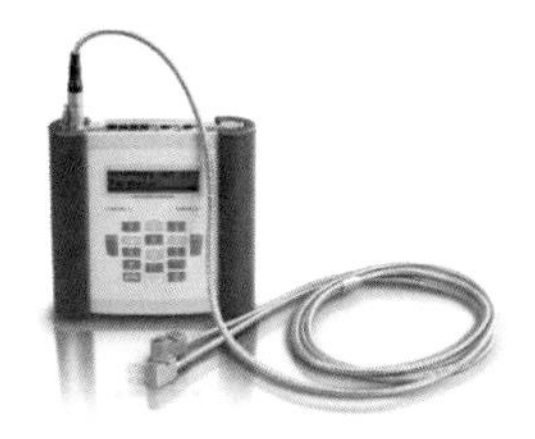

Figure 16-5 Gas Flow Meter

- How much energy am I using right now?
- How much have I used this week/month/year?
- What are my biggest energy users?
- How does my use compare to others? (with Smart Grid)

Figure 16-6 Where is Energy Used?

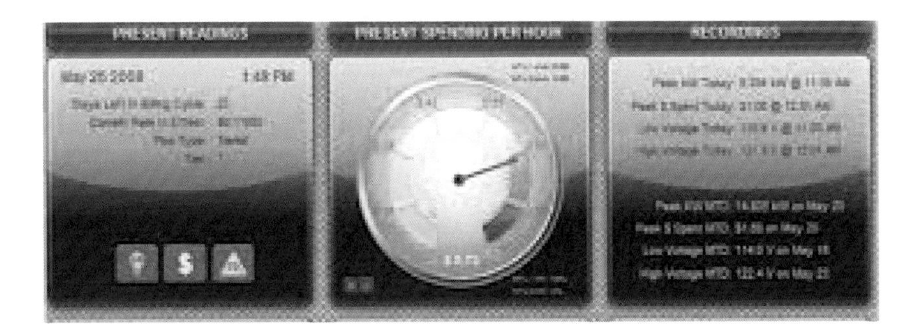

Figure 16-7 Energy Use Over Time

ENERGY MANAGEMENT

Energy monitoring provides the user with real time and historical usage and cost data so that they can proactively control their usage. But before making the decision to move to active energy management there are some things to consider:

- If the utility does not offer incentive rate plans (or dis-incentive rate plans) then there may be no point to control appliances.
- Or they may offer incentive rate plans to curb peak demand, but the user does not have enough consumption or generate significant enough peak usage to justify appliance control.

Figure 16-8 Energy Use Dashboard

Many energy monitoring device companies make claims (such as saving 20%). However, this requires perpetual attention to the monitor to achieve those savings over the long term. We all know how likely that is! Energy Management makes most sense when there is an economic incentive offered by the utility to curb your peak demand use. This applies mainly to homes that have multiple Electric HVAC systems, electric water heaters, electric floor heat, electric clothes dryers, pools, etc. By preventing multiple appliances from operating simultaneously, you significantly reduce your peak demand, and the Utility rewards you with lower peak and off-peak rates, when compared to the flat rate structures. You need to do the ROI analysis up front to determine if the cost of the Energy management system and installation is justified. If the decision is made to take on full-blown energy management, there are several ways the system can interact with the subsystems of the home:

- Manage thermostats and lights based on occupancy
- Manage lights by the time of day using astronomical clocks
- Utilize daylight harvesting technologies
- Manage house temperature using shades
- Use controllable power strips to kill loads at certain times of the day
- Use power conditioners with IP-controllable outlets
- Communicate directly with smart appliances
- Use "away" and "green" modes to address multiple systems

SUMMARY

Whether you are new to the industry or a veteran of home technologies, this emerging sector holds a great deal of promise for your business and career. Manufacturers, especially control system makers, are all creating products and platforms to address this growing segment of the industry. You are encouraged to learn as much as possible about both the monitoring and management of energy, in order to take advantage of this opportunity.

Questions

1. An electricity billing rate which increases based on the amount the customer uses is known as a _____.
   a. Time Of Use (TOU) rate
   b. Tiered rate
   c. Flat rate
   d. Demand rate

2. A "phantom load" describes a situation where:
   a. The user turns on a device remotely
   b. Electricity is being used but cannot be measured
   c. Electricity is being used even though the device is turned off
   d. A device operates without using any electricity at all

3. The "Prius Effect" is when:
   a. A person uses less energy simply by being aware of their usage
   b. Fossil fuels and electricity work together
   c. A user is unaware of energy use because they cannot hear it
   d. A client wants energy savings solely because it is the "right thing to do"

4. The first step toward controlling energy use is _______.
   a. Dimming lights
   b. Adjusting the thermostat
   c. Knowing how it is being used
   d. Being willing to make some sacrifices

5. A Kilowatt Hour is a measure of what?
   a. Current
   b. Power
   c. Energy
   d. Time

Chapter 17
SECURITY, ALARM, SURVEILLANCE, AND ACCESS CONTROL

OVERVIEW

Security, alarm, surveillance and access control systems, while unique and individual, are often thought of as part of a larger integrated system. In fact, they are likely to be controlled in some way by a central control system in larger residential installations, even when various components are from different manufacturers.

As the network becomes the backbone for integrating all low-voltage systems, we will see more and more centralized control and monitoring, as well as control and access remotely via mobile devices.

Value to the Client

Each of these related systems serves the homeowner in a different way, but they all share a common added value – peace of mind.

The home owner is provided with peace of mind knowing that that personal safety as well as property protection is being addressed. They receive the benefit of early detection of a security breach or fire emergency by the system which will then notify a monitored facility and/or the home owner themselves.

- A security system lets the occupants know if there is an intruder, and communicates with authorities if there is a problem. Fire and smoke detection can provide early warning of an emergency and make the appropriate call to first responders. Sensors can also detect Carbon Monoxide, high water, and a number of other conditions.
- Homes without security systems are about 3 times as likely to be broken into.
- Control over door locks allows the client to physically lock and unlock the home from anywhere in the world.
- Surveillance cameras allow the homeowner to observe what is happening in and around the home from remote locations.
- Systems of this type can result in discounts on homeowners insurance.

Including These Systems in a Business Model

Access control and camera systems can be integrated into a larger system with great success and not a great deal of difficulty. The added value to the client is substantial. But burglar and fire alarm carry an additional level of liability because they involve life safety. In nearly all parts of the US as well as many other regions, installing security and alarm systems requires special licensure and involves a completely different level of responsibility that other low voltage systems like audio, video, control, etc.

Some integrators make the decision to either get the needed licensure, or add a security division with its own business identity which becomes a security company on its own. Other companies choose to partner with a security and alarm company who is good to work with, and whose services compliment the work done by the integrator.

One of the most attractive aspects of security and alarm is the Recurring Monthly Revenue (RMR) which is generated by the fee paid by the customer to the central station monitoring company, which also generates a small revenue stream to the security company. Companies with hundreds or even thousands of these subscriptions see substantial monthly income as long as the customers retain the monitoring service. It should be noted that other RMR opportunities are also becoming popular, such as maintenance contracts, network management agreements, and home health services.

SECURITY AND ALARM SYSTEMS

Wired vs. Wireless

Before discussing the individual devices and components, let's take a general look at infrastructure. As with all electronic systems, the wireless age is officially here. But there are always advantages to a hard-wired system.

Hardwired

In a wired security and/or fire alarm system, all sensors and detectors are homerun wired back to the main control panel. The cabling usually provides power to the sensor(s) if needed and also provides a signaling link to the control panel. This is usually done in a new construction environment. The advantage to this is a more reliable system, since EMI problems can be minimized, RF range limitations are avoided, and since the sensors receive power from the control panel they are smaller and do not need batteries. The problem is that hard wired systems are more expensive to install and usually cannot be moved once installed. Retro fitting a hard wired system into an existing home can be quite challenging.

Surveillance and access are often hardwired too. Cameras and microphones can be wired to speakers and video processors and monitors. Electronic strikes and keypads are wired to power supplies and/or control panels.

Wireless

A wireless system functions the same way as a wired system, only with "invisible" wires. Each device uses battery power and has a transmitter that communicates to a receiver that can then pass the information on to the control panel. They use the same main components as a hard wired system, but since a battery and transmitter are also included, the size of each device may be a little larger. There can be some additional programming involved with wireless systems. This programming mainly involves "learning" the transmitting devices to the receiving unit.

Addressable and Analog Addressable

Addressable and analog addressable systems are capable of having detector/sensor units that have a unique location and thus their exact location can be determined/reported by the controller. These are the most intelligent of systems, and often used in life safety systems such as fire alarm. These type of systems require more programming which can be accomplished via the control panel or some other dedicated programming unit, or a laptop interface in which proprietary software is loaded onto the technician's laptop for the purpose of programming of such a system.

Combining Wired and Wireless

There are obviously advantages to both wired and wireless connections and in larger systems of all kinds, the best solution is often a combination of both. In a new home where wiring is easy there still may be some call for wireless devices. In a retrofit project where most communication is wireless, some wired connections may be possible and advantageous.

What the System Does

Since most installed security and alarm systems deal with both intrusion and smoke/fire detection, we will discuss them together. The basic functionality can described pretty simply:

- Detecting entry (or attempted entry), smoke, heat etc. in a defined area
- Alerting the occupants
- Sending a status or event notification
  - » to a central monitoring station
  - » Via text, email or other method

System Components

Security and alarm systems are often combined into the same system. The following system components and locations listed will cover this type of system.

- Central control panel
  - » Main processor
  - » Power supply
  - » Battery backup
- Input Devices
  - » Control point (user interface, such as a keypad)
  - » Detection devices
- Output devices
  - » Sounders, horns, strobes
  - » Communication: via telephone, cellular, internet
- Wired or RF infrastructure
- Power

INPUT DEVICES

- Motion Detectors
- Glass Break Detectors
- Volumetric Detectors
- Magnetic Contacts
- Photoelectric Beams
- Wireless Transmitters
- Keypad Inputs
  - *Arm/Disarm
  - *Panic/Medical
  - *Voice

ALARM CONTROL PANEL

Transformer
Battery
RJ31X Jack & Cord

OUTPUT DEVICES

- Speaker / Voice Driver
- Siren / Siren Driver
- Lights
  - *Lighting Control Integration
  - *Dedicated Strobe Lights
- Keypad
  - *Sounder
  - *Two Way Voice
  - *LED Status
  - *LCD Readout
- Communicator (built in Dialer)
  - *Digital
  - *Cellular
  - *Radio
  - *Internet

Figure 17-1 Central Control Panel

Central Control Panel

The central control panel is just that. It is the brains of the system that provides power to (and receives information from) the devices installed throughout the home. It will usually be installed in a central location so that if wired, the wiring runs are not too long. If wireless, it will be able to receive information from the transmitting devices throughout the home. It receives and processes information and then acts according to its programming to alert occupants, communicate with the outside world, or with other systems in the home. In order to ensure continued operation the main processor usually has battery backup.

Input Devices

Input devices include the user interface and all of the various types of sensors and detectors throughout the home which are there to relay information to the central control panel. A very simple system may include just perimeter doors, smoke alarms, and one or two other devices to detect a breaking window or other situation. More comprehensive systems include many more input devices to cover many different things.

User Interface / Keypad - This is where the user interacts with the system. Keypads, and touch screens are commonly used for this application. The main functions of the user interface are:

- Arming/disarming the system
- Bypassing a particular zone temporarily
- Allowing for a "panic" alarm to be initiated
- Indicating system status

Detection Devices - In security, two broad groupings of detectors or input devices or sensors are perimeter type devices and interior type devices. Fire alarm systems typically cover just the interior of a residence. Here is a list of the most common types of detection devices:

- Magnetic and contact sensors
  - » Cost effective
  - » Used for doors, windows, garage doors
  - » Require no additional power
- Passive Infrared (PIR) motion detectors
  - » Looks for heat signature
  - » Various detection patterns available
- Dual technology motion detectors
  - » Combine microwave and PIR
  - » Both must be triggered to detect movement
  - » More reliable than PIR
- Photoelectric beams
  - » Uses send and receive units
  - » Sends signal when beam is interrupted
  - » Good for large, straight line applications
- Glass break detectors
  - » Today most are the acoustical type
  - » Sense a specific frequency of sound
  - » Will protect an entire room
- Pressure sensors
  - » Under carpet
- Smoke detectors
  - » Ionization type
    - ◊ Positive and negative plate
    - ◊ Causes Oxygen and Nitrogen to shed electrons
    - ◊ Smoke neutralizes these ions and causes change in current flow
  - » Photoelectric
    - ◊ Light beam and photoelectric cell
    - ◊ Smoke blocks beam, triggers alarm
- Heat detectors
  - » Used in areas where smoke detectors may give false alarm (kitchen, bar)
  - » Heat detectors sense temperature
  - » Rate-of-rise sense rapid change in temperature
- Other sensors
  - » Gas detectors – sense a natural gas leak
  - » Carbon Monoxide sensors – sense CO in the air (from faulty furnace, water
  - » Water flow sensors– detect a leak or faucet left on
  - » Water level sensors – detect flood in basement
  - » Pilot light sensors – monitor pilot light on gas fueled appliances

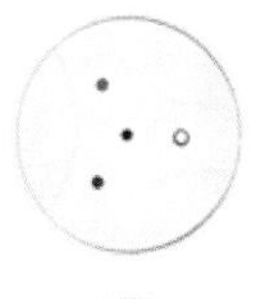

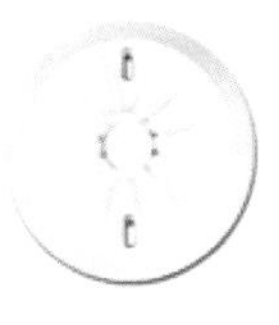

Zones and System Wiring Basics

It is important to have a fundamental knowledge of how a security system is set up and wired. If you are involved in the design or installation of these systems you will pursue a great deal more training. In this section we will just provide the fundamentals.

Zones

Every individual input device (sensor) has the ability to send a signal to the main process, but in practice we usually group similar sensors into "zones". This allows more sensors to be managed by the processor, and allows quicker interpretation of an alarm on the user interface. A zone is a circuit which reports to the main processor, and is organized by sensor type or location in the home. The user then has the ability to arm or disarm certain zones, for instance outside doors or motion detectors. The homeowner would want the perimeter doors armed at night but not the motion detectors, which would be activated if the occupants got up and walked around the house. Some zones should be programmed to "always on", such as those related to life safety (fire, smoke, carbon monoxide)

There are three types of zones:

- Hardwired – a zone input with a physical connection to the control panel
- Wireless – a zone input originating in a self-contained device with an onboard wireless transmitter and battery power
- Soft zone – a zone input that does not represent a physical location, but rather represents a device like a portable "panic button".

Cable Types

The cable types used in security and alarm wiring are all relatively simple. To review from Chapter 6, many cables are described according to the gauge of the conductors, and the number of conductors, in a simple notation such as 18/2 (2 conductors, each 18 gauge), or 22/4 (4 conductors, each 22 gauge). In the case of twisted pair cable, like Cat6, the designation is sometimes 24/4 or 24/4P (4 pairs of 24 gauge conductors). Some other abbreviations you should make sure you know:

- SOL – solid conductors
- STR – stranded conductors
- UTP – unshielded twisted pair
- STP – shielded twisted pair
- FPLR, FPLP – Cable rated for fire alarm systems, for riser and plenum installation (NEC ratings for vertical runs and installation in plenum areas, where HVAC return air is present)
- An overview of cable types commonly used in security and alarm system installation:
- 22/2 SOL or STR - Contacts
- 22/4 SOL or STR - Keypads, motion detectors, glass break and connection to telephone
- 18/2 SOL or STR – Power, audible devices
- 18/2 or 18/4 FPLR, FPLP – Fire alarm circuits

Wiring Basics

Sensing devices are provided current by the control panel and the processor monitors their status. Some devices are "normally closed", which means when in their normal state they pass the current. Others are "normally open", meaning like a switch that is turned off they do not pass the current. Some can be configured either way, as needed.

Within the system design, some zones may be just one device and others may have multiple devices. These devices within a zone may be wired in series or parallel, and we will see how this works.

Many circuits include a resistor of a known value as part of the path the current must flow through. This allows the processor to "see" a specific resistance (in Ohms). This this is done because a short circuit cause by something like a stray screw might make a normally closed zone look fine when in reality the "closed" status is being caused by something other than the sensors being closed. This type of circuit, with the resister at the "end of the line" is known as "supervised". The recommended circuit type for burglar alarm zones such as doors is a supervised normally closed circuit. When more than one of this type of sensor is used together in a zone, they are wired in series, so that if any one of them opens the whole circuit is open.

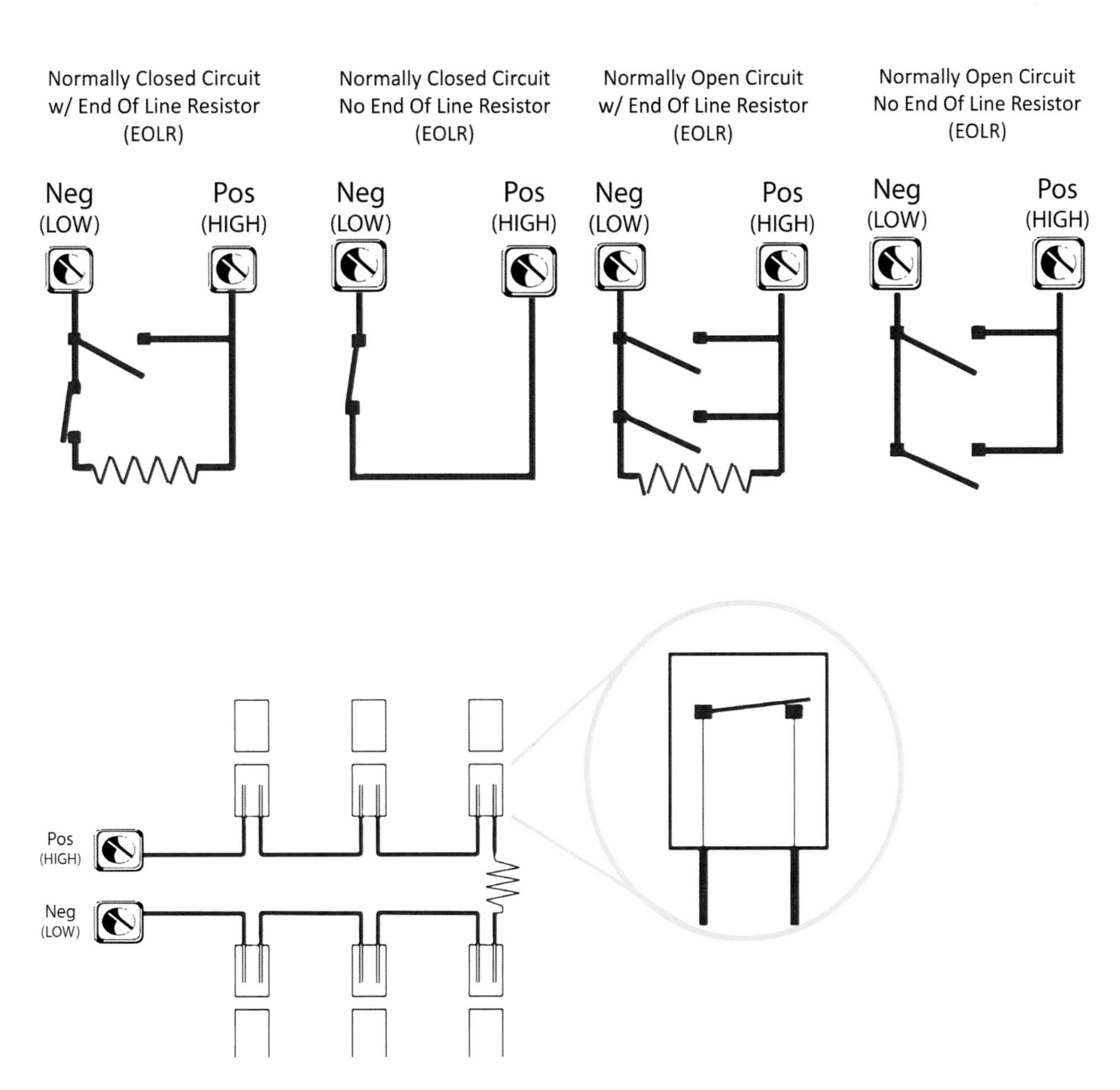

Figure 17-2 Series Circuit

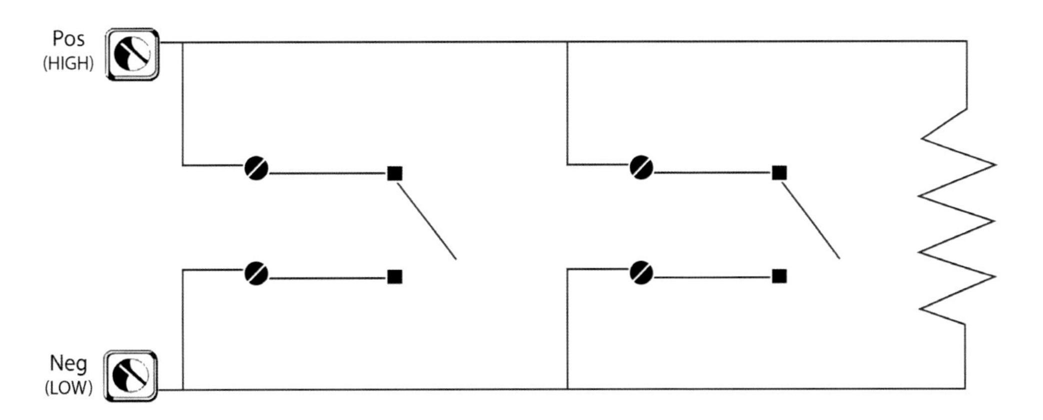

Figure 17-3 Parallel Circuit

Output Devices

This part of the system lets people within the secured area know that there is an alarm state, either a breached area or fire/smoke status. They can also provide an alert to a home owner that some other situation exists such as a low battery or power issue. Often referred to as Notification Appliances, these devices include bells, buzzers, horns, sirens, loudspeakers, chimes, strobes and more. If used in the security application, sirens and horns have the added value of creating noise that can confuse and deter intruders. In a fire alarm situation buzzers and strobes are common. The following provides a breakdown of various types of annunciation devices their primary use and typical mounting locations.

Visual - Strobes, etc.

The most common type of visual alarm indicator are strobes, to alert occupants of the alarm status. They are mainly used on the interior in an area where they will not compete with other bright lighting. If more than one strobe is used and more than one can be seen in a location, they should be synchronized to avoid adverse effects on certain individuals. They are generally required to be clear or white light. Wall and ceiling mount options are available and are often combined with some audible annunciation device in addition to the strobe.

Audible - Bells, Buzzers, Horns, Chimes, and Sirens

Bells that use a solenoid to cause a clapper to repeatedly strike a gong can cause electrical noise problems with the system they are connected unless isolated and/or filtered. They also use a high amount of current to operate. Motorized bells are preferable because they are often louder, less electrically noisy, and use less current than the solenoid type. They are to be mounted in areas of the home where they can be heard in any area of the home.

Horns usually use even more power than bells but are less electrically noisy. The buzzer is similar to the horn in that a continuously vibrating element is responsible for making the sound but they use less current.

Chimes and sirens are basically loudspeakers, and draw relatively low levels of current. Some annunciators are capable of producing voice messages. If more than one siren is used, they should be synchronized.

Communication

System communications is the aspect of the security system that notifies people off-premise. Notifying people on-premises with visual and audible signals as in the previous section is important, but protection of life and property also requires off-site notification and response (dispatch of relevant authorities).

A monitoring company that receives this communication and acts according to what kind of alarm it is, is known as a Central Station. Their services are paid for on a subscription basis.

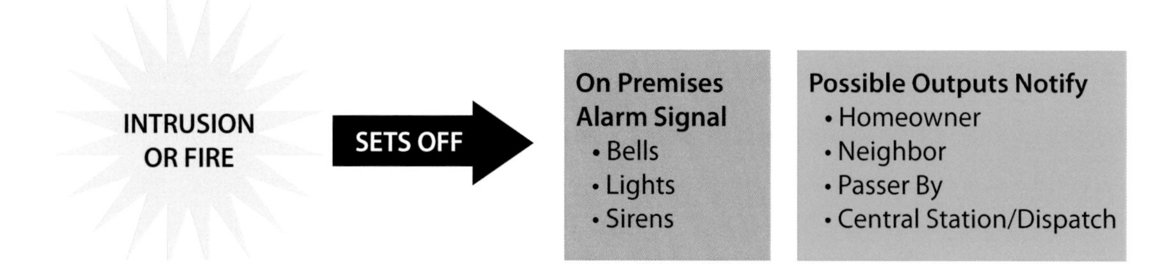

The signal to the Central Station generally includes a user ID, the type of alarm, and zone information. This communication may come from the home's security/alarm system in a number of different ways:

- Digital communicator (dialer) – uses existing phone lines
- Long range radio transmitter – uses RF transmission, often a backup to phone line
- Cellular transmitter – uses cell phone network, becoming more common with fewer land lines
- Internet transmitter – also often a backup, can include video

We can now envision the entire process, from beginning to end:

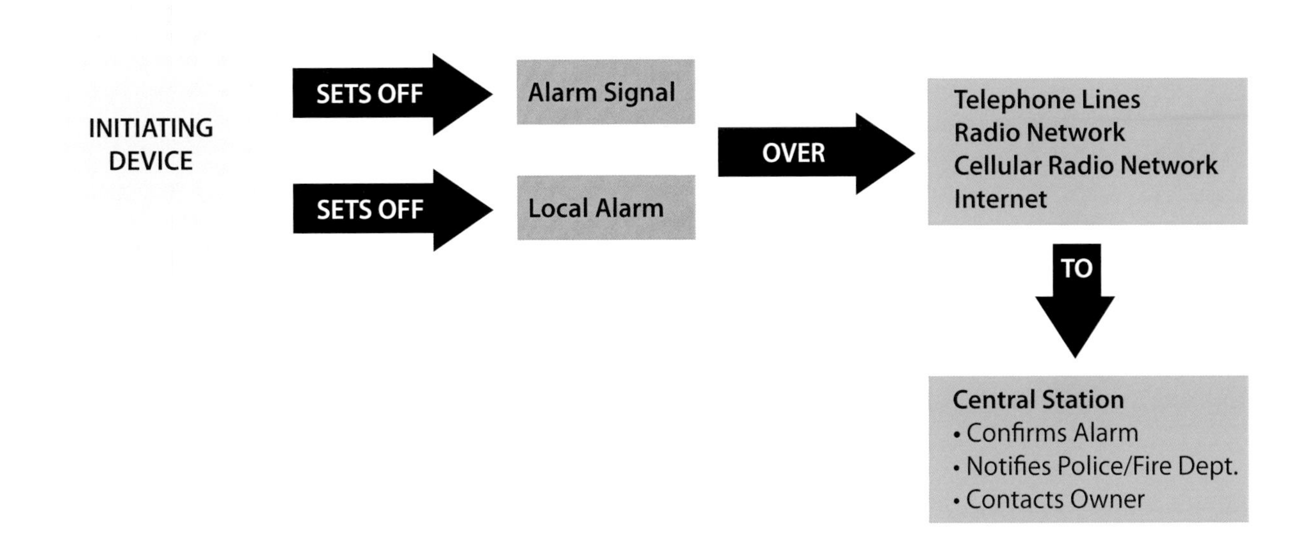

ACCESS CONTROL

Access control is a subsystem that provides the homeowner with enhanced control over who enters the home or property. An access system that is integrated with a security system should allow easy entry for family and guests while presenting obstacles for entry for intruders. In such a system, two things are needed. Hardware devices allow locks and gates to be locked and unlocked. Software can add intelligence to the system enhancing user friendliness of such a system and adding even more control options.

The hardware has the ability to "read" a keyed code entered, or coded credentials to identify the person attempting to gain access. Coded credentials include:

- Swipe cards
- Key Fobs
- Proximity Cards
- Smart Cards

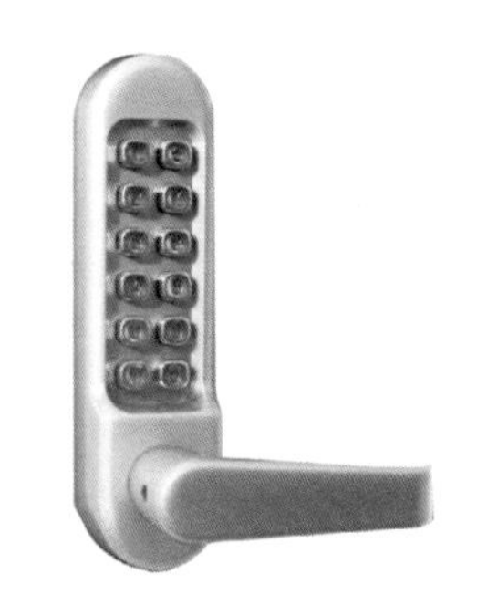

A more advanced type of reader is the biometric reader, which recognizes unique physical characteristics of individuals, such as fingerprints or retinal scans.

Locking devices including garage door openers and gate openers are the hardware that secures the entry points. There are a wide variety of these devices that range in difficulty to install.

A Cipher Lock is a keypad interfaced to an electric lockset such that entry of the correct code can allow entry into the residence. These can be battery powered or powered by a low voltage electrical connection.

Electronic strikes and bolts can be installed into the door frame to prevent access. Since these devices use a lot of current to operate, they are usually hard wired to a low voltage source. There are also electromagnetic locks that can be used in areas where free egress in case of emergency is important. Garage door openers and gate openers also prevent overt entry into a secured area of a residence.

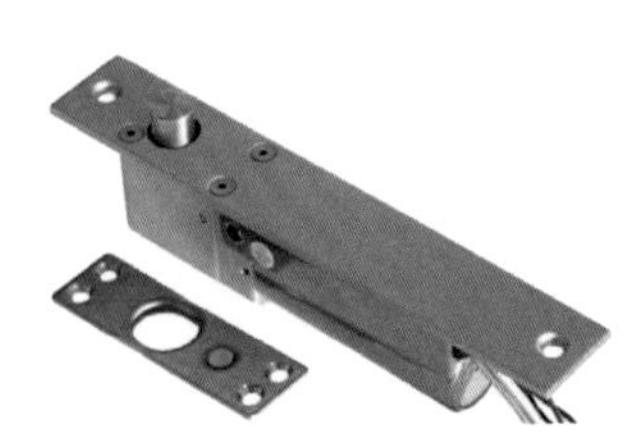

All of the physical devices mentioned above need a micro-processor system and software to enhance the user friendliness of the system. This software component of the security system is known as access control. The available options for the home owner are numerous with the addition of such a capability. Various entry points throughout the home can be set to allow entry to specified users at certain times. The times and locations of entry for particular users can be tracked. Locks can be controlled remotely via mobile devices. This information can be used by the home owner to optimize the system operation or just keep tabs on who comes into the home and when.

CCTV AND SURVEILLANCE

A surveillance system can operate independently or be truly integrated into a main control system. Cameras may be viewed on televisions or computers in the home, or remotely via the internet. We will discuss the key components in this sub-system.

Cameras

Video cameras are often used to allow viewing from anywhere, and enhance the operation of a security system. Cameras can be strategically placed throughout the protected residence both indoors and out to identify visitors, spot an intruder, or simply check in on a pet or how a particular space on the premises is being used.

Cameras can be outfitted in housings that enable movement of the camera so that one can look around an area of interest. Most cameras will be placed in a set position such as the front doorstep or the gate entry area in the driveway.

Analog Video Cameras

The traditional surveillance system starts with analog video cameras. An analog camera requires an external power supply and outputs an analog composite baseband video signal. Sometimes the cable to an analog camera is a combination cable which includes a coaxial cable for video and a two conductor cable to provide power, called "Siamese" cable.

Analog cameras can be viewed directly on a monitor or modulated and combined with the CATV distribution system for viewing any TV by tuning to the correct channel. In a multi-camera system they are usually connected to a Digital Video Recorder (DVR) so that recorded video information from all cameras can be stored and viewed at a later time, or even accessed remotely via the internet. Many DVRs have circuitry that allows them to interface to an integrated security system such that motion detected by a camera or cameras can trigger some event such as an alarm in the system.

IP Cameras

Internet Protocol (IP) cameras operate on the network and do not have analog video outputs. The IP camera was invented in 1996 and has become increasingly popular ever since. The video captured by such a unit is converted to a digital format and then transmitted over an IP network where it can be viewed directly, or to a PC server with video management software to enable recording and playback options. This opens a world of options for the home owner in that video images can be sent and retrieved over the internet so one does not have to be on the premises in order to view the video images. The real value this gives to the home owner is the ability to be" in the home" even while at work or on vacation. The user no longer has to wonder what is happening at home, they no longer have to just hope all is well, they can verify it for themselves.

A variety of IP camera types are available. Fixed View cameras allow only one viewing position while some cameras offer PTZ (Pan, Tilt, and Zoom functions) so that the user can control the camera to look around in an area. Indoor/outdoor housings can be added as well as domed and hidden types that blend in with the existing environment. Cameras placed in public view can deter would be intruders from threatening the home owners safety and/or belongings. Hidden cameras placed strategically throughout the home can act as motion sensors that trigger other events in a security alarm system from simply recording for later viewing to initiating an alarm to sending email or text notifications of a possible breach. IP cameras can easily be integrated into main controls systems, allowing images to be seen on touch panels as well as mobile devices which access the system.

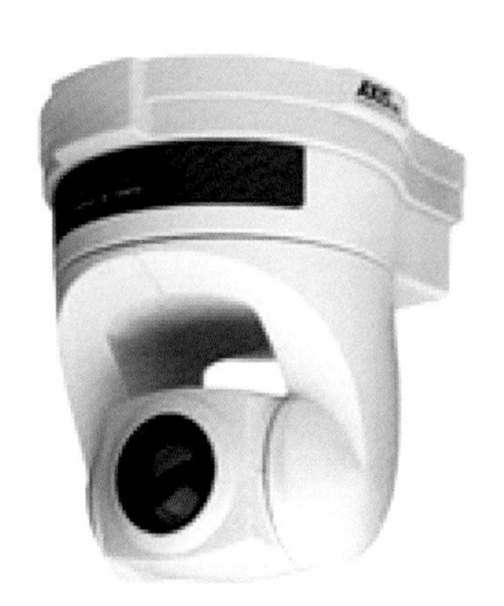

SUMMARY

This chapter has provided a very basic overview of these systems. Each has its own specific technologies and design/install considerations. In many situations, home technology professionals may only be interfacing with a security/alarm system, rather than actually installing it. Cameras and access control, however, are frequently specified and installed as part of a larger integrated system. More advanced training is offered by manufacturers, as well as the Electronic Security Association (ESA).

Questions

1. Which is NOT an input device in a security system?
   a. Annunciator
   b. Motion detector
   c. Photoelectric beam
   d. Keypad

2. Which input device requires no additional power to operate?
   a. Photoelectric beam
   b. Pass Infrared (PIR) motion detector
   c. Magnetic contact
   d. Glass break sensor

3. Which communication method is quickly becoming the primary method of central station communication?
   a. Dialer
   b. Long range radio
   c. Cellular
   d. Internet
4. An electric lockset which includes a keypad, to control who can open the door, is called:
   a. A cipher lock
   b. A combo lock
   c. A biometric lock
   d. An IP lock

5. An IP camera may have the capability to:
   a. Serve as a motion detector, triggering a message or recorder
   b. Pan, Tilt, and Zoom to better view the area
   c. Send images to mobile devices, TVs and touchpanels
   d. All of the above

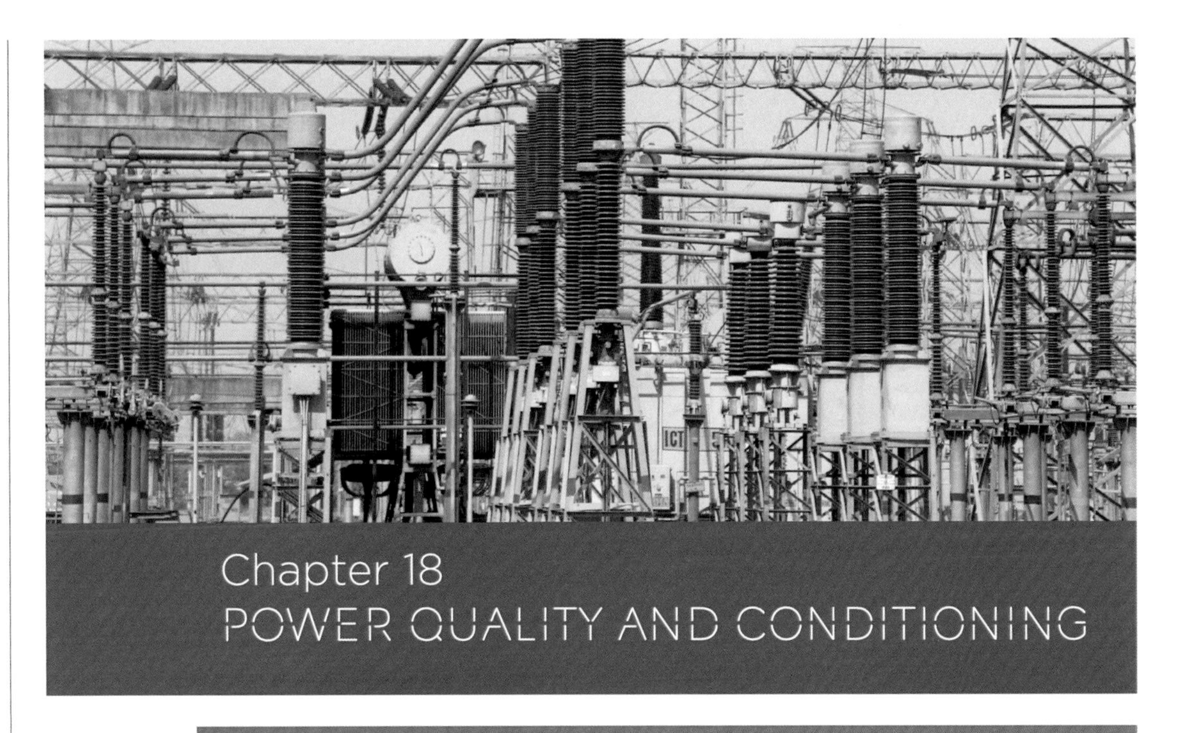

Chapter 18
POWER QUALITY AND CONDITIONING

OVERVIEW

Globally, a challenge common to all home technology professionals is the quality of the electrical power our electronic systems are connected to. As the number and complexity of electronic systems has grown along with the increasing demands on the electrical utilities the ability of these utilities to supply the needed electrical power has been compromised.

NOISE AND OTHER ANOMOLIES

As a result of these compromises today's sensitive electronic equipment is subject to severe and numerous undesired types of noise. This electrical noise is any kind of unwanted signal that causes undesirable effects in the circuits of the systems in which it occurs. This noise can be generated from internal sources within the system, or it can be generated from external sources through conduction or radiation. It is basically any unwanted distortion to the power signal. This noise disrupts electronic systems.

This noise can be caused by the following:

- Other electronic devices
- Control and switching circuits
- Arcing of contacts
- Motor and inductive loads
- Solid state rectifiers
- Power supply switching
- Electronic ballasts

Improper grounding and reversed polarity can also make electrical system noise worse when it does not conduct noise away from the power system or when it allows additional noise to enter the system.

The most common types of electrical noise are:

- **Over-voltages** - Above normal voltages for longer than one minute
- Under-voltages - Below normal voltages for longer than one minute
- **Brownouts** - Overstressed power grids, lasting from a few minutes to several hours
- **Surges** - Voltage increases between 10% and 35% that last anywhere from 15 milliseconds to several minutes
- **Spikes** - More severe types of surges that last for a few milliseconds but with short pulses of energy with voltages of 6 kv or greater
- **Sags** - Brief drops in voltage between 10% and 90% tat last from .5 to cycles to 1 minute

"Transients" is a term often used to denote both surges and spikes. Transients are also described as an undesirable but momentary event. Any time an unexpected change in current or voltage happens a transient can occur.

Electrical noise (or transients) can cause serious problems in electronic systems and equipment. This damage may not be immediately evident but can accumulate over time. In an effort to understand how these transients can damage equipment, consider electrical current flowing through wire like water flowing through a hose. If the hose gets kinks or knots in it that cause the water to build up too much pressure it will burst. Similarly if too much current flows through the lines it can harm the equipment it powers. It can get too hot causing damage to sensitive components in the equipment.

Premature failure of electronic equipment, lost or corrupted data and unexpected operation of equipment, stuck or fouled contacts in relays are signs that the equipment is being damaged by transients.

In addition to things like air conditioners, electric heaters, microwaves, copy machines, lighting systems and any other electrical equipment powering on and off. Generally the greater the load the electrical equipment is on the system the greater the chances that it will cause transients like spikes, surges or sags.

Geography can also be a factor in determining how vulnerable the electrical system is to transients and how susceptible your equipment will be to transient damage. Geography affects the chance of experiencing spikes due to lighting strikes. Some areas of the world are more susceptible than other areas. Being located in large metropolitan areas with rapid growth and outdated electrical systems are also contributing factors. In these areas brownouts, under-voltage and overvoltage situations are common.

Knowing what kinds of events are occurring is important to knowing which solutions should be employed. With short term events such as surges and spikes tools like Transient Voltage Surge Suppressors (TVSS) can be used to protect the connected equipment. With longer term events such as under-voltage, overvoltage and sags voltage regulators are needed to ensure that the voltage to the connected equipment remains stable and safe.

When the undesired events have greater or longer duration more aggressive protection means are needed. Brownouts, under-voltage and sags are best dealt with by using an Uninterruptible Power Supply (UPS) with battery back-up. Many UPS systems have the ability to protect the connected equipment from not only long term voltage fluctuations and brownouts but many have TVSS protection built in.

SOLUTIONS

With the cost of equipment and the losses associated with failed equipment and the resultant down time proper transient protection should be used in every electronic system.

Surge Protectors

Keeping voltages from exceeding desired levels is accomplished in most case with a component know as a Metal Oxide Varistor (MOV). These solid state devices are used in most consumer surge protectors. They have a variable resistance to electrical flow depending on the voltage present in the circuit. As the voltage level in the circuit exceeds the rated clamping voltage the MOV begins to work. When functioning fully an MOV can reduce the voltage to a safe level. In North America the most aggressive MOVs will have a clamping voltage of 200 volts. Globally, expect to see MOV based surge protectors operating at 330 volts for the best solutions.

Surge Protector ratings are as follows:

- Clamping voltage: 330 volts is best but surge protectors clamping at 400 volts and 500 volts are common. Look for the lowest clamping voltage for best protection.
- Energy Absorption/Dissipation: The higher the number the better the level of protection. Many manufacturers rate their performance in Joules. One way to define a Joule is the amount of work required to produce one watt of energy for one second. Quality surge protectors have ratings from 1575 Joules to 2700 Joules some inexpensive consumer products offer protection at about 1000 Joules. A few manufacturers rate their products using the classification known as "Nominal Discharge Current Rating". In order to comply with UL 1449 this rating a rating of 3kA or higher is required. During testing a surge protector must survive 15 events of this current level to be approved.
- Response Time: This measures the length of time it takes for the MOVs to start protecting the circuit. The longer the time, the more exposure to the surge the circuit will experience. Times of 1 nanosecond or less offer the highest level of protection.

Surge protectors must carry the approval a governmental recognized testing laboratory. These include:

- Underwriters Laboratory, much of the world including North America
- Canadian Standards Association, (CSA), Canada and North America
- Electrical Testing Laboratory, (ETL), globally many countries accept this
- TUV Product Service, (TUV), based in Germany but recognized globally

There are two classes of surge protectors. The first, "Series-operated/Connected", employ the protective devices in series with the load, in other words they are part of the power pathway, when the surge protector fails the power to the load is interrupted. In order to restore power to the protected loads replacement of the surge protective device may be required. These are rare in residential installations.

The second class of surge protectors is known as "Parallel-Connected". These devices are not part of the power pathway to the load. They provide a safety valve diverting unwanted electrical noise back to the source over the ground or neutral before it can damage the protected load. Since they are not part of the conductive pathway their failure will not disconnect the protected load from the power source. Unfortunately the failure of the surge protector may not be noted and the connected loads will be at risk of damage or failure due to the lost protection. These surge protectors often take the form of cord connected outlet strips which sit behind or under the protected equipment. They may have status LEDs showing their current state but if they are not readily accessible or visible loss of protection will not be noticed.

In many cases surge protectors will be incorporated into uninterruptable power supplies.

Uninterruptable Power Supplies (UPS)

There are 5 main types of UPS systems available in the market place. As residential and small commercial integrators we commonly use two types.

The most common type used in residential applications is the **"Standby UPS".**

This type consists of some level of surge protection and filtering, a transfer switch, battery charger, battery and an inverter. While power is present the back-up is in standby mode. In standby mode the battery charge is maintained at full by the charger. It is not until the power is lost that the transfer switch connects the battery backup to the load. During normal operation the load is protected from transients and the power might be filtered but is not until normal voltage levels drop or are lost that the backup comes on line. The length of back up is dependent on the size of the load and amount of battery storage in the UPS. Generally these UPS systems are meant to carry the load thru short duration outages and brown outs. These UPSs allow the load to remain stable during the many small short term power glitches that occur. They also allow for an orderly shut-down of the load if the loss of power of longer duration. When power is restored the back-up goes off line and the battery charger returns the battery to a fully charged state.

The second most commonly used UPS is the "Line Interactive" UPS". This design is used in larger residential systems and is often found in small business applications. This type consists of a transfer switch which is located at the beginning of the system. The switch only opens during an outage. As the name might indicate the battery and inverter are always part of the circuit. During normal operation the charger/inverter maintain the battery at full charge. When the power is interrupted the battery supplies power to the load thru the inverter. The design of this type allows the capacity and back up time to be increased by the addition of more or larger batteries. These UPSs are typically larger and more costly than a "Standby UPS", they could be as large as 5kVA or as small as .5kVA. More sophisticated "Line Interactive UPSs" incorporate some form of voltage regulation, typically a variable tap transformer, to maintain voltage levels during brown outs and sags. Adding this technology will extend the life of the batteries.

The remaining types of UPSs are "Standby Ferro UPS", "Double Conversion On-line UPS" and "Delta Conversion On-line UPS". These three are usually found in large commercial and industrial installations, they are large, expensive and generally not found in residential or small commercial systems.

Voltage Regulators

Voltage regulation deals with long term variations in voltage. Wide fluctuations in voltage are common throughout the world. As demands on the power utilities continue to increase, the problems of Sags, Brownouts and Under-voltage will only get worse. In countries where growth

outpaces the infrastructure and generating capabilities of the serving utilities, the need for these solutions will only increase, which provides new opportunities for home technology professionals. While a UPS can help with these problems voltage regulation is a better solution for these problems. Many commercial large UPSs also incorporate voltage regulation in their systems. To compensate for power inconsistencies voltage regulators are used to equalize voltage between the supply and the connected equipment. They operate to maintain a steady rate of voltage in order to minimize any damage from voltage irregularities and to prevent annoying pops and clicks in the audio system. In order to compensate for voltage fluctuations many voltage regulators employ multi-tap transformers. As the supplied voltage changes the transformer taps change to maintain a stable voltage level at the equipment side of the regulator. In order to create a stable voltage the regulator draws greater current from the power source.

Power Sequencers

Many power conditioning systems also include the ability to determine the order and schedule of connected equipment power up. This feature can protect equipment such as loudspeakers by preventing "pops" in the system that might be caused when an upstream component is powered on last. The same applies to powering down; power amps are turned off first, so nothing is amplified during the process. There are also other components which perform more predictably when powered up in the right sequence, such as modems and routers. When a power distribution center of this kind is network enabled, rebooting is possible from a remote location. This can help avoid a costly service call.

Servicing Power Conditioning Equipment

Transient voltage surge protectors, uninterruptible power supplies, and other power quality and conditioning equipment require service or preventative maintenance at times. These devices are often located in difficult to see or service spaces. The phrase "out of sight out of mind" is unfortunately often applicable, when it comes to surge protective devices. Many times, these devices will have prevented serious damage to connected equipment, but suffered catastrophic failure themselves. Since most of the protective circuits can fail while still allowing the connected equipment to be powered the next time an event occurs, there will be no protection in place and the connected equipment may suffer damage or failure in the future as a result. Some protective equipment may be equipped with status LEDs indicating their current state but if these are not visible, or checked regularly, the operator may not be aware of their possible failed state.

Uninterruptable Power Supplies also contain surge protection circuitry that can over time fail. Additionally, UPSs contain batteries, which have a limited life span and may not provide back up during an unexpected outage. Typically these batteries have a 5 year service life. This service life can be shorter if the UPS is located in a hot location or if the electrical utility has difficulty providing reliable power. UPS batteries often are designed for fewer than 200 discharge cycles. Temperature, age and a large number of discharge cycles will shorten the life of the batteries resulting in unreliable performance.

Power quality and conditioning equipment should be included in any routine maintenance schedule.

TYPE	OUTLET	COUNTRIES, REGIONS (PARTIAL)
A		USA, CANADA, MEXICO, CENTRAL AMERICA, COLUMBIA, PERU
B		USA, CANADA, MEXICO, LATIN AMERICA
C		(EUROPLUG) COMPATIBLE WITH MANY OTHERS, AND USED IN EUROPE, SOUTH AMERICA, AND AFRICA
D	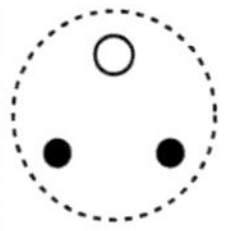	INDIA
E	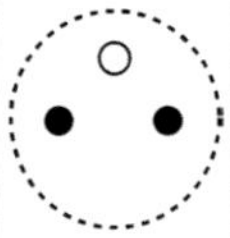	FRANCE, POLAND, BELGIUM , CZECH REPUBLIC, SLOVAKIA
F	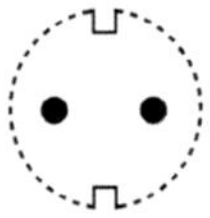	GERMANY, AFGHANISTAN, NETHERLANDS
G		UK, IRELAND, SINGAPORE, SAUDI ARABIA
H		ISRAEL
I		AUSTRALIA, NEW ZEALAND, CHINA, ARGENTINA
J		SWITZERLAND
K		DENMARK
L	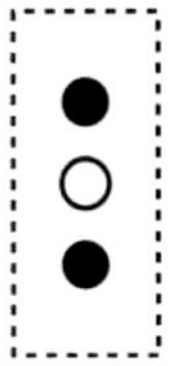	ITALY
M		SOUTH AFRICA
N	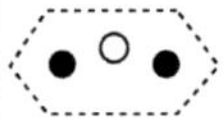	SOUTH AFRICA, BRAZIL

There are a wide variety of power types and configurations used in different parts of the world. It is important that you fully understand the power standards in every region you do projects. While is it sometimes possible to operate small devices like phone chargers and laptop computers on different voltages and frequencies using just an adapter, all audio, video and control equipment must be provided exactly the power type they are designed for.

Regardless of the location, operating voltage or frequency of the power system supplying the equipment the same problems and solutions apply.

Globally almost every electrical power system is Alternating Current (AC). The rules for dealing with AC power are the same regardless of voltage levels and operating frequency. Alternating current systems used in residential settings consist of a "hot" wire also known as the "current carrying conductor" a "neutral wire", also known as the "grounded current carrying conductor", and a grounding conductor. In cases of large appliances or other heavy loads two current carrying conductors maybe used to supply sufficient power to the load. In this case a neutral conductor may not be present but the grounding conductor will still be required.

Voltage and frequency vary from region to region and in many cases from country to country. In a few cases you can find voltage and frequency variations in the same country. In cases where voltages are similar but the frequency varies equipment made for a specific frequency may not function satisfactorily.

Motor driven equipment such as clocks, fans, turntables, CD and DVD players may not run at the proper speeds due to the variation in the power system frequency. In an increasingly global marketplace some equipment is being designed with the ability to operate correctly at either 50 or 60 Hertz. It is up to the home technology professional to ensure that the selected equipment is designed to operate on the voltage and frequency of the power supply.

In North America including Canada, the US and Mexico power systems operate at 60 Hz but in Mexico expect to see 127 volts AC rather than the 120 volts AC supplied in the US and Canada. In Europe including Easter Europe, the United Kingdom, The Republic of Ireland, Russia and the European areas of the former Soviet Union the frequency will be 50 Hz. Voltages of 220, 230 and 240 volts AC will be found and may vary from country to country. Most Middle Eastern countries have 50 Hz and 220 volts AC. In South America both 50 and 60 Hz will be seen along with voltages of 110 and 220 volts AC. In Africa the frequency will be 50 Hz and voltages of 220 and 230 AC are used. China has standardized on 50 Hz and 220 volts AC. India is also 50 Hz but operates at 230 volts AC. Japan is an interesting case, while the entire country uses a 100 volt AC power supply but Eastern Japan has a 50 Hz power system and Western Japan uses a 60 Hz power supply.

BEST PRACTICES

Power conditioning is critical to the service life of the electronic equipment in the home, and the satisfaction of the client. A multi-layered system of protection will offer the best solution to dealing with the challenges of electrical power systems. The first layer of defense in a total system solution begins with whole house surge protection at the service entrance. This will keep the worst of the external noise out of the residence but will not totally protect the electronic systems from internally generated noise. Surge protection can also be installed in each electrical load center alongside the overcurrent protection, (circuit breakers). These protectors will have an impact on much of the internally generated noise. Finally a complete power conditioning and protection system includes protection at the equipment location. Every layer should include transient voltage surge protection, then as required for more complex problems the final layer should include voltage regulation and where needed back up.

Every electronic component in the system should have some level of protection.

Video projectors that require an orderly shutdown and a cooling period to protect the life of the lamp should include battery back-up. Equipment such as DVRs (Digital Video Recorders), set-top boxes, and servers should also include a high level of protection including battery back-up. If the client relies on voice over IP for telephone service the network equipment, then modems, routers, switches, etc. necessary to connect to the internet should also be backed up. Any life safety systems such as security, smoke/fire and CO (Carbon Monoxide) detectors should also be backed up to ensure an "always on" state.

Amplifiers, and most source components should be protected from transients, but do not need to be backed up.

SAFETY, GROUNDING, AND OTHER CONSIDERATIONS

As with any electrical system, care to ensure the safety of those who come into contact with the system is essential. The importance of proper grounding cannot be overstated. Not only will a properly grounded electrical system offer protection from electrical shock and hazard to the user, but it will also perform better.

Without a solidly grounded system the most sophisticated surge protectors will not function correctly. If no properly designed and installed low impedance grounded return pathway exists, there will be no way for unwanted voltage transients to be dealt with. The simplest, lowest cost, first step in dealing with all these issues is to ensure the systems are properly grounded.

A final note concerning power quality and protection: There are some devices that rely on power line carrier technologies to pass data and control signals, and communicate with other equipment. These devices may not function (or might function intermittently) if connected to the electrical system through a surge protector or other power protection equipment. In particular Ethernet over power line may not function reliably when connected to a surge protector. The power conditioning equipment may treat the desired data signals as noise and remove them from the power line. The same could be true for some power line carrier lighting control. Some manufacturers recommend connecting their equipment directly to the power sources with the surge protection in line. In these cases a whole home surge protector is even more important to protect the installed electronic systems.

Questions

1. An increase in voltage of 10-35% which lasts up to a few minutes is called a:
   a. Spike
   b. Surge
   c. Over-voltage
   d. Sag

2. A voltage regulator keeps voltage constant by
   a. Filtering out changes
   b. Adding voltage from another electrical source
   c. Saving overages to use when voltage drops

3. Drawing different amounts of current to compensate for change in output voltage A good surge protector should have a response time of:
   a. 1 second
   b. 1 millisecond
   c. 1 nanosecond
   d. 1 minute or less

4. Uninterruptable Power Supplies have a battery which _______.
   a. Need to have water added annually
   b. Have a 5 year service life
   c. Is usually not replaceable
   d. Can be picked up at any convenience store

5. Virtually all domestic electrical services in the world are:
   a. 230 volts
   b. Balanced systems
   c. AC
   d. Grounded

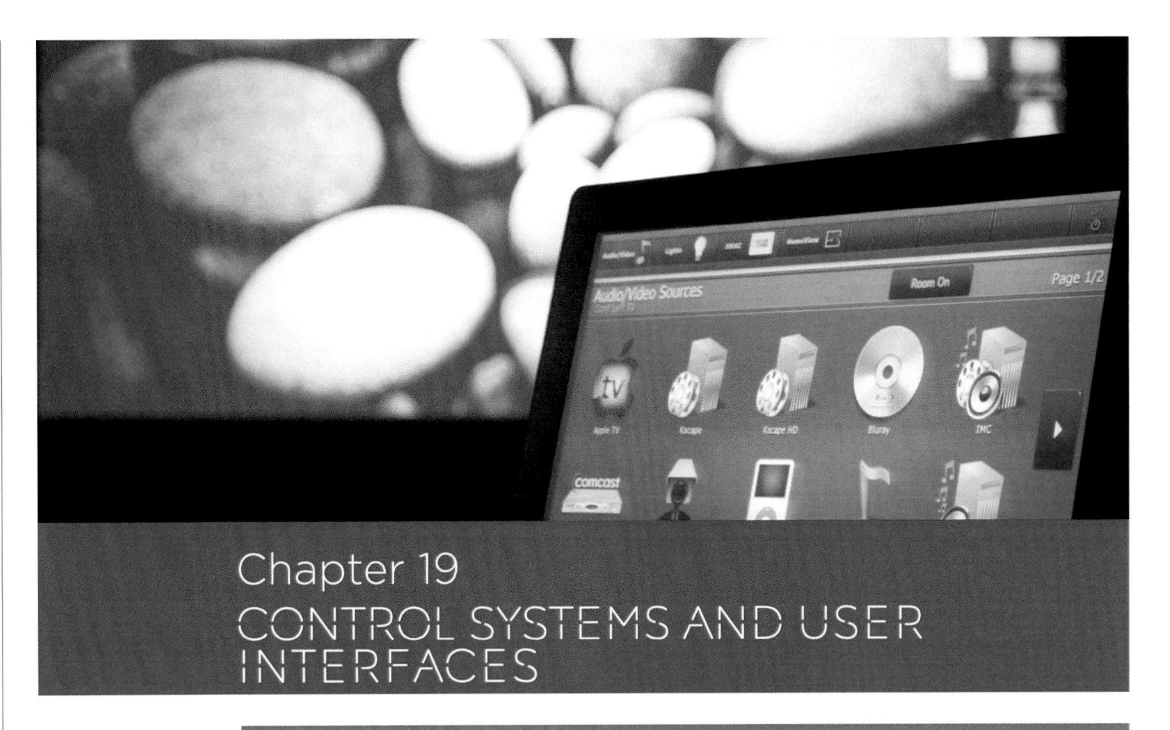

Chapter 19
CONTROL SYSTEMS AND USER INTERFACES

HISTORY OF CONTROL

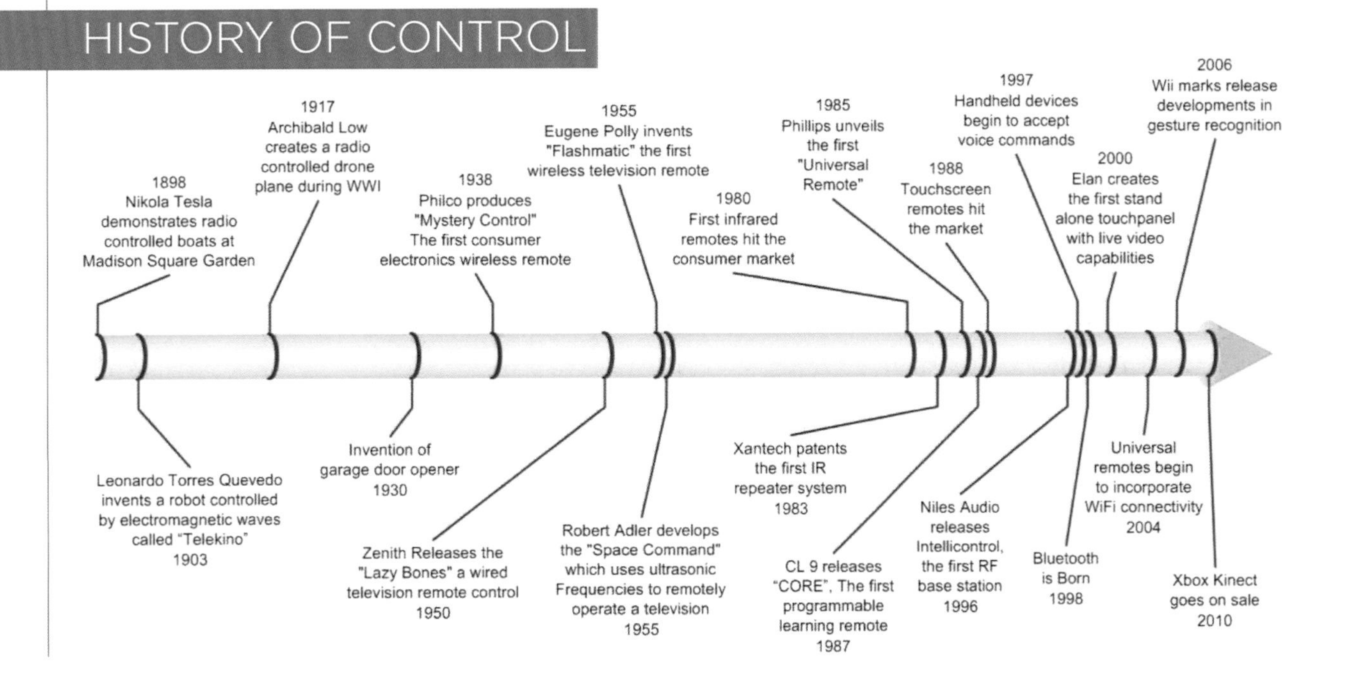

Early Radio Control

Nikola Tesla's demonstration of radio controlled boats in 1898 is shown as the starting point. In 1917 the successful creation of a radio controlled plane proved the viability of the technology and from then on development of remote control devices for military purposes became a strong focus for engineers from around the world, especially during World War II.

The first remote controls used for consumer applications appeared during the 1930's, but had limited scope and high costs.

Early Consumer Control

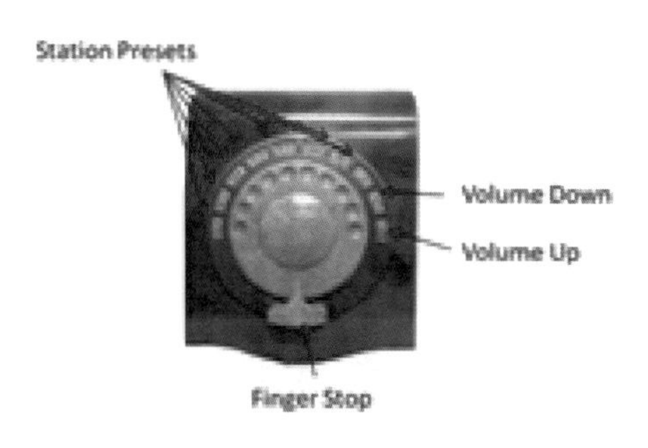

The invention of the garage door opener has been documented as one of the earliest remote controls developed for consumer use. According to a 1931 issue of Popular Science, two inventors on opposite sides of the country applied for extremely similar patents at virtually the same time, unbeknownst to each other. The first wireless control developed for consumer electronics use was the Philco "Mystery Control". Released in 1939, it was designed to operate Philco radios wirelessly through the use of RF. Modeled after a rotary telephone, the user could spin the dial in order to choose one of 8 preset radio stations as well as control the volume up or down.

Evolution of the TV Remote

The evolution of the television remote control began when Zenith unveiled the "Lazy Bones" in 1950, and even though it was not a wireless remote, it gained considerable consumer popularity. The next developments in the consumer market also came courtesy of Zenith, by way of the somewhat problem ridden "Flashmatic", followed closely by the much more reliable "Space Command".

IR In the Beginning

IR, which stands for "Infrared", was developed from military applications. The technology became available to the consumer market around 1980 replacing the technology of the ultrasonic "Space Command" remotes which were in use prior for a quarter of a century.

As IR remotes gained popularity, the necessity to consolidate the number of remotes used to control consumer devices became more and more apparent. Thus in 1987, the first programmable, learning remote control was brought to market. Although the new product got rave revues from the technically savvy, due to its cumbersome programming design which created a necessity for a very high level of expertise, it failed to catch a wide audience and eventually died.

Through the 1990s and early 2000s many companies continued to develop the technologies used for control applications including the invention of RF base stations, Voice Command applications, Wi-Fi integration, Bluetooth communication technologies, and gesture recognition for control. One recent development in the control industry was the release of the Xbox Kinect, which signifies breakthroughs in both voice and gesture recognition technologies that hold limitless possibilities for the future.

The IR remote control is made available with almost every consumer electronic device. Some TV's, A/V Receivers, and Blu-ray players come with a pre-programmed IR remote that will be discussed later.

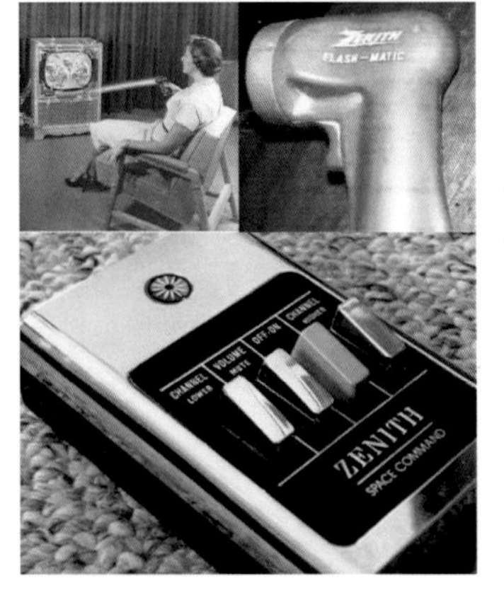

Figure 19-1 Early TV Remotes

INTRODUCTION TO USER INTERFACES

A user interface is the means by which a user accomplishes a desired task. In our everyday lives, we encounter all sorts of user interfaces, although we may not think of them as such. Additionally it is any device that triggers an event based on human interaction. The goal of the user interface is to simplify the user's interactive experience with the system, make common functions simple, and make complex functions possible. One of the most brilliantly conceived interface devices still in use today is the light switch. Additional examples of commonly used interfaces are:

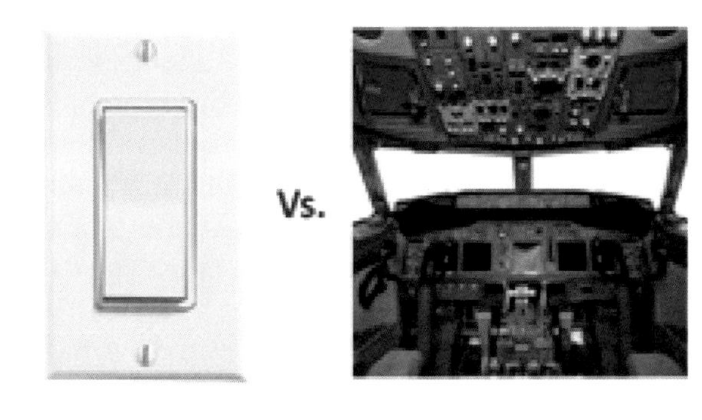

- **Door handles** – enable users to open and close doors
- **Climate controls in cars** – enable users to adjust the temperature
- **Computer keyboards** – enable users to interact with computers
- Consider your experience with these and other simple user interfaces. Think of how frustrating it is when a user interface is confusing and you cannot get it to do what you want it to do. Whether the interface is simple or complex, users expect it to do what it looks like, or what their previous experience suggests it should do.

Who Will Be Interacting with the Interface?

It is to be expected that there will be a wide range of users utilizing many types of interfaces. It is important that we identify the skill level of the user technologically and functionally. The best system in the world will do nothing but collect dust if the user interface that enables the user to take advantage of all the features is too confusing. Skill levels will include:

- **The Power User** – Wants every function available on all controllable components in the system
- **The Intermediate User** – May feel comfortable following one or more steps to accomplish a task
- **The Novice** – Knows very little and wants to accomplish simple tasks such watch TV with a minimal amount of interaction or buttons pressed

User Interfaces in Home Integration and Entertainment

In the home integration/entertainment business, user interface typically means a remote-control, touch-screen, keypad, or mobile device. It is sometimes easy to forget that users want the same thing from their touch-screens that they want from their door handles. They want them to open doors without having to figure out how they work. The difference is that these door handles open doors to music, movies, lighting, environment controls, etc.

Providing a Good User Interface

The best system in the world will do nothing but collect dust if the user interface that enables the user to take advantage of all the features is too confusing. If the user cannot make the movie play, then it does not matter how good the movie would have looked or sounded if it had played. It is essential that the user interface fit the people that are going to be using it.

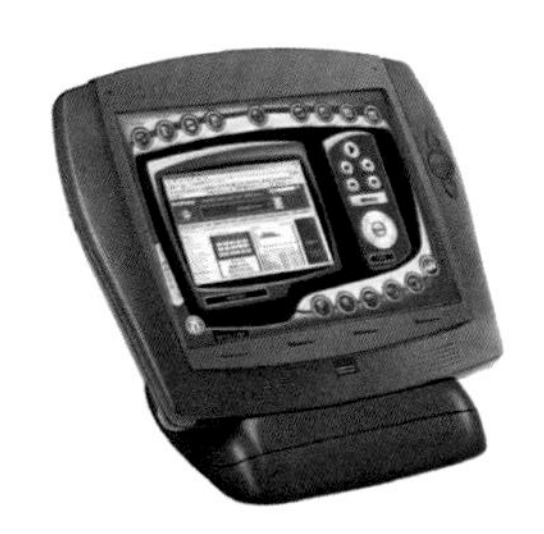

Perhaps the most important added value a system integrator brings to the client is the ability to make complex systems simple to operate. Every user should have an easy, understandable path to any desired function in a system.

Here are some characteristics of a good user interface.

- It is appropriate for the application and location in which it is to be used. A great UI in the wrong location is a bad UI.
- It is transparent to the operation of the device. In other words, the user is interested in turning on lights, not manipulating the UI. Therefore, the UI should make the user's intended operation as easy as possible.
- It is intuitive to use, requiring little or no learning curve to perform basic functions. Users are not willing to expend effort to relearn how to do simple things, so the UI should make common functions simple.
- It allows users to do complex tasks. Users are willing to spend some time to learn complex tasks and are willing to put in commensurate effort to the payoff they get by accomplishing their task.

Types of User Interfaces

Hand-held Remote Controls

Basic handheld IR remotes are devices that transmit serial data via an infrared carrier signal. These are a staple of the industry and there are many variations. We will examine the pros and cons of IR remotes in general, as well as dedicated (one-device) and programmable IR remotes. All IR remotes share the following pros and cons.

Pros

- **Ubiquitous** – These devices are well known by the users and familiarity is a big part of acceptance of the UI (User Interface).
- **Fast** – IR does not add significant latency delays.
- **Inexpensive** – IR is the least expensive of remote UI technologies.
- **Long lasting** – Unless the model used has backlit displays or other high power devices, the batteries last months or years without requiring recharging.

Cons

- Limited to line of site – Users must point the remote at the component or an IR repeater and people can block the signal as they move around.
- Susceptible to plasma TV and other interference – Next generation TVs, florescent lights and other light-emitting devices can interfere with the signal. Sometimes this problem can be solved with filters, but in some cases, it is unfixable.
- Ineffective in sunlight – Range is greatly reduced in direct and indirect sunlight.
- Complicated – Manufacturers do work to simplify them. However, the trade-off between reliable discreet commands and button reduction using toggling functions can complicate the UI.

Dedicated, one-device, remotes are the typical remotes that come in the box with a DVD player, tuner, CD or other source. These work great when users only have one device. However, the proliferation of remotes as users add devices leads to a pile of remotes on the coffee table. In addition, each remote has a unique UI so there is no continuity from remote to remote. Spending five minutes rummaging through a confusing pile of remotes just to watch the news is no one's idea of fun.

Dedicated remotes have the following pros and cons, in addition to all of those listed for IR remotes in general.

Pros

- Included with source component
- Preloaded with the proper IR codes

- Optimized by the manufacturer for the specific device
- Very effective - for one device

Cons
- Pile up very quickly with multiple entertainment devices
- Different user interfaces (each remote has its own)

Low-end learning/programmable remotes sometimes come as standard equipment with a cable box, TV or receiver. These remotes can consolidate the functions of the pile of dedicated remote-controls, which eliminates the pile of remotes on the table, but brings a new set of problems.

Pros
- **Inexpensive** – Sometimes included free or can be purchased inexpensively
- **Efficient** – Reduce the pile of dedicated remotes to one single remote

Cons
- Not optimized for individual components – Some buttons are improperly labeled for a specific device's function, or they are missing
- Confusing – Complex functions are difficult or impossible for users to figure out due to mislabeled buttons
- Not as usable as a good, dedicated remote – Users are not going to want to replace the TiVo remote with a less usable and more complex multi-remote

High-end learning/programmable remotes add a display, a menu driven interface and macros to help improve the UI. They are programmed on a computer, often using "wizards" to simplify the process. The result is a dependable control system that can be customized to the user's needs.

Pros
- Navigation is easier
- Buttons are (can be) properly labeled
- Macros (discussed later in this chapter) can be used to simplify complex actions

Cons
- The battery requires a charging cradle or must be changed
- If signal is IR, there can be issues with line-of-sight and macro dropouts
- Toggling (discussed later in this chapter) commands frustrates users. Sources with toggling commands get out of sync

RF/IR Hybrid Remotes

Infrared/radio frequency (IR/RF) hybrid learning/programmable remotes use radio frequency signals instead of IR to send commands from the remote-control to a receiver located near the source equipment. This central processor stores the actual commands, sometimes both IR and RS232 or even has the ability to control devices via contact closure. RF eliminates the line-of-sight issues that IR remotes have, but does have its own issues. This type of system will be discussed later in the chapter.

Pros
- There is no need to aim the remote. People movement will not block signal.

Cons

- User frustration – Sources with toggling commands can still get out of sync, even though it is less likely.
- RF signals can sometimes bleed from house to house

Keypads

Keypads can be IR, or proprietary to a specific control system.

Stand-alone keypads are often dedicated to specific sub-systems. For example, the security system may use keypads at the entrances. Users cannot access any other sub-system controls from these keypads. Each sub-system would have its own keypad. These keypads share the general IR remote pros and cons previously mentioned. They also have a few unique pros and cons.

Pros

- **Convenience** – Like light switches, they are always in the same place.
- **Simplicity** – They do not need lots of buttons.
- Wired systems are very stable and dependable, and need no batteries (power is supplied via the cabling)

Cons

- Limited expansion – They have a limited capacity for expansion, which makes them less desirable for complex systems.
- Clutter – Multiple keypads can cause clutter.

System keypads use a network (typically wired) to connect the keypads to other devices, such as multi-zone audio controllers or a system automation controllers. From these devices, users can access and control multiple sub-systems. For example, users could control the audio, lighting, security and HVAC sub-systems from one keypad.

Pros

- **Reliable** – They will do what they are told without much potential for error in the transmission of the command because their connection to the systems is more stable.
- **Easy to use** – Simple buttons with titles
- Provide meta-data – Information such as a current song can be presented on an integrated display (if the keypad is display- and meta-data capable).
- **Scalable** – System keypads are more scalable than stand-alone keypads.

Cons

- **Expensive** – They can be more expensive.
- **Installation and Programming** – They can be more difficult to install properly.

Touchscreens

Like stand-alone keypads, **stand-alone touch-screens** typically only control one sub-system. Touch-screens add a graphical user interface (GUI) for control of IR and optional RS232 controllable devices. The GUI is an advantage in that it is flexible, but a disadvantage is that it is also complex. This type of interface may also incorporate video display capabilities (music server interface or door camera). As more functional systems emerge, the stand-alone IR touch panel is being replaced by more sophisticated devices.

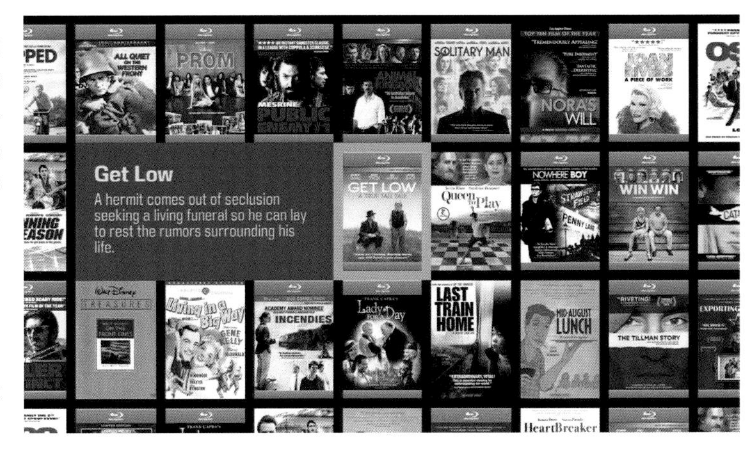

A **system touch-screen** is very similar to a system keypad. It also shares the advantages and disadvantages of the GUI with the stand-alone touch-screen. System touchscreens are very versatile and often can be customized to meet the customer's needs.

Pros

- Reliable control
- Meta-data
- Ability to browse content such as artist or genre list
- Color graphics
- Integration of multiple-controlled systems

Cons

- More Expensive
- Can be more difficult to integrate/program
- Difficult to do simple things if there are no (or not enough) hard buttons
- May seem complex to nontechnical users

Wireless versions of both stand-alone and system touch-screens exist and share the same pros and cons as their tethered counterparts. They require battery management, which is a drawback. They do have the advantage of being mobile, but may have a limited range.

TV/Display Onscreen GUI's

A Graphical User Interface (GUI) is what everyone is used to using on today's computers and tablets. In many ways, video (TV) display-based GUIs are similar to touch-screens in terms of flexible interaction. However, the input is far more limited. The user needs to navigate the environment using up/down/left/right type commands. Many systems and content providers are now utilizing on-screen interfaces, and most customers are comfortable using them. Applications include; cable TV and satellite, content services like Netflix, multi-purpose services like Apple TV and Google TV, proprietary media servers, and even the new "Smart TV" features where the internet access and applications are built directly into flat panel televisions.

Computer and Mobile Control

As different devices perform similar functions, and the lines between the various services and delivery methods continue to blur, the industry will see more diverse functionality in computers, laptops, phones and tablets. In the case of mobile devices, this is accomplished with "apps" (applications) that are usually proprietary to a single control system or device. For instance a user may have a tablet with apps that work with the automated lighting and the distributed audio system, and also use that same device to bring their personal music or movie collection to the media room system wirelessly. This discipline is complex and constantly evolving – but not going away.

These two control device options may be interfaced with any of the protocols discussed in this chapter by using the right "bridge" devices.

In regards to communication a protocol is a set of instructions for a device or rules for communication between devices. To follow are some of the most common protocols in the A/V and automation control world.

Infrared (IR)

Infrared light has a wavelength just beyond the visual spectrum of human sight, so even though we don't see it, it behaves much as visual spectrum light. It travels in a straight line, is blocked by solid objects, such as walls and cabinet doors. IR is the most popular method of remote control. The handheld device is pointed at an IR receiver (either in the actual device or installed as part of an IR distribution system), the remoter sends commands over a carrier frequency, and the sensor (eye) picks them up. In the case of a purpose-built IR distribution system, the eye picks up the signal and sends it to an active distribution block over a physical cable such as UTP (Unshielded Twisted Pair Category 5). A minimum of 3 conductors are needed (power, data, and ground). The distribution block then passes the signal on to the devices via UTP and an emitter for each device to be controlled.

Radio Frequency (RF)

RF signals, like radio frequencies in the air, do not require direct line of sight to the receiver. Some systems use RF from the handheld controller to a main processor, which then has the ability to send out IR, RS232, or contact closure signals to devices downstream. RF transmits radio waves using binary code that uses 1s and 0s, or positive and negative voltage. This range up to 100 feet is different because of different technologies. RF is used when IR is not an option or when it is faulty. This happens a lot with clients who do not want to make sure they're pointing the remote an IR receiver every time. If an RF transmitter is added to the system clients do not have to worry about that. The two RF frequencies used the most are 418 MHz and 433 MHz. If you are in a crowded neighborhood, you should stay away from 418 MHz. There are more things around the 418 range that will cause interference. Most devices will give you an option between 418 and 433. 433 will be more reliable but the downside is the distance is a little less. It is similar to going from 2.4 GHz with Wi-Fi to 5 GHz. 5 GHz is much more reliable but the distance goes down.

Serial (RS232, 485)

Serial control is highly dependable, carried only over cable, and often provides two-way communication. In telecommunications Recommended Standard 232 (RS-232) is a protocol standard for serial binary data interchange. This standard includes the physical, electrical and functional characteristics of the interface. RS-232 is the standard American format for serial data transmission over cable (that is, from a computer terminal to a modem).

Looking into front of connectors

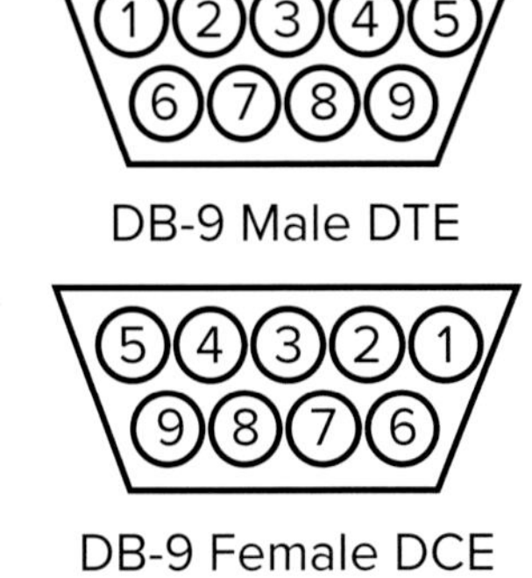

1	CD	carrier detect
2	RD	receive data
3	TD	transmit data
4	DTR	data terminal ready
5	GND	signal ground
6	DSR	data set ready
7	RTS	request to send
8	CTS	clear to send
9	RI	ring indicator

Figure 19-2 DB-9 Connector

RS-232 transmission usually uses a 9-pin DB-9 connector, but in many settings not all of the conductors are used. RS-232 typically connects between data terminal equipment (DTE) and data communications equipment (DCE).

Internet Protocol (IP)

IP is the universal language of the internet and the home network, and is quickly becoming more important in the integration industry. IP is not proprietary; it is the same for all devices and in all parts of the world. This language is used on the Local Area Network (LAN) as well as the Wide Area Network (WAN). There are actually many different Internet protocols, but Transmission Control Protocol (TCP) and Internet Protocol (IP) are the most recognizable and widely used. IP control is quickly replacing RS232 in today's residential electronic devices. IP has multiple advantages over RS232, and because Ethernet and category cables are already in most clients homes, it is extremely easy to install systems with the customer's existing infrastructure. For more information on IP and networking, see Chapter 13.

Wi-Fi is widely used in wireless networking. Its primary use is to allow an electronic device (primarily a computer) to exchange data or connect to the internet. It works on IEEE802.11 standards. It is being used today for control of Audio/Video and Automation applications. Two popular frequencies that Wi-Fi works in are 2.4 GHz and 5 GHz bandwidth.

Other Protocols

Newer wireless communication methods are finding their way into the remote control/touch-screen industry. In many cases they allow the device to provide functions not previously found on hand held controllers (streaming video, web browsing, email, etc.) You should be familiar with all of them.

ZigBee is a standards-based wireless technology based on an IEEE 802.15 standard. It is used in applications that only require a low data rate which make it an excellent choice for transmitting control data. It creates a "Mesh" network which relays data from one device (node) to the next in the network.

Z-Wave is an RF wireless technology that allows household products, like lights, door locks and thermostats to "talk" to each other securely and can be accessed and controlled using phone, tablet or pc. Z-Wave also supports "Mesh" networks. It operates in the sub-1GHz bandwidth as opposed to many devices such as cordless telephones that operate in the 2.4 GHz bandwidth.

Bluetooth is another wireless technology used for transmitting data. It works in the 2.4 GHz bandwidth. It was originally intended to be an alternative to RS-232 (Serial Protocol) data cables. Bluetooth has many applications in addition to control including hands-free headsets, telephones, media players and robotics systems. It is easy to implement but has limited range.

Low Voltage/Contact Closure

Electronic devices such as relays are sometimes used to trigger events, for instance controlling a sprinkler system. A small amount of voltage can also be applied to a relay to raise and lower motorized shades or projector screens. In some cases simply closing a circuit will trigger action by another device.

Sensors

In many situations, it is necessary to know the status of a device or component in order for a control system to know which command needs to be sent. Some examples include the status of a garage door (up or down) or the On/Off status of a TV. Various sensors are made to provide feedback on such things as the presence of current flow, light, or moisture.

We have discussed the various types of interface devices commonly used, and the communication protocols that are available. Now we will dig just a little deeper into how these all work together in a whole house control system, as well as touch on some specifics of how IR can be routed and distributed.

Single-Room System

The simplest control system example is a one-room solution where a single controller addresses multiple components. This is essentially replacing an array of factory remotes with one programmed device which simplifies the operation of the system, usually one of these situations:

- TV with a set-top box of some type and a DVD or Blu-ray player
- Media Room system (similar to above but with a surround receiver & speakers
- Home Theater (other sources, display options and/or lighting added)

In all of these cases everything being controlled is in the same room. The control system may be all IR but more likely will be an RF/IR hybrid.

IR Distribution

When some components are in a cabinet, closet, or a different room, the IR signal must be converted to an electrical signal and distributed, and then converted back to an IR beam.

This is call an IR repeating system, or IR distribution. A common IR repeating system includes the following hardware. An IR receiver, a connecting block, a power supply, and IR emitters.

The IR Receiver does just as its name suggests. It receives the IR signal, converts the infrared light to voltage and passes it on to a connecting block and then to one or more emitters, which are placed directly onto the equipment to be controlled. There are many types of IR receivers including, stand alone, in-wall, in-ceiling and sometimes integrated into in-wall touch panels or keypads.

The connecting block acts as a splitter for multiple emitters to connect to, and sometimes allows for the integration of more than one IR receiver. Additional connecting blocks may be added to the system to allow the system to be scaled. There are multiple types of connecting blocks including amplified and programmable/zoneable. An amplified connecting block is a large connecting block that provides power for more IR emitters. The typical IR repeater system allows up to four dual emitters as a maximum. An amplified block will drive more emitters over longer cable runs.

The IR emitter converts the voltage back to IR light pulses and is typically placed directly over the equipment's IR receiving window. This is where the user would normally aim the basic factory IR remote to control the device. Types of emitters include single head, dual head, and direct connect where the equipment would have a special jack that the 3.5mm connector (most often used connector size) is plugged into. Most IR Emitters are supplied with a 3.5mm connector.

IR systems require a power supply/source such as a basic "wall wart" or sometimes a dedicated box.

RF/IR Hybrid

In an RF/IR hybrid system the basic IR Receiver is replaced by an RF base station. This allows the remote control to use RF rather than IR, which means direct line of sight is no longer neces-

sary. Emitters are connected directly to the base station. More advanced base stations include support for all of the protocols mentioned in this chapter. All base stations utilize a CPU (Central Processing Unit). Advanced base stations employ a more powerful CPU. These CPU's can execute millions of instructions/ commands per second referred to as "MIPS" (Million Instructions Per Second). The advanced type of base stations are generally integrated into automation systems but can be used for a multitude of applications.

Bridge Devices

There are many devices that bridge IR and IP control. These devices allow protocols such as IR, Serial, and Low Voltage to be integrated with a system that uses IP as its primary means of communication/control. Not all components allow for IP control, although more and more A/V and automation devices are starting to include IP integration. One example of this kind of device connects directly to the home network like any other network enabled device. A control device such as a smart phone or tablet can then communicate with the bridge over Wi-Fi via the network router, which then sends the signal over Wi-Fi or the wired network to the bridge device which is hard wired to the equipment to be controlled, either by IR or serial communication.

Centralized Control Systems

When multiple sub-systems are to be controlled, a main processor is used to store all programming and communicate with all devices and systems in the home. These main processors have a great deal of computing power and the ability to communicate via virtually any of the protocols described earlier. The manufacturers who produce these sophisticated control systems typically offer every imaginable type of control devices (hand-held, in-wall, mobile apps, etc.) and all of the hardware needed to control any type of electronic device or subsystem (HVAC, security, lighting, and much more).

The configuration and programming of these higher-end systems is a specialized field requiring an advanced skillset, and a full understanding of all components and systems in the home. Experienced control system designer/programmers are some of the most highly paid, and sought after, professionals in the industry.

Control Cabling and Connectors

A cable is a transmission line. Its purpose is to move information or a signal from one point to another without significant alteration. Cables are used to transmit signals or information in the form of electrical energy or light energy in all low-voltage systems. In ALL cases, remember that any wiring installed inside a wall must meet local or regional code requirements. This includes any wiring associated with control systems and IR distribution.

IR - Emitters will usually have short cable attached to them. However the attached cable is not always long enough and may be extended using twisted pair cable such as Cat5e. Generally three pairs are used.

- Power (orange pair)
- Signal (blue pair)
- Ground (green pair)

For long runs we suggest just two pairs:

- Solid orange for power
- Solid blue for signal
- Blue/white and orange/white together for ground

The most often used connector for IR devices is a 3.5mm mini-phone type connector (some companies have used 2.5mm). There are times when a DB9 connector may used, and in some instances an RJ45 plug/jack may also be used.

Serial Control may be converted using a DB9 to RJ45 adapter and run over Unshielded Twisted Pair (UTP) cable.

Proprietary Systems may have special cable configurations. Follow the manufacturer's recommendations.

MACROS AND OTHER CONCEPTS

Discrete Codes

Discrete codes are commands that separate similar functions. For instance the On/Off command (referred to as a "Toggle Command") normally associated with the "Power" button on a remote can be separated into to two distinct or discrete commands. On and Off. For instance if TV is powered on and the "Power" command is sent, it will turn the TV off. Conversely if the TV is powered off and the "Power" command is sent it will turn the TV on. Hence the term, "Toggle". However if a TV is already powered on and the "On" command is sent the TV will do nothing. If the goal is to turn on multiple components in a system it is beneficial send only the On command. This ensures that if a device is already on that it will not be turned off putting it out of sync with the rest of the system. Devices with multiple inputs may also have an associated discrete command for each input providing easy access.

Macros

A **Macro** is a set of instructions or a series of commands that is represented in an abbreviated format. In the case of control systems, this means one button push initiates a series of individual commands, in sequence.

One of the problems with off-the-shelf universal remotes is that the user must remember all the steps in sequence. When using macro commands this problem is eliminated. The user issues one command and the rest goes on behind the scenes. This is the perfect example of "making the complex simple", a concept which is central to the systems integration industry.

For instance, a single command (Watch TV, Play DVD, Listen to Radio) can then activate a complex series of commands. The user is not confused by all of these individual commands, and finds the system much easier to use.

Effective macros are repeatable no matter what state the component is in before the macro is issued. The use of discrete On and Off commands allows the proper power command to be issued and ensures, whether the device is on or off, it will be set to the right state after the macro is complete. Specifying components with discrete Power, Input, and Mode commands allows these to be included in the macro, and ensures the proper settings for each component. If a device gets out of sync, either by a missed IR code or by someone pressing a button on the component, the macro will reset the system to where it should be. This can be done by re-issuing the macro.

Workarounds

When components do not have discrete on and off commands, power sensing devices can be used to tell the control system the state of the component. For instance this type of device can sense if a DVD player is powered on by sensing the voltage from a video output, so they control system knows not to re-issue a power command.

SUMMARY

At the very heart of what our industry does is the ability to enhance the homeowner's lifestyle by using technology to make life easier, safer, more convenient, and more entertaining. Without well engineered control systems all of this would be too complicated to operate, therefore delivering only frustration rather than benefit.

It is important to understand the difference between integration and automation. Integration is when multiple components or systems work together, and are therefore more easy to control. This can be in a single room solution or for the whole house. Automation takes this one step further and allows actions to be taken by the technology WITHOUT human interaction. For instance a home with integrates entertainment, lighting, security and HVAC will not only allow control of these diverse systems from one interface (integration), it will also allow programming of actions requiring no direct control, many of which can conserve energy, protect property, or even save lives. An common example; the alarm system senses smoke and triggers the annunciators in the home while also alerting first responders. But it is also programmed to send a command to turn off all air handlers in the home, to prevent smoke from being spread throughout the living space. Lighting in this home may be easily controlled by the user, but it may also have the ability to control itself and give the appearance of an occupied home, when set in "vacation" mode.

In recent years, control and integration has become the largest revenue generating segment among CEDIA member companies. A solid understanding of these concepts will be critical to advancing our career.

Questions

1. An IR signal carried over a cable requires a minimum of ___ conductors?
   a. 2
   b. 3
   c. 4
   d. 8

2. The most important advantage of an RF remote system is:
   a. Batteries last longer
   b. It doesn't affect nearby homes
   c. It does not require direct line of sight
   d. It can store and issue macro commands

3. In order to address devices requiring IR or RF commands from an IP based control system, a ________ is needed.
   a. Router
   b. Hub
   c. Bridge
   d. Connecting block

4. A "macro" ___________.
   a. Allows a remote to communicate with several different devices at once
   b. Is a set of instructions, in sequence, initiated by one user interaction
   c. Sends the IR signal out in all directions equally
   d. Is a program that can be used moved from one device to another, regardless of manufacturer

5. Serial control:
   a. Includes RS-232 and RS-485
   b. Is carried only over cable
   c. Is highly dependable
   d. All of the above

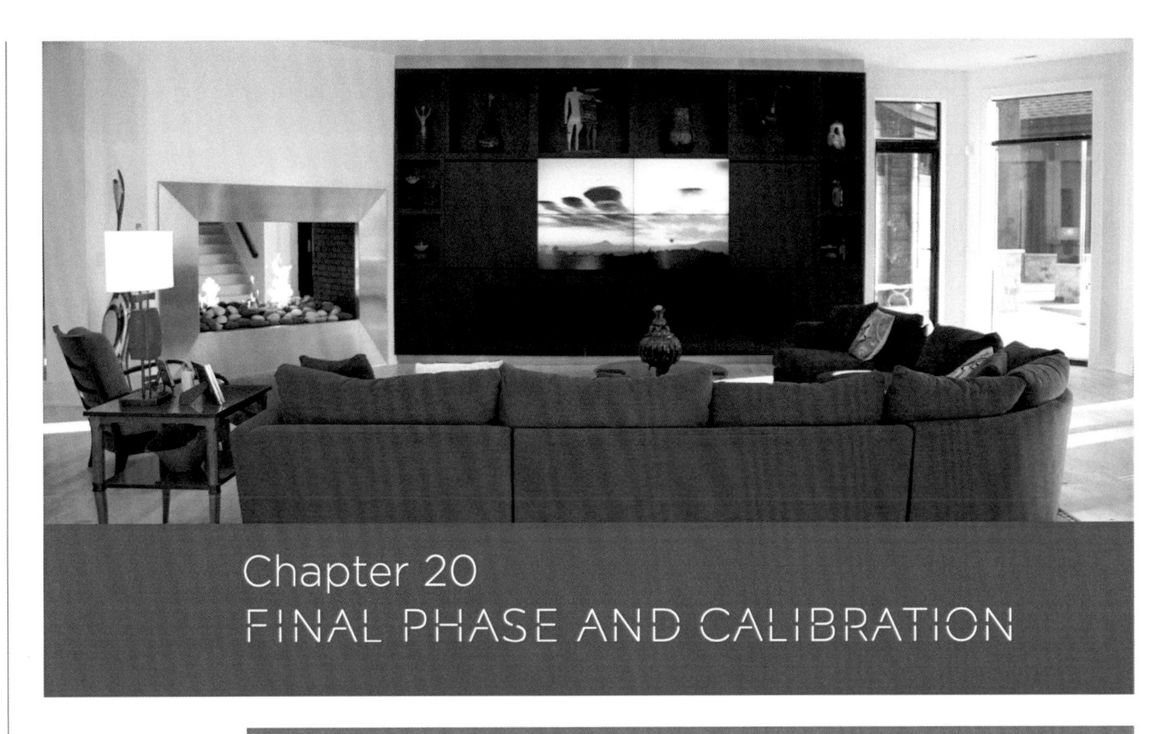

Chapter 20
FINAL PHASE AND CALIBRATION

OVERVIEW

Welcome to the final phase of a systems integration project, and the final chapter of this Study Guide! By this time you have been introduced to a wide spectrum of technologies and practices that make up the residential systems industry. In this chapter we will discuss what takes place during the final installation of equipment in the home, and how we ensure that the system will be enjoyed and appreciated by the client.

What Happens in This Phase?

During the Trim-out Phase, the platework, architectural speakers, and other permanently attached equipment were installed. In the final phase, all remaining components, displays, control devices, etc. are installed, tested, configured, and calibrated. In addition, the client will be trained in the operation of the system to ensure their long term satisfaction.

Working in a Finished Home

The final phase of a new construction project, like many smaller jobs, often takes place in the home while the family is living there. The other contractors are gone, and the keys to the home may have already changed hands, from the builder to the owner. The family may have moved in, or they may be in the process of moving in. This means all personnel must take special care to respect the people and property in the home, and exercise the highest level of professionalism. All surfaces (wall, floor, countertops, etc.) should be protected from damage or dirt. Great care should be taken regarding footprints and smudges that could be difficult to remove. We want to have the least possible impact on the home and its occupants.

INSTALLATION, TESTING, AND TROUBLESHOOTING

Preparation

One of the best ways to minimize the impact of this phase on the homeowner is to do as much of it as possible offsite.

- Load, wire, and configure equipment racks
- Program and test control systems
- Verify proper operation of audio and video equipment
- Configure and test the network components

When equipment racks are loaded and tested at the shop, all troubleshooting can take place without impacting the client. If a rack is too heavy and cumbersome to transport to the jobsite, some heavier but less complex components (power amps, power conditioners, etc.) can be removed to reduce the weight, and then reinstalled onsite.

Network hardware can be configured and tested completely, and then when installed the additional work should be minimal (assuming all infrastructure has been tested and verified in the Trim-out Phase).

Powering up audio and video components in advance can prevent surprises and delays during the final installation. This is also a chance to verify that all control communications are working as planned.

In some cases it may be appropriate to invite the client by the shop to see the preparation, programming and prewiring that is being done. This provides an opportunity to explain why it is done this way, and how this procedure will minimize the intrusion in their home during the final installation.

Installation

Once onsite, the actual installation of equipment takes place. Design documents should provide all the necessary details about location, wiring, etc. and everything should be done according to industry best practices and company policy. All wiring should be neatly dressed and professionally labeled.

If any actual installation varies from the work order or drawings, these variations must be documented so that final "as-built" documents are accurate.

Testing

Every system should be tested for proper functionality. These systems may include:

- Distributed audio
- Video distribution
- Theater and media room systems
- CCTV
- Lighting
- Interface with Security and HVAC
- Network
- Telephone
- Control system and mobile device control

Troubleshooting

Any problem that arises should be systematically diagnosed and corrected. Remember to document any changes from the original design documents. If most of the testing, programming, and configuration is done in advance, there should be minimal troubleshooting required.

CALIBRATION

Distributed Audio

Distribute audio systems should be calibrated properly to provide the appropriate listening levels for each zone. If speakers are specified and located appropriately this will be a simple procedure. The goal is even coverage and ample level to be heard above background noise and in some cases loud conversations. Each zone may be calibrated a little differently depending on its intended use. Use the anticipated maximum background noise level as a benchmark. Some general guidelines include:

- Getting balanced levels at all locations in a room. Goal is ≤ 6dB variance.
- Check this at 12 – 15 dB above anticipated ambient noise level.
- Set up for 75 dB playback at listening locations, higher in some situations
- Equalization may be called for, especially in unusual rooms and outdoors.
- Make sure all sources are equal in volume.
- Document all settings for future reference.

Home Theater / Media Room Audio

In order to get the best possible experience from a surround sound system, as discussed in Chapter 12, the system must be properly set up and calibrated. Along with proper room and system design, calibration will allow the user to get the best performance the equipment is capable of. As we covered earlier, the goals of a surround system are clarity, focus, envelopment, smooth response, and dynamics.

The final step in achieving these goals is the professional calibration of the surround system. Every surround system must be calibrated. This should be included in every project estimate or quote. Regardless of the quality of the audio equipment, the room plays a critical role in what the listener hears. Audio calibration compensates for both the equipment and the room, to deliver the most accurate sound to the listener.

Our goal in this book is to cover the fundamentals that should be understood by everyone in the industry. So we will not delve deep into the process of calibration, but rather provide an overview of the main steps, and what they contribute to the performance of the system.

- Verify performance - make sure all channels, inputs, and speakers are functioning
- Verify speaker phase, location, and aiming
  - » "In phase" means all speakers are moving in the same direction (in or out) when sent the same signal. Two speakers which are "out of phase" will work against each other, because sound waves with opposite polarity will cancel each other out, especially at lower frequencies
  - » Full range front speakers should be aimed at the primary listening position so the listener is hearing "on-axis" sound reproduction
- Set channel delays – done by measuring the distance to each speaker and entering values into the processor's setup menu
  - » A sound played through all speakers at once should reach the listener at exactly the same time. Since speakers are not often actually arranged equidistant from the listener, digital delay is used to make this adjustment, creating the illusion that all speakers are "virtually" the same distance from the listener
- Set channel levels – done by playing pink noise through each channel, one at a time, and measured on an SPL (Sound Pressure Level) meter
  - » All speakers should be perceived at the same sound pressure level. Since distance and room characteristics affect this, we calibrate the signal sent to that amplifier channel to balance all levels
- Set channel equalization – done by playing pink noise, one channel at time, and viewed on an RTA (Real Time Analyzer)
  - » All full range speakers in the system should have similar tonal characteristics, but the room affects what the listener hears so when necessary we may adjust certain frequency ranges to make them all closer to smooth response at the listening position

The final calibration tool should always be your ears. The technician should have some audio content they use regularly, which reveals flaws in a system, and is very familiar to them. Listening to this content can provide clues as to any shortcomings in the system that might require special attention. This is also good for "before and after" comparisons.

Also, there may be special circumstances that require special calibration. For instance a client with hearing loss may have difficulty understanding dialog when the system is calibrated properly. This may require equalizing the center channel for enhanced performance in the speech range, to ensure that the client fully enjoys the system.

CEDIA and other associations offer comprehensive audio calibration training and it is recommended that experience technicians know this procedure.

Video Displays

Every video display needs to be calibrated. They are often shipped with brightness and color settings that are nowhere near reality, in order to stand out in brightly lit retail environment. Basic calibration can be done very quickly and vastly improve the performance of a display. Calibration not only helps deliver a more accurate and pleasant image, it also results in a reduction in energy consumption and extended operating life for the display. This value proposition should always be explained and basic calibration should always be specified. In theaters and media rooms, a complete calibration process should be done, including gray scale calibration, which requires more specialized equipment and training. Some displays allow for separate calibration of different inputs (sources) and separate calibrations for different presets, such as "cinema", "game", "night" or "day". The capabilities of the display and the calibration plan should be determined in advance.

As with audio calibration, we will provide an overview of the calibration steps and what they do. More advanced training is available for ESTs who wish to learn this important skill.

Basic calibration can be accomplished using a pattern generator or test disc and a blue filter. The test pattern is displayed and settings are adjusted on the TV to get the desired results. Color settings are done by viewing a special pattern developed by the Society of Motion Picture and Television Engineers (SMPTE), while looking through a special blue filter which allows only blue to be seen in the pattern. This is the minimum calibration that should be performed on all displays. This is a skill that any experienced technician should have, and use regularly.

- Set ambient light to the level that will be used for the setting being calibrated
- Set all display settings to default and turn off any automatic features
- Set Black Level using the Brightness control (B&W PLUGE pattern)
- Set White Level using the Contrast control (B&W PLUGE pattern)
- Set Color Level using the Color control (SMPTE Color Bar pattern and blue filter)
- Set Hue using Hue or Tint control (SMPTE Color Bar pattern and blue filter)
- Set Sharpness using the Sharpness control (Resolution pattern)

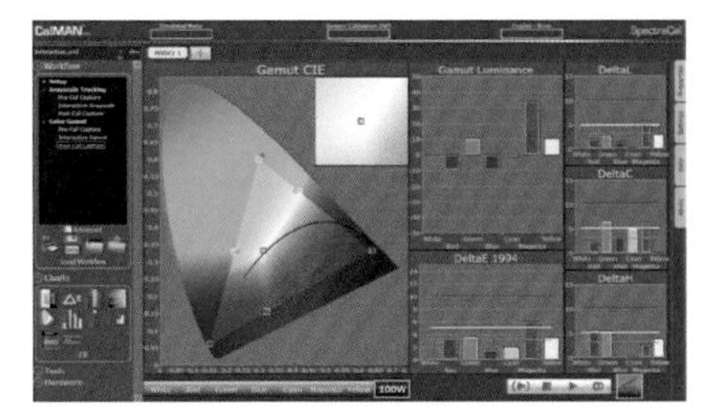

The final step in calibration is Gray Scale calibration, which requires more advance equipment and training, along with calibration software. This step ensures that all levels of gray between black and white contain a constant mix of red, blue, and green. This complete calibration should always be used for critical displays such as theaters and media rooms.

Basic calibration can be accomplished using a pattern generator or test disc and a blue filter. The test pattern is displayed and settings are adjusted on the TV to get the desired results. Brightness and contrast settings are done using one of many possible patterns with black and white areas. Color settings are done by viewing a special pattern developed by the Society of Motion Picture and Television Engineers (SMPTE), while looking through a special blue filter which allows only blue to be seen in the pattern. A color image of the SMPTE pattern can be seen on the back cover of this book. This is the minimum calibration that should be performed on all displays. It a skillset that any experienced technician should have, and use regularly.

DOCUMENTATION AND CLEANUP

Updating Documentation

Everything in the project file needs to reflect the way the installation was actually executed. There will always be changes, and when someone comes back to service or upgrade the system they must be looking at documentation and drawings that are absolutely correct.

Cleanup

If the home is occupied, it must always be left completely clean. At the end of each day, the jobsite must be completely cleaned up and free of boxes, tools and any sign that work was being done. It should look as good, or better, than when you arrived.

Storing Important Items

Most integrations projects involve a control system which replaces factory supplied remotes. Most flat panel displays are purchased with pedestal type stands, which may not be used because the display is wall-mounted or motorized. These items, however, are still very valuable to the client and must be safely stored. Sometimes a TV is taken off to college, or moved to another location and the pedestal and remote are then needed. Other components may also be removed when a system is upgraded, and they need to have their accessories with them to be repurposed elsewhere. Your company should have an established policy for these items, and all of the Owner's Manuals that come with the equipment. One popular option is to use a plastic storage bin with the company name and contact info on it. All remotes and manuals can be stored in this bin, in a location which is documented in the file. That way the owner or another technician knows where to find them when needed. TV pedestals should be tagged as to the make and model TV they go with and all hardware attached.

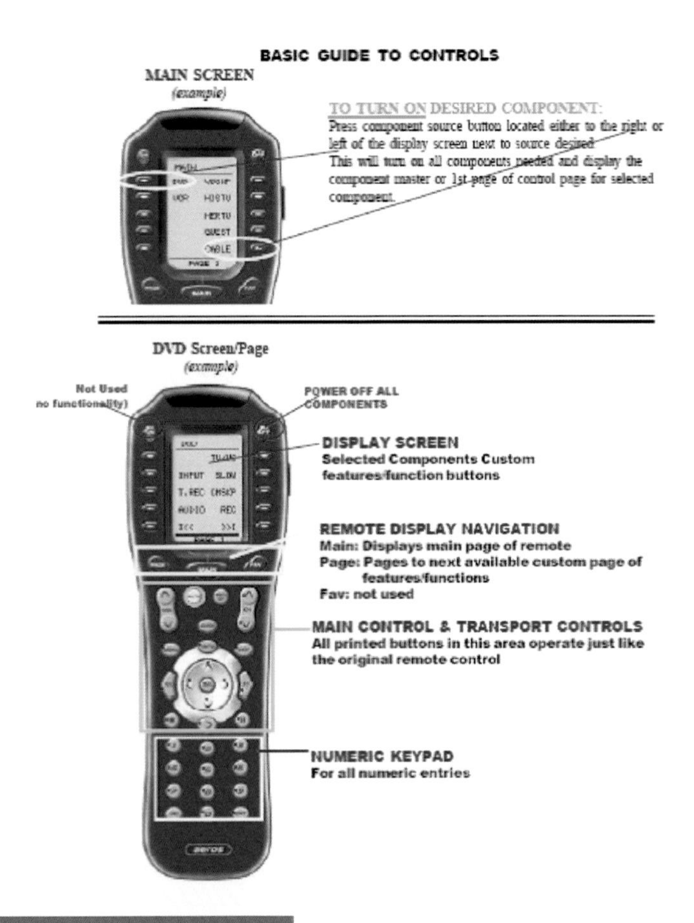

CLIENT ORIENTATION

No integrated system project can be considered a success unless the users are comfortable using it. Remember the goal is to enhance their lifestyle by simplifying complex systems and making them easy to use. This philosophy should be at the core of the entire design process and be reflected in the way that all of the user interfaces look and function. The client should already be familiar with the types of user interfaces that are used in the system, because in the initial sales and design meetings they were shown examples of various devices. In the case of touch panels, mockups should have been produced and shared with the client during the design phase.

The key to successful client orientation is taking the time to make sure they really do understand how to make things happen and are not afraid of the system.

Here are some general guidelines:

- Start out with the more simple control devices and functions
- Let them actually control the system and navigate the interfaces
- Avoid using technical terms, model numbers, etc. – Explain in simple terms
- Encourage them to ask questions
- Make sure they understand before moving on

User Guide

A good tool to help the end user learn the system is the User Guide, or "cheat sheet". An example would be a laminated sheet that stays in the media room and describes, step-by-step, how to do the most common tasks like "Watch TV" or "Play Blu-ray". If the steps are clearly noted, the client can learn on their own and become proficient at accessing the functions they want.

Followup

Most successful integrators have a policy of scheduling a follow up visit at a set time interval after final installation. This ensures that the customer is completely satisfied and provides an opportunity to continue the orientation as needed, answer questions, and even identify programming changes that might make the system more user-friendly. Have them log any questions or concerns they have, to be discussed during the followup visit. This creates an atmosphere of participation rather than the client feeling overwhelmed.

Questions

1. Equipment racks should be:
   a. Loaded and tested at the shop, to minimize the impact on the homeowner
   b. Installed and loaded onsite, to ensure proper interaction with installed components
   c. Completely sealed and covered, to prevent client access
   d. Made of wood, to prevent ground loops

2. Left, right, and center speakers should be aimed at the primary listening position so that:
   a. The listener gets the highest possible SPL
   b. They sound "in phase"
   c. The listener is hearing "on-axis" response from each speaker
   d. The listener gets optimum bass response

3. The contrast control on a video display adjusts the_______.
   a. Black level
   b. White level
   c. Sharpness
   d. Color intensity

4. ____________ should be stored onsite in a container with your company information on it.
   a. TV pedestals
   b. Factory remotes
   c. User manuals
   d. All of the above

5. Distributed audio within a room should vary in SPL by no more than _____.
   a. 6 dB
   b. 3 dB
   c. 12 dB
   d. 10 watts

APPENDIX

CEDIA ESC JOB TASK ANALYSIS
CEDIA MEMBER CODE OF ETHICS
APPLICATION NOTES
- CEA/CEDIA CEB-22
- CEA/CEDIA CEB-23
- CEA/CEDIA CEB-24

ITU RECOMMENDATIONS FOR SURROUND PRODUCTION
5.1 SURROUND SPEAKER LOCATIONS
7.1 SURROUND SPEAKER LOCATIONS
LARGE ROOM SPEAKER LOCATIONS
SURROUND SPEAKER ELEVATION
EXAMPLE CABLE AND LOCATION SCHEDULE
ADDITIONAL RESOURCES

JOB TASK ANALYSIS (EXAM BLUEPRINT)

International (English)

This document outlines the body of knowledge on which the certification exam is built. Within each domain, the content which needs to be mastered is the fundamental knowledge and terminology, not advanced techniques or problem solving. A comprehensive overview of this material can be found in the Fundamentals of Residential Electronic Systems book, as well as in CEDIA eCourses and other resources. To identify all of the resources available, see the ESC Certification Exam Prep Resources document.

Domain  Industry Overview and Fundamentals

KNOWLEDGE AREA 1
INTRODUCTION TO THE INDUSTRY

1. Definition and scope of the industry; history, trends, most commonly installed sub-systems
2. CEDIA history, mission, and ethical standards
3. Company types: integrators, retailers, distributors, manufacturer reps, manufacturers, specialty designers
4. Project stakeholders; clients, architects, interior designers, builders, other trades
5. Career paths

KNOWLEDGE AREA 2
JOBSITE AND BUSINESS PROFESSIONALISM

1. Personal/professional behavior and appearance
2. Project documentation

KNOWLEDGE AREA 3
INDUSTRY-RELATED MATH

1. Applied jobsite mathematics
2. Mathematic conversions (fraction/decimal and metric/standard)

KNOWLEDGE AREA 4
FUNDAMENTALS OF JOBSITE SAFETY, CODES, AND STANDARDS

1. General safety practices, basic first aid, and emergency procedures
2. Understanding of applicable codes, standards, and recommended practices and their role and importance
3. Proper use and care of tools

KNOWLEDGE AREA 5
THE BUSINESS OF RESIDENTIAL TECHNOLOGIES

1. Small-business fundamentals
2. Project management fundamentals
3. Servicing high-end clientele
4. Sales, service, recurring revenue, and monetizing design and engineering

Domain  Infrastructure

KNOWLEDGE AREA 1
PRE-WIRE PHASE

1. Construction methods and materials: wood frame, metal stud, and concrete
2. Cable/wire types and applications
3. Cabling topologies, service entries, wiring for the future
4. Device placement and cabling practices/labeling

KNOWLEDGE AREA 2
TRIM-OUT PHASE

1. Termination methods and plate-work
2. Termination tools and test equipment
3. Cable identification (labeling)

Domain  Equipment Installation

KNOWLEDGE AREA 1
RACKS AND CABINETS

1. Cabinets and equipment racking systems
2. Basic cable/interconnect management

KNOWLEDGE AREA 2
EQUIPMENT MOUNTING

1. Mounting hardware (brackets, lags, toggle bolts, drywall anchors, etc.)
2. Measurement tools and techniques
3. Proper installation techniques
4. Retrofit and safety considerations

Domain  Sub-Systems Overview

KNOWLEDGE AREA 1
AUDIO

1. Basic audio device recognition (receiver, amplifier, speaker types, etc.)
2. Basic audio terminology (frequency, wavelength, tweeter, crossover, etc.)
3. Audio signals and interconnects (analog and digital)
4. Fundamentals of multi-room audio
5. The role of room acoustics and sound isolation issues in audio performance

KNOWLEDGE AREA 2
VIDEO

1. Basic video device recognition (sources, display technologies)
2. Basic video terminology (pixel, resolution, brightness, etc.)
3. Video signals and interconnects (analog and digital, HDMI)

KNOWLEDGE AREA 3
HOME THEATER/MEDIA ROOM

1. Design and performance goals
2. System components and their function
3. Basic layout and configuration, recommended practices

KNOWLEDGE AREA 4
NETWORKING

1. Basic network device recognition (router, switch, access point, etc.)
2. Basic data and networking terminology (bits, bytes, bandwidth, etc.)
3. Basic network topologies (wired and wireless)

KNOWLEDGE AREA 5
OTHER SUB-SYSTEMS

1. Automated lighting components and operation
2. Motorized devices: shades, lifts, mounts, etc.
3. Energy monitoring and management
4. Other devices and sub-systems (security, HVAC, telephone, etc.)

KNOWLEDGE AREA 6
SYSTEMS CONTROL

1. Basic control device recognition (remote, keypad, processor, etc.)
2. Basic control systems terminology (discrete code, toggle command, macro, etc.)
3. Basic control protocols (IR, RF, RS232, IP)

KNOWLEDGE AREA 7
POWER QUALITY AND MANAGEMENT

1. Power types and quality issues
2. Basic device recognition (surge protector, battery backup, power conditioner, etc.)
3. Basic electricity and electrical distribution

KNOWLEDGE AREA 8
FINAL SYSTEM CALIBRATION, TESTING, AND COMMISSIONING

1. Fundamental goals and techniques of audio and video calibration
2. Final system setup and client orientation

CEDIA
MEMBER CODE OF ETHICS

This code of ethics is designed to ensure quality of service, responsiveness, responsible business practices, proper legal and ethical conduct, and overall excellence.

EACH MEMBER OF CEDIA SHALL AGREE TO ADHERE TO THE FOLLOWING:

- Provide to all persons **truthful and accurate information** with respect to the professional performance of duties.
- Maintain the **highest standards of personal conduct** to bring credit to the custom electronic and design industry.
- Promote and encourage the **highest level of ethics** within the profession.
- Recognize and discharge by responsibility, to **uphold all laws and regulations** relating to CEDIA policies and activities.
- **Strive for excellence** in all aspects of the industry.
- Use only **legal and ethical** means in all industry activities.
- **Protect the public against fraud and unfair practices** and attempt to eliminate from CEDIA all practices which bring discredit to the profession.
- **Use written contracts clearly stating all charges, services, products and other essential information.** To the extent that the customer does not own the software codes for all programming implemented by the member, contracts should address the parties' rights and responsibilities as to such codes.
- **Demonstrate respect** for every professional within the industry by clearly stating and consistently performing at or above the standards acceptable to the industry.
- Make a commitment to **increase professional growth and knowledge** by attending educational programs recommended, but not limited to, those prescribed by CEDIA.
- **Contribute knowledge** to professional meetings and journals to raise the consciousness of the industry.
- **Maintain the highest standards of safety** and any other responsibilities.
- When providing services or products, **maintain in full force adequate or appropriate insurance.**
- **Cooperate with professional colleagues,** suppliers and employees to provide the highest quality service.
- **Extend these same professional commitments** to all those persons supervised or employed.
- **Subscribe to CEDIA's Principles of Conduct and Ethics** and abide by the CEDIA Bylaws.

Home Theater Recommended Practice: Audio Design

These points represent only the key elements of CEB-22 and are intended only to be a quick reference. For a complete understanding of the Recommended Practice, refer to the full original document. Also see CEDIA courses ESD131, ESD232, ESD301, ESD302, and ESD303

The Goal of Home Theater Audio is to faithfully reproduce the audio content of the source material, whether it is cinema, television, or a video game, by adhering to motion picture industry standards and making well informed decisions when dealing with design compromises.

General Performance Objectives

1. No audible coloration of the sound
2. A smooth and balanced front soundstage
3. A surround system that provides convincing envelopment
4. Intelligible dialog
5. No perceivable distortion
6. Performance consistent across entire seating area, best at primary seating
7. Accurate low frequency performance which is consistent seat to seat
8. Inaudible background noise
9. Sound isolation from adjacent spaces; both egress and ingress

Design Fundamentals

1. The room size and shape, image size, sight lines, speaker location, and seating configuration are all part of an overall system which must be designed simultaneously for best results.
2. There should be one prime seating location, optimized for both audio and video. The best seating location is on the center meridian of the room.
3. Other seating locations should all provide good overall performance. The number of seats must be appropriate for the room size.
4. Loudspeaker type and location, acoustical treatment, and room décor should be designed to complement one another and deliver the stated objectives.

Loudspeaker Objectives and Placement

1. All speakers should have a similar timbral signature (same manufacturer/series)
2. L, C, R at approximately seated ear height (no more than 15° above)
   a. If L,C,R are higher, aim to cover smoothly
3. An unobstructed path from all full range speakers to all listeners
   a. Especially critical for Center speaker
4. Center speaker on center line of room
5. Speakers mounted in baffles or cabinets should be decoupled and damped to reduce resonances.
6. L and R speakers at an angle of 45-60° to each other, not too far from screen. L and R speakers toed in toward primary listening position.
7. Surround speaker placement for optimal envelopment and accuracy
   a. In 5.1 configuration, Surround Speakers at 110-120° off center line (A)
   b. In 7.1 configuration, Surround Speakers at 60-100°,
      Rear Speakers at 135-150°. (B)
   c. In larger room, multiple Surrounds to ensure envelopment. (C)

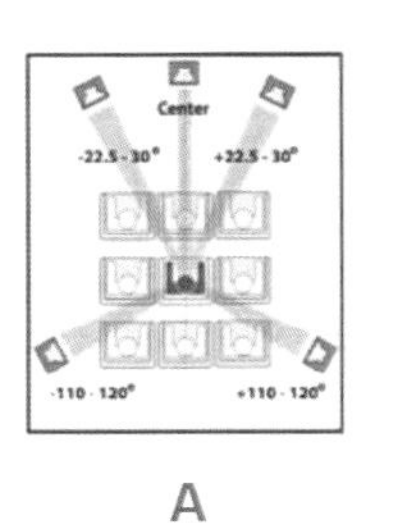

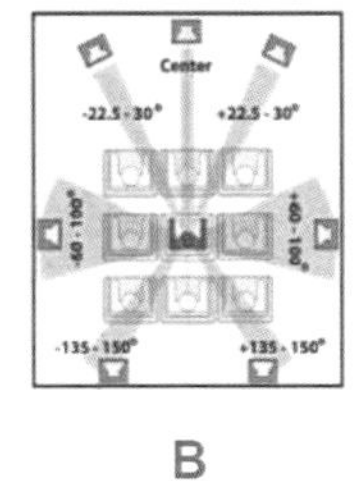

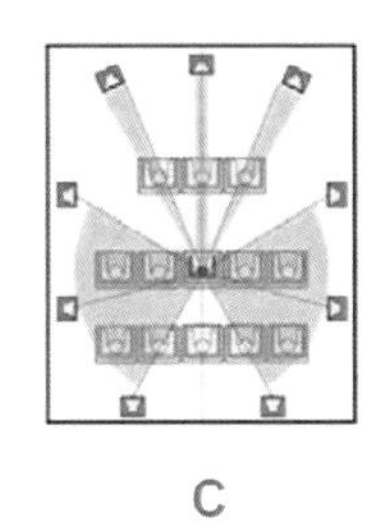

A B C

Subwoofer Guidelines

1. For reduced seat-to-seat variation, multiple subwoofers are ALWAYS better than just one.
2. In a closed rectangular room, using 2 or 4 subwoofers in certain configurations will yield predictable modal behavior, allowing for changes in dimension or seating to provide even response at several seating locations.(D)

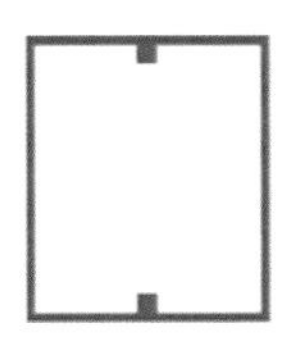 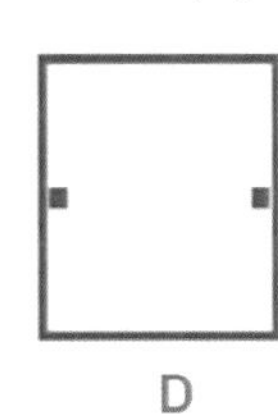 

D

Seating Guidelines

1. Prime location at geometric center of speaker layout
2. Use center line for seating, not for an aisle (E)
3. No seating too close to any speakers
4. Avoid seating in 2nd order modal null points of room - 1/4 in from each wall (F)

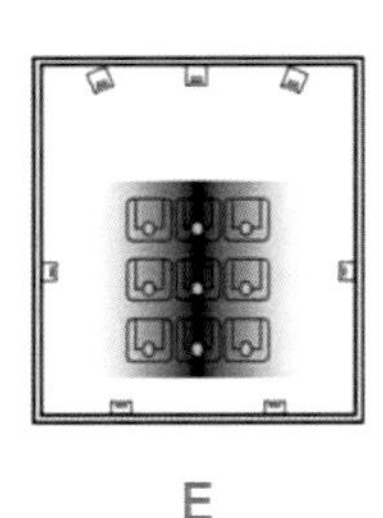

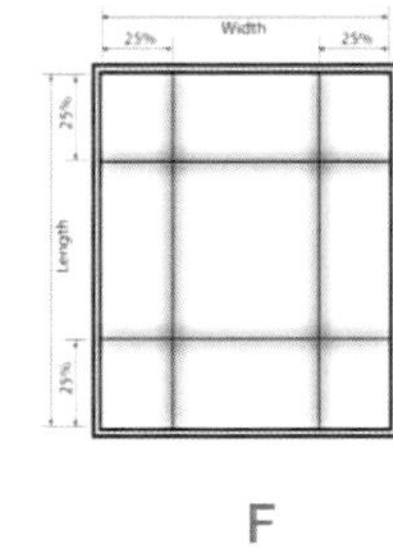

E F

Acoustical Treatment

1. Reverberation Times (RT) at 500Hz should be in the range of .2 - .5 seconds.
2. Shorter RT may make room too “dead” and adversely affect the experience.
3. A typical room can achieve good sound with;
   a. Wall to wall carpet
   b. A combination of absorption (3”minimum) and diffusion, not to exceed 50% of wall area, placed to control flutter echoes between parallel walls near listening area, and control first reflections (optional)
4. All materials should be Class A fire rated.
5. There should be no audible rattles in the room.

Background Noise should be ≤ 30dB/A using RC mk II criteria. Ideal is ≤ 20dB/A.

1. Utilize oversized HVAC ducts
2. Specify low noise lighting dimmers
3. Acoustically treat nearby plumbing
4. Isolate HVAC equipment, lift pumps, etc.

Home Theater Recommended Practice: Video Design

These points represent only some of the key elements of CEB-23 and are intended to be a quick reference. For a complete understanding of the Recommended Practice, refer to the full original document.

The Goal of Home Theater Video is to faithfully reproduce the image which was painstakingly crafted by the director and to allow the viewer to "suspend disbelief" and become fully immersed in the program material. This is best accomplished by designing, installing, and calibrating the system to be consistent with established industry standards.

RECOMMENDATIONS:

General Design Considerations

1. Entire system to meet or exceed local codes
2. As energy efficient as practicable, while meeting performance goals
3. Signal processing and scaling taking place only the one optimum place in signal path

Room Layout & Installation Guidelines

1. Primary viewing position
    a. 3 X Image Height (Figure B)
    b. On the horizontal centerline of the screen
    c. Vertical angle ≤ 5° from orthogonal axis (dead center) (Figure C)
2. All other viewing positions
    a. Vertical angle ≤ 15° from orthogonal axis
3. Projector location - as far back in the room as possible (within throw range)
    a. Minimal fan noise
    b. Use of "sweet spot" of lens
    c. More accurate image geometry (minimal pincushion/barrel distortion)
    d. More efficient use of angular reflective screen (gain greater than 1.0)
4. Room lighting and décor
    a. No external light leaking into a dedicated theater
    b. Entry and egress lighting subdued and controllable by viewer
    c. All lighting must meet local codes
    d. Minimum of three lighting scenes
        i. Entry/Exit - approx. 50% dimmed
        ii. Watching - all off except path and exit
        iii. Cleaning/Task - All on
    e. Wall and ceiling finishes
        i. Dark and neutral
        ii. Non-reflective as possible
5. Source location
    a. Convenient for users
    b. Accessible for servicing
    c. Minimal cable run lengths

Projector

1. Fan noise
    a. Overall background noise at primary position ≤ 30dB/A (CEB-22)
    b. Fan can be a major contributor to this noise
    c. Maximum recommended projector noise (dBA) shown in Figure A

16 Hz	-	55dB
31.5 Hz	-	48dB
63 Hz	-	43 dB
125 Hz	-	30 dB
250 Hz	-	22 dB

A

2. Thermal issues
    a. Operating temperature should be no more than 10°F (5.55°C) higher than ambient room temperature
    b. Ventilation should remove hot air and draw in room air
    c. Uninterrupted Power Supply (UPS) recommended to ensure proper cool-down of lamp

Power & Cabling

1. Dedicated AC circuit for theater video system, on same phase as rest of theater (US)
2. HDMI or DVI is preferred interconnect
3. Interconnect attenuation ≤ 3dB (1dB ideal)
4. Strain relief where cabling connects to components
5. All cabling labeled and secured with Velcro™ when possible
6. No zip-ties or mechanical fasteners
7. Service loops where appropriate
8. Flexible conduit (minimum 1.5") from equipment to projector for future use

Image Performance Objectives

1. Luminance
   a. Minimum 14 foot lamberts (1 candela/ft^2 or 3.425 candela/m^2)
2. Contrast Ratio
   a. Sequential Contrast Method = ≥ 2,000:1 (minimum ≥1,200:1)
   b. Intra-Frame Contrast Method = ≥ 150:1
3. Calibration - for all inputs and viewing modes permitted by display device
4. Gray Scale - calibrated across entire IRE range
5. Chroma Uniformity = ≤ 20% variance between any two points on image
6. All calibration and measurements documented

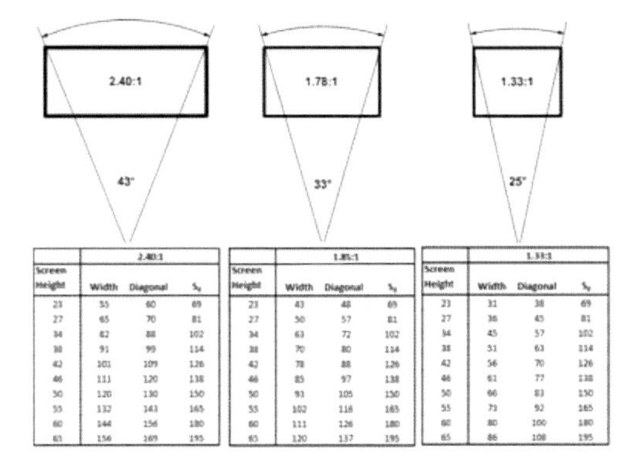

	2.40:1		
Screen Height	Width	Diagonal	S_v
23	55	60	69
27	65	70	81
34	82	88	102
38	91	99	114
42	101	109	126
46	111	120	138
50	120	130	150
55	132	143	165
60	144	156	180
65	156	169	195

	1.85:1		
Screen Height	Width	Diagonal	S_v
23	43	48	69
27	50	57	81
34	63	72	102
38	70	80	114
42	78	88	126
46	85	97	138
50	93	105	150
55	102	116	165
60	111	126	180
65	120	137	195

	1.33:1		
Screen Height	Width	Diagonal	S_v
23	31	38	69
27	36	45	81
34	45	57	102
38	51	63	114
42	56	70	126
46	61	77	138
50	66	83	150
55	73	92	165
60	80	100	180
65	86	108	195

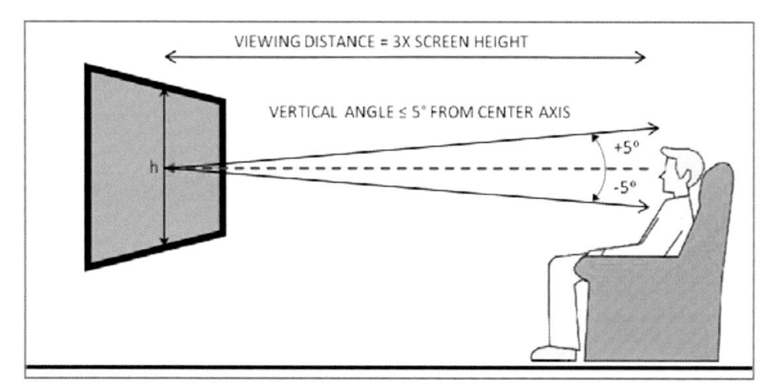

Home Theater Recommended Practice: HVAC

These points represent only some of the key elements of CEB-24 and are intended to be a quick reference. For a complete understanding of the Recommended Practice, refer to the full original document.

The Three Goals of Home Theater HVAC Design

1. Occupant Comfort
2. Acceptable Equipment Environment
3. Acceptable Noise Floor

MAX 85°F

MAX 75°F

RECOMMENDATIONS:

Ambient Room Conditions

1. Temperature: 68°-75°F (20°-23.9°C)
2. Relative Humidity: 30%-60%

Note; Lower humidity causes static electricity, higher can cause condensation in equipment.

Equipment Conditions

1. Operating Temperature: ≤ 85°F (29.4°C)
2. Fresh Air Inlet to Rack/Cabinet: ≤ 75°F

Note A; 99.6% of all power (electricity) going into a rack is converted to heat.
Note B; for every 10°F above 85°F, reliability of equipment drops 50%.

Noise Floor (at primary listening position)

≤ 30dB/A

1. Noise Generated by HVAC: RC(N) < 25
2. Total Noise: ≤ 30dB/A (also noted in CEB-22)

Noise Egress to Adjacent Spaces via HVAC

1. Maximum ≤ 30dB/A - measured 1m from any duct in the home

HVAC Recommendations

HVAC DESIGN

1. Air handler located far from theater
2. Ductwork of Non-metallic Construction (fiberglass)
3. Radiused elbows vs. mitered elbows
4. Oversized ductwork to reduce velocity
5. Dampers located far from terminal outlets
6. Ductwork with minimum 1" Wall Thickness / 3lb/ft^2 material
7. Multiple Gradual 90° Bends

CEDIA
ITU RECOMMENDATIONS FOR SURROUND PRODUCTION

ITU-R BS.775-2 (2006)

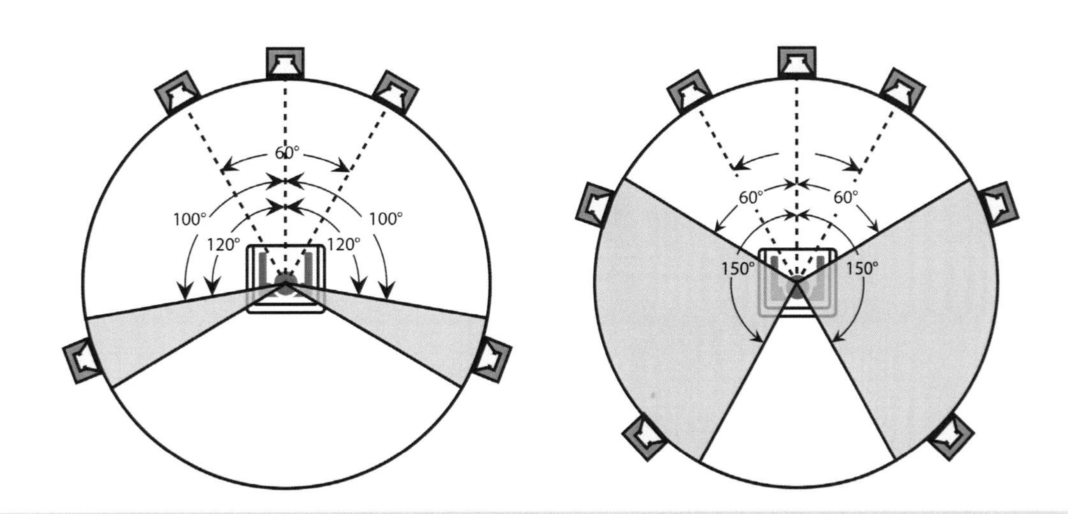

ABOUT

The International Telecommunications Union has set the standards for multichannel playback as it applies to professionals, mainly in the broadcast side of the industry. Obviously the radially located, equidistant, loudspeakers are quite impractical in homes. Only in a very large room are these free-standing loudspeakers able to perform well. However, it provides a reference standard to be used by some professionals, university research departments and a handful of consumers.

5.1 SURROUND SPEAKER LOCATIONS

This diagram is available for download in high resolution

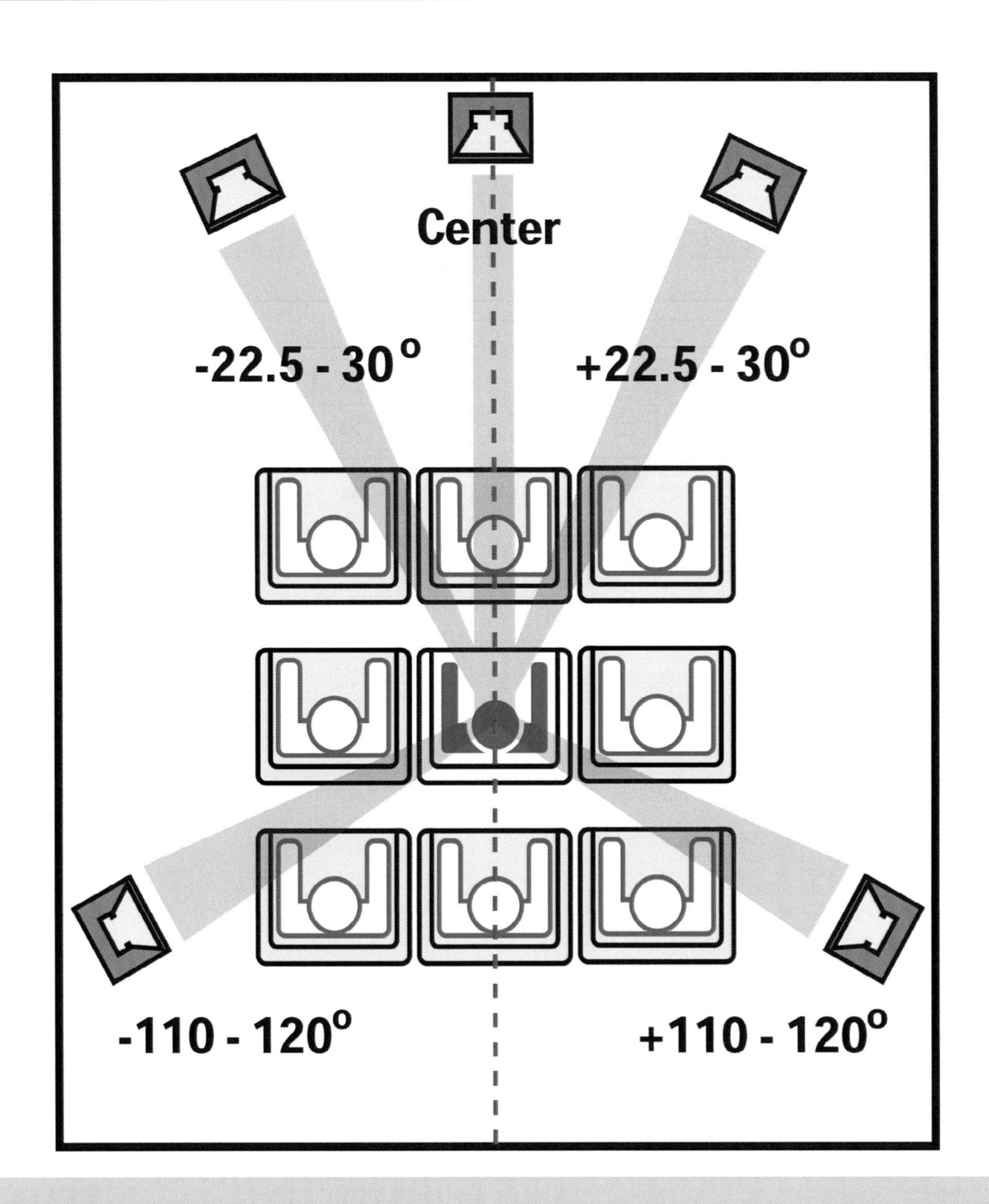

ABOUT

In a basic 5.1 surround sound layout:

- The Center speaker should be on the center line,
- The total angle between the Left and Right speakers should be between 45 and 60 degrees, and
- The Surround speakers should be just behind the prime listening position, at 110-120 degrees off the center line.

CEA/CEDIA CEB-22

CEDIA
7.1 SURROUND SPEAKER LOCATIONS

This diagram is available for download in high resolution

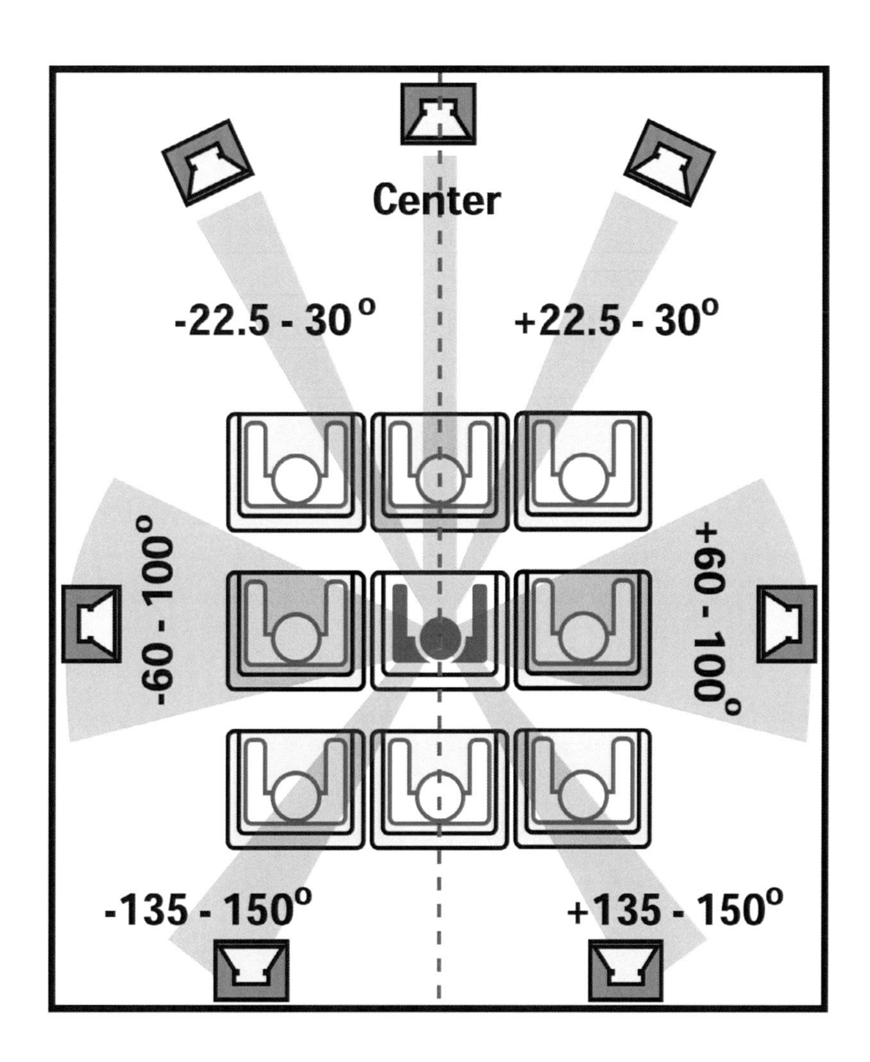

ABOUT

In a 7.1 surround sound configuration:

- The Center speaker should be on the center line,
- The total angle between the Left and Right speakers should be between 45 and 60 degrees,
- The Surround speakers should be between 60 and 100 degrees off center, and
- The Rear speakers should be between 135 and 150 degrees off the center line.

CEA/CEDIA CEB-22

LARGE ROOM SPEAKER LOCATIONS

This diagram is available for download in high resolution

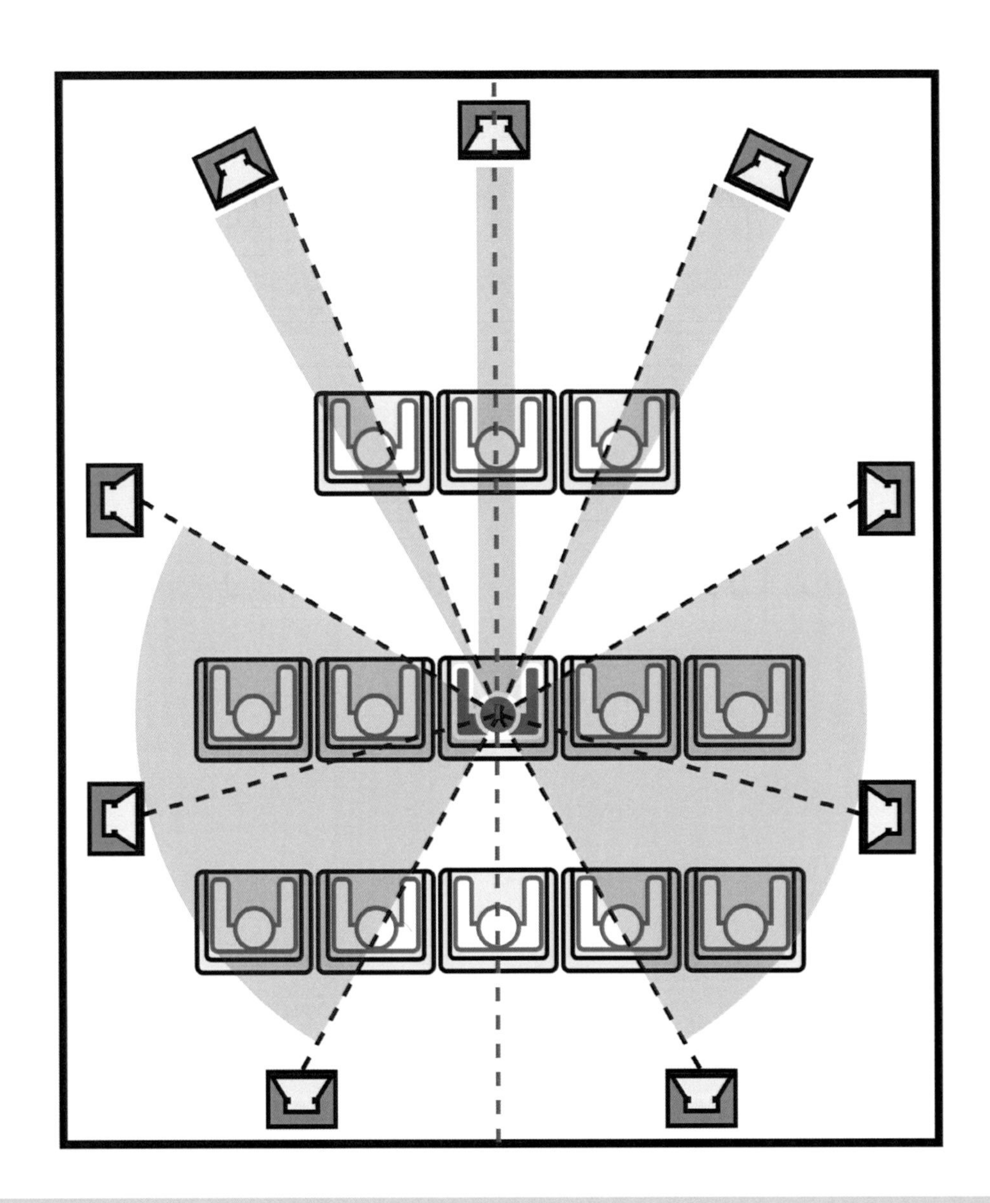

ABOUT

In larger rooms, the goal is to create convincing envelopment for all listeners. This may be accomplished with multiple surround speakers on each side. The locations are still optimized for the prime listening location.

CEA/CEDIA CEB-22

CEDIA
SURROUND SPEAKER ELEVATION

This chart is available for download in high resolution

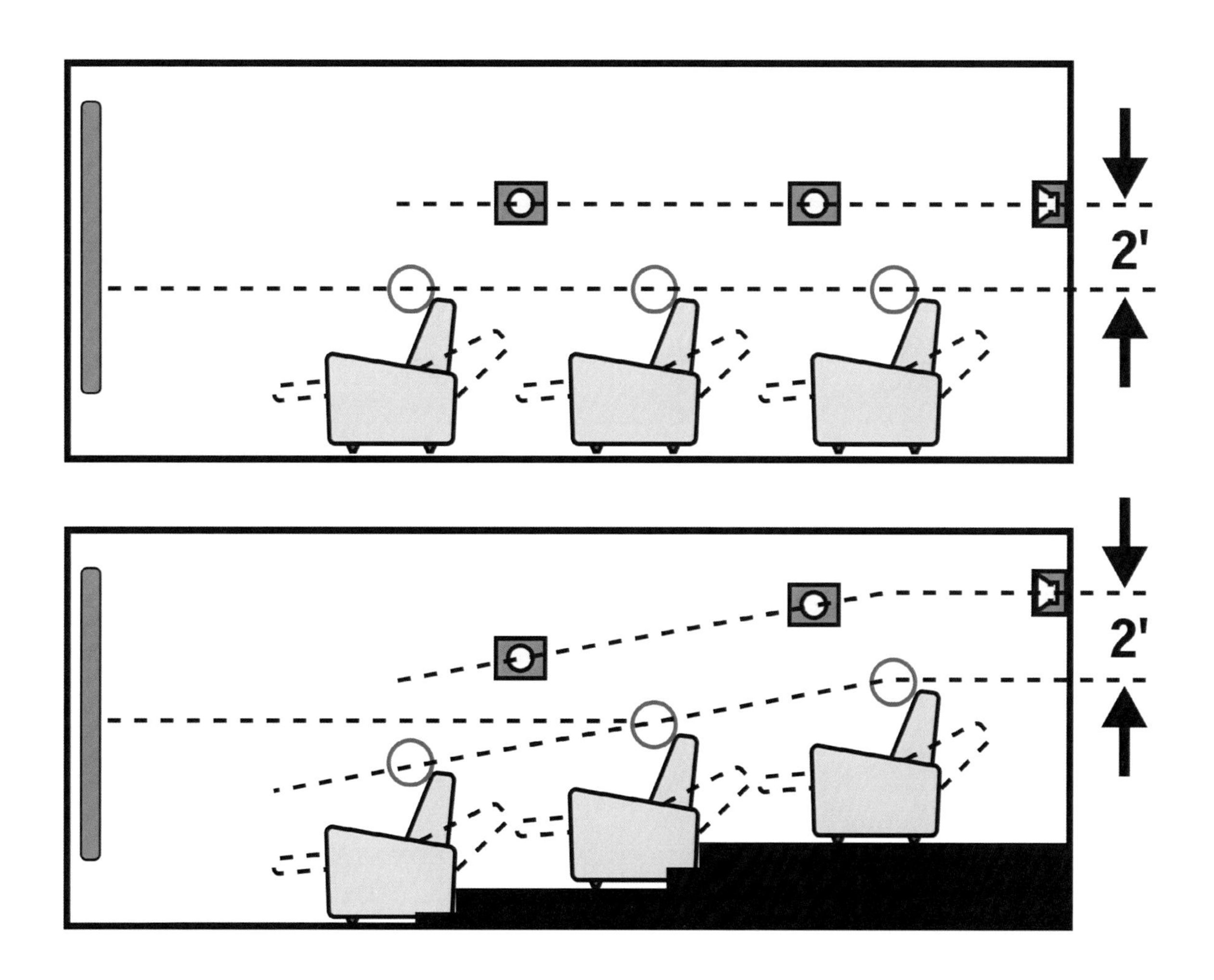

ABOUT

To ensure smooth envelopment, it is recommended that surround speakers be located at least 2 feet above the listeners' ears.

When seating is raked, as shown here, this 2 foot minimum must still be observed.

CEA/CEDIA CEB-22

CEDIA

EXAMPLE CABLE AND LOCATION SCHEDULE

CABLE AND ROUGH-IN SCHEDULE
WALKER PROJECT

Job: NEW CONSTRUCTION / WALKER   Job ID: ____
Location: HIDDEN LAKES LOT 99
Prewired by: JTG   Date: ____
Trim-out by: JLR   Date: ____

Prewire Cable Schedule / Prewire Location Schedule

ROOM	ID	Type	From	Term	TO	Term	Trim	Pull	Trim	Test	Estimated Run Length						NOTES	ROOM	Location	1G	2G	Other	NOTES
											C5	RG6	RG59	16/4	16/2	CAM							
KIT	1	CAT5e	DC	VM	WALL	RJ45	RJ45				50						PLATE ABOVE COUNTER	KIT	A	1G			
NOOK	2	RG6	DC	CM	WALL	F	F					80					HIGH IN CORNER	NOOK	B	1G			
MBR	3	CAT5e	DC	VM	WALL	RJ45	RJ45				100						BED WALL	MBR	C	1G			
MBR	4	CAT5e	DC	DM	WALL	RJ45	RJ45				100							MBR	D	1G			
MBR	5	RG6	DC	CM	WALL	F	F					100					OUTLET HEIGHT	MBR	E	1G			
BR2	6	CAT5e	DC	VM	WALL	RJ45	RJ45				60						BED WALL	BR2	F	1G			
BR2	7	CAT5e	DC	DM	WALL	RJ45	RJ45				60							BR2	G	1G			
BR2	8	RG6	DC	CM	WALL	F	F					60					TV LOCATION	BR2	H	1G			
BR3	9	CAT5e	DC	VM	WALL	RJ45	RJ45				60						BED WALL	BR3	I	1G			
BR3	10	CAT5e	DC	DM	WALL	RJ45	RJ45				60							BR3	J	1G			
BR3	11	RG6	DC	CM	WALL	F	F					60					TV LOCATION	BR3	K	1G			
GRT RM	12	CAT5e	DC	VM	WALL	RJ45	RJ45				100						PHONE & DATA SAME BOX	GRT RM	L	1G			
GRT RM	13	CAT5e	DC	DM	WALL	RJ45	RJ45				100						PHONE & DATA SAME BOX	GRT RM	L	1G			
GRT RM	14	RG6	DC	CM	WALL	F	F					120					OUTLET HEIGHT CBL TV	GRT RM	M		2G		
GRT RM	16	RG59	AV	RCA	WALL	F	RCA						80				HD COMPNT B	GRT RM	M				
GRT RM	17	RG59	AV	RCA	WALL	F	RCA						80				HD COMPNT G	GRT RM	M				
GRT RM	18	RG59	AV	RCA	WALL	F	RCA						80				AUDIO L	GRT RM	M				
GRT RM	19	RG59	AV	RCA	WALL	F	RCA						80				AUDIO R	GRT RM	M				
GRT RM	20	RG59	AV	RCA	WALL	F	RCA						80				COMPST VIDEO	GRT RM	M				
GRT RM	21	CAT5e	AV	RJ45	WALL	RJ45	RJ45				100						CONTROL	GRT RM	M				
GRT RM	15	RG59	AV	RCA	WALL	F	RCA						80				HD COMPNT R	GRT RM	N	1G			
AV CLOSET	22	RG6	DC	F	WALL	F	F					80					CBL TV - PLATE AT SHELF HEIGHT	AV CLOSET	N				
AV CLOSET	23	RG6	DC	F	WALL	F	F					80					SAT TV - PLATE AT SHELF HEIGHT	AV CLOSET	N				
AV CLOSET	24	CAT5e	VM	RJ45	WALL	RJ45	RJ45				80						PHONE FOR AV CLOSET	AV CLOSET	N				
TP SRVC	25	CAT5e	DC	RJ45	DEMARC	N/A	N/A				30						PHONE SERVICE, TO EXTERIOR	TP SRVC	O			N/A	
CBL SRVC	26	RG6	DC	R	DEMARC	N/A	N/A					30					CABLE SERVICE, TO EXTERIOR	CBL SRVC	O			N/A	
SAT PW	27	RG6	DC	N/A	LACE	N/A	N/A					40					LACE IN WALL, NO TERM	SAT PW	P			N/A	
SAT PW	28	RG6	DC	N/A	LACE	N/A	N/A					40					LACE IN WALL, NO TERM	SAT PW	P			N/A	
SAT PW	29	RG6	DC	N/A	LACE	N/A	N/A					40					LACE IN WALL, NO TERM	SAT PW	P			N/A	
SAT PW	30	RG6	DC	N/A	LACE	N/A	N/A					40					LACE IN WALL, NO TERM	SAT PW	P			N/A	
SAT PW	31	RG6	DC	N/A	LACE	N/A	N/A					40					LACE IN WALL, NO TERM	SAT PW	P			N/A	
SAT PW	32	RG6	DC	N/A	LACE	N/A	N/A					40					LACE IN WALL, NO TERM	SAT PW	P			N/A	
CAM PW	33	SIA	DC	N/A	NW	N/A	N/A									120	CAM PREWIRE ONLY, BLANK CVR	CAM PW	Q	1G			STAINLESS STEEL BLANK COVER
CAM PW	34	SIA	DC	N/A	NE	N/A	N/A									120	CAM PREWIRE ONLY, BLANK CVR	CAM PW	R	1G			STAINLESS STEEL BLANK COVER
CAM PW	35	SIA	DC	N/A	SE	N/A	N/A									120	CAM PREWIRE ONLY, BLANK CVR	CAM PW	S	1G			STAINLESS STEEL BLANK COVER
CAM PW	36	SIA	DC	N/A	FRNT DR	N/A	N/A									120	CAM PREWIRE ONLY, BLANK CVR	CAM PW	T	1G			STAINLESS STEEL BLANK COVER
ATTIC	37	RG6	DC	N/A	LOOP	N/A	N/A					100					SPARE IN ATTIC	ATTIC	U	1G			SERVICE LOOP - ACCESSIBLE
ATTIC	38	CAT5e	DC	N/A	LOOP	N/A	N/A				100						SPARE IN ATTIC	ATTIC	U				SERVICE LOOP - ACCESSIBLE
ATTIC	39	RG6	AV	N/A	LOOP	N/A	N/A					120					OFF AIR ANTENNA IN ATTIC	ATTIC	U				SERVICE LOOP - ACCESSIBLE
BONUS	40	RG6	DC	CM	EQ	F	F					120					CBL/SAT TV TO BONUS EQUIP CAB	BONUS	V	1G			REAR WALL OF EQUIPMENT CABINET
BONUS	41	RG6	DC	CM	EQ	F	F					120					CBL/SAT TV TO BONUS EQUIP CAB	BONUS	V				REAR WALL OF EQUIPMENT CABINET
BONUS	44	CAT5e	DC	VM	EQ	RJ45	RJ45				120						EQUIP CAB	BONUS	V				REAR WALL OF EQUIPMENT CABINET
BONUS	45	CAT5e	DC	DM	EQ	RJ45	RJ45				120						EQUIP CAB	BONUS	V				REAR WALL OF EQUIPMENT CABINET
BONUS	42	CAT5e	DC	VM	WALL	RJ45	RJ45				120						DESK CORNER	BONUS	W		2G		
BONUS	43	CAT5e	DC	DM	WALL	RJ45	RJ45				120						DESK CORNER	BONUS	W				
BONUS	46	RG59	EQ	RCA	PROJ	F	RCA						60				HD COMPNT R	BONUS	Y	1G			CEILING ABOVE PROJECTOR
BONUS	47	RG59	EQ	RCA	PROJ	F	RCA						60				HD COMPNT B	BONUS	Y				CEILING ABOVE PROJECTOR
BONUS	48	RG59	EQ	RCA	PROJ	F	RCA						60				HD COMPNT G	BONUS	Y				CEILING ABOVE PROJECTOR
BONUS	49	RG59	EQ	RCA	PROJ	F	RCA						60				COMPST VIDEO	BONUS	Y				CEILING ABOVE PROJECTOR
BONUS	50	CAT5e	EQ	RJ45	PROJ	RJ45	RJ45				60						CONTROL	BONUS	Y				CEILING ABOVE PROJECTOR
BONUS	51	CAT5e	EQ	RJ45	SCREEN	N/A	N/A				50						CONTROL PREWIRE ONLY	BONUS	Z	1G			ZIGZAG AFC
BONUS	52	16/2	EQ	BP	DORMER	PB	BP								40		BINDING POST / WALL - LEFT	BONUS	AA		2G		IN WALL BELOW DORMER WINDOW
BONUS	53	16/2	EQ	BP	DORMER	PB	BP								40		BINDING POST / WALL - CENTER	BONUS	AA				IN WALL BELOW DORMER WINDOW
BONUS	54	16/2	EQ	BP	DORMER	PB	BP								40		BINDING POST / WALL - RIGHT	BONUS	AA				IN WALL BELOW DORMER WINDOW
BONUS	55	16/2	EQ	BP	REAR	N/A	N/A								40		LACE IN CEILING FOR SURR L	BONUS	BB			N/A	ZIGZAG AFC
BONUS	56	16/2	EQ	BP	REAR	N/A	N/A								40		LACE IN CEILING FOR SURR R	BONUS	CC			N/A	ZIGZAG AFC
BONUS	57	CAT5e	EQ	RJ45	LIGHTS	N/A	N/A				60						PREWIRE LIGHTING CONTROL	BONUS	DD			N/A	TAKE TO LIGHT SWITCH BOX
KIT/NOOK	58	16/4	AV	N/A	VC	N/A	N/A							100			PW VOLUME CONTROL LOCATION	KIT/NOOK	EE		2G		SHARE WITH DECK VC
DECK	67	16/4	AV	N/A	VC	N/A	N/A							100			PW VOLUME CONTROL LOCATION	DECK	EE				SHARE WITH KT/NOOK VC
KIT/NOOK	59	16/2	VC	N/A	SPKR L	N/A	N/A								30		PW & LACE LEFT CLNG SPKR	KIT/NOOK	FF			NA	ZIGZAG AFC
KIT/NOOK	60	16/2	VC	N/A	SPKR L	N/A	N/A								30		PW & LACE LEFT CLNG SPKR	KIT/NOOK	GG			N/A	ZIGZAG AFC
DINING	61	16/4	AV	N/A	VC	N/A	N/A							100			PW VOLUME CONTROL LOCATION	DINING	HH	1G			
DINING	62	16/2	VC	N/A	SPKR L	N/A	N/A								30		PW & LACE LEFT CLNG SPKR	DINING	II			N/A	ZIGZAG AFC
DINING	63	16/2	VC	N/A	SPKR L	N/A	N/A								30		PW & LACE LEFT CLNG SPKR	DINING	JJ			N/A	ZIGZAG AFC
MBR	64	16/4	AV	N/A	VC	N/A	N/A							100			PW VOLUME CONTROL LOCATION	MBR	KK	1G			
MBR	65	16/2	VC	N/A	SPKR L	N/A	N/A								30		PW & LACE LEFT CLNG SPKR	MBR	LL			N/A	ZIGZAG AFC
MBR	66	16/2	VC	N/A	SPKR L	N/A	N/A								30		PW & LACE LEFT CLNG SPKR	MBR	MM			N/A	ZIGZAG AFC
DECK	68	16/2	VC	N/A	SPKR L	N/A	N/A								30		PW & LACE LEFT CLNG SPKR	DECK	NN			N/A	ZIGZAG AFC
DECK	69	16/2	VC	N/A	SPKR L	N/A	N/A								30		PW & LACE LEFT CLNG SPKR	DECK	OO			N/A	ZIGZAG AFC
								Cable:			C5	RG6	RG59	16/4	16/2	CAM							
								Total Length:			1650	1310	720	400	440	480			69	24	4	19	
																			TOTALS				

ABBREVIATIONS
DC = Distribution Center
F = F connector
VC = Volume Control
SP = Speaker
CM = Coax Module
VM = Voice Module
DM = Data Module
SM = Speaker Module
MOD = Modulator Module
AV = A/V Closet (music and cable TV)

PW = Prewire
N/A = Leave UnTerminated
BP = Binding Post

ADDITIONAL RESOURCES

CEDIA Training: available online, at EXPO, and at local training events – a wide variety of topics at all learning levels - including technical, business, and emerging trends. www.cedia.net/train

CEDIA Resource Library – over 100 documents, drawings and other resources to improve the quality of your projects www.cedia.net/train/resources

CEDIA Advanced Residential Electronic Systems – this book covers the industry at a more advanced level, and supports CEDIA's ESC-T, ESC-D, and ESC-N certifications. A valuable resource for any integrator, and anyone seeking CEDIA Professional Certification. Available in print or eBook.

Standards and Recommended Practices – technical documents intended to help establish functional goals and standardized practices in the industry.

CEA/CEDIA CEB-22	Home Theater Recommended Practices: Audio Design
CEA/CEDIA CEB-23	Home Theater Recommended Practices: Video Design
CEA/CEDIA CEB-24	Home Theater HVAC
CEA CEB-29	Installation of Smart Grid Device
ANSI J-STD-710	Architectural Drawing Symbols Standard
ANSI/CEA/CEDIA 2030-A	Multi-room Audio Cabling
ANSI/CEA/CEDIA 863B	Connection Color Codes for Home Theater Systems
ANSI/CEA/CEDIA 897	F-Connector Color Coding for the Home
EIA/TIA 570-C	Residential Telecommunications Infrastructure Standard
ITU 775-1	Multi-channel Speaker Configuration
UL-1678	Mounting A/V Devices

CEDIA White Papers and Other Reference Documents – More than 40 topic-specific publications aimed directly at the residential industry

Technical White Papers

Dimming CFLs
Dimming LED Lamps
HDMI Fundamentals White Paper Compilation
Mobile Devices Pt.1 – The Business Case
Mobile Devices Pt.2 – Ten Steps to Creating a Robust Wireless Environment
Mobile Devices Pt.3 – Using Mobile Devices as a Control Platform
Network Security Best Practices
Safety Tips and Regulations
Selecting Display Size Based on Room Size and Seating
Telepresence Opportunities
Ten Steps to Creating a Robust Wireless Environment
Troubleshooting for the Home Technology Professional
Ultra HD-4K Industry White Paper
Wireless Audio Solutions: Technologies and Best Practices

Business White Papers

Inbound Marketing
Increasing Profits: How to Best Utilize Subcontractors
Increasing Profits: How to Create Job Cost Reports
Increasing Profits: How to Measure and Track Employee Productivitiy
Managing Cash Flow
Markup vs. Margin

Other Documents

CEDIA Recommended Wiring Guidelines (US)
CEDIA Recommended Wiring Guidelines (Europe, Middle-East, Africa)

Publications

US and Canada

CE Pro – www.cepro.com
Residential Systems – www.residentialsystems.com
Technology Integrator – www.technologyintegrator.net

UK and Europe

Inside CI - www.insideci.co.uk
Hiddenwires - http://hiddenwires.co.uk
SVI - www.svimag.com
Essential Install – www.essentialinstallive.eu/magazine

Mexico

HomeTech Magazine - http://musitech.com.mx/hometech/

Brazil

Home Theater and Casa Digital - www.hometheater.com.br
AVI Latinoamerica - www.avilatinoamerica.com

Columbia

AXXIS - http://revistaaxxis.com.co

Australia

Connected Home – www.trade.connectedhometechnology.com

GLOSSARY

110 punch-down tool	hand tool used to terminate twisted pair cable to modular jacks or a 110 punch block
2.4 GHz	an ISM (Industrial, Scientific and Medical) radio band used for wireless communications
3D Television	television display or projector which delivers a three-dimensional image using either active or passive glasses or by autostereoscopy (no glasses needed).
5 GHz	an ISM (Industrial, Scientific and Medical) radio band used for wireless communications
70-Volt system	type of speaker distribution system that uses transformers at the output of an amplifier and at each speaker in order to provide a constant voltage of 70.7 volts. Typically used in large commercial installations. (also knowns as a constant-voltage system)
A/D Converter	converts an analog signal to a digital signal.
Absorber	anything that absorbs sound and thus reduces sound reflections in a room; can be a device intended for this purpose, or ordinary objects and people in the room.
A-BUS	audio distribution technology where a matrix selects the sourcs and sends signal to a keypad, which includes a small power amplifier which drives speakers in the room
AC	see alternating current.
AAC	Advanced Audio Coding - a common compression algorithm (ipods, etc)
Accent lighting	lighting that adds interest to a room or landscape by drawing attention to an object or architectural detail.
Access control	technology which allows verified users to gain access via doors and gates and allows control from remote locations
Access point (AP)	a wireless tranceiver connecting wired Ethernet to a wireless 802.11x network
Acoustically transparent (AT) screen	video projection screen made with fabric mesh or vinyl with small perforations to allow sound to pass through
Active Loudspeaker	a loudspeaker which has the power amplifier(s) built in. In home theater, usually a subwoofer

Additive color process	photographic process in which the desired colors are produced by combining appropriate proportions of three primary colors. This the method used to create a color image in video.
Advanced Television Systems Committee (ATSC)	standards committee that addresses digital television broadcasts
Alarm	a condition or state of a security or fire alarm system when triggered by a the state of a detection device. The alarm state may trigger a number of different actions, such as audible sound, phone call, lockdown, HVAC functions, or lighting functions.
Alternating Current (AC)	an electric current that changes its amplitude and reverses its direction of flow (polarity) in a sinusoidal cycle at a frequency that is measured in Hertz (Hz).
Ambient lighting	overall illumination in a room or space. May include both natural light and artifical light.
American Wire Gauge (AWG)	a system of defining the size of copper wire based on its cross-section, used primarily in US and Canada since 1857, also known as Brown & Sharpe Wire Gauge
AM/FM	Analog radio broadcast technologies (Amplitude Modulation/Frequency Modulation)
Ammeter	an instrument used to detect and to measure electrical current.
Ampere (Amp, or A)	a quantity measurement combined with a time measurement to define current, the number of electrons flowing through a circuit in a given time
Amplifier	component which increases the amplitude of an analog signal (audio or video)
Amplitude	the magnatude of an audio signal, or sound in the air
Analog	a signal in which the waveform is the same as the original sound wave
Analog to Digital converter	a device which changes an analog signal to a digital signal
Analog video signal	video signal that creates a picture by taking the form of continuous waves that are displayed via a series of successive scans
Anamorphic lens	a lens used in the process of modifying a wide image to occupy a narrower frame, or to reverse this process to recover the wide image
Annunciator	visual or audio device used to notify users of various happenings such as fires and and intrusions
AP	see Access point.
App	abbreviation of "application". A small software application, optimized for use on mobile devices and smart phones. Usually focuses on one specific task, calculation, game, or product.

Architect	person responsible for the design and oversight of a building's construction.
Architectural drawings	set of drawings created by architects and engineers that tell how a building should be constructed. These can include floor plans, site plans, framing plans, electrical plans, plumbing plans, elevations, sectional drawings, and presentation drawings
As-built drawings	set of plans updated to show how the project was actually built. These drawings are the result of the field-marked, red line drawings.
Aspect ratio	relationship of screen width to screen height
ATSC	see Advanced Television Systems Committee.
Attenuation	decrease in the value of the power of a signal or sound. Commonly measured in decibels (dB).
Audio Calibration	the process of evaluating, verifying, and tuning an audio system to optimize performance according to established sound quality goals.
Audio power amplifier	an audio component that receives a line level signal and increases its amplitude to drive loudspeakers
Audio/Video Receiver	a component which selects sources, processes audio and video, tunes radio stations, and provides amplified output to drive speakers
Authority Having Jurisdiction	The individual who has the final say on enforcement of codes and regulations. May be a city inspector, or even a building superintendant
A/V floor plan	architectural drawing showing layout and location of all A/V devices
A-weighting	see dBA-weighting
Back box	casings for in-wall and in-celing speakers that provide benefits such as: increased performance, enhanced sound isolation, and reduced vibration
Backup communication	any redundant means of communication in a security system which provides a link to the central station. May be a phone line, cellular telephone, internet or two-way radio.
Balance sheet	financial document that shows a company's assets and liabilities
Balun	devices used to send a signal over twisted pair cable by converting an unbalanced signal to balanced, then reversing the process at the destination
Banana plug	a two conductor connector used for speaker connections
Bandwidth	the range of frequencies a device is capable of carrying. In audio, the ideal bandwidth is 20Hz-20kHz. In networking bandwidth is measured in MHz.
Base 2	see Binary Numbering System

Bass trap	device or structure used to absorb bass frequencies, lessening the effect of standing waves; also called low-frequency absorber. Most operate at frequencies above 80Hz.
Bend radius	the curvature of a bend in an installed cable. The minimum bend radius is the smallest recommended curvature, beyond which performance may be compromised
Binding post	a simple connector used mainly for speaker leads
Bill-of-Materials (BOM)	document that results from cost estimating process which lists quantity and cost of each resource needed for a project.
Binary numbering system	a mathematical representation of counting sequentially in which all numbers are comprised of two unique values: 0 and 1.
Biometric reader	a device that identifies a user by unique physical traits such as fingerprint or retinal scan
Bit	shorthand abbreviation for binary digit. Represented by a lower case 'b'
Black level	a video signal level that represents the blackest intensity.
Block diagram	diagram used to show circuits or flow by breaking the information into smaller sections or blocks to show the relationship or flow between components.
Blueprint	scale drawings of space layout, construction, and specifications needed to erect a building or home.
Bluetooth	wireless radio data transfer specification working on the 2.4Ghz band that allows for PAN (Personal Area Network) communication between devices.
Blu-ray disc	recording format that has the ability to store several hours of high definition video signals (720p/60, 1080i/60, 1080p/24, or 1080p/60) on a disc the same size and shape as a conventional DVD, but with up to 25 GB of capacity for a single-layer disc, and 50 GB of capacity on a dual-layer disc.
British Thermal Unit (BTU)	measure of thermal energy used to heat and cool in HVAC systems; heat required to raise a one pound-mass of water (usually around one pint) 1° Fahrenheit (F). One Watt = 3.41 BTU.
Broadband Internet	Internet access that is usually provided over a cable or DSL connection.
Brownout	a temporary drop or "sag" in voltage
BTU	see British Thermal Unit.
Business plan	document that spells out a company's expected course of action for a specified period, usually including a detailed listing and analysis of risks and uncertainties.
Bus topology	network architecture layout based on a single main communications line (trunk) to which nodes are attached; also known as linear bus. An example would be RS485.

Byte	series of 8 bits. Represented by an upper case 'B'
Cable	two or more wires inside a common jacket, designed for a specific application
Cable schedule	document showing every cable run in a project, including its start point, destination, cable type, use, and termination type. Should also have a provision to check off or initial when each cable is run, terminated, and tested
Capacitance	a measure of the amount of electric charge either stored or separated for a given electric potential.
Capacitor	device which resists changes in voltage in an electric circuit.
Card reader	a device used to provide access by means of a magnetic strip on an identification card
Cathode Ray Tube (CRT) technology	"legacy" video display utilizing an electron-beam gun at the back of a vacuum enclosure.
CEC	specific feature on HDMI-enabled equipment that allows communication between HDMI-connected devices via simple commands (like "play") that control other devices without any special programming.
Central station	a company which monitors subscribers' alarm systems and calls approprite authorities as needed
CFL	See Compact Flourescent Light
Change order (CO)	document used to record changes to an existing agreement; becomes part of the contract when agreed upon by all parties involved.
CIE chromaticity diagram	a method for visualizing what our eyes perceive as the visible light color space within a 2D realm without concern for luminous intensity. Created by the International Commission on Illumination in 1931.
Cipher lock	a device which combines a door handle and lock with some way to identify the user and allow access only authorized individuals
Circuit	a complete loop through which current can flow.
Circuit breaker	a switch designed to protect an electrical circuit from damage caused by overload or short circuit. Unlike a fuse, a circuit breaker operates automatically and can be reset.
CMR	See Common Mode Rejection
Coaxial cable	a type of cable used for satellite, CATV, CCTV systems, comprised of an insulated central conducting wire, a dielectric barrier, a second "coaxial" conductor, usually braided, and usually some additional sheilding. Coaxial cable is durable, withstands linear stress well, and is capable of carrying a lot of data. The most common types are RG-6 lSeries 6) and RG-59.

Codec	abbreviation for compressor/ decompressor or encode/decode; a technology for compressing and decompressing digital data.
Color gamut	complete subset of colors.
Color space	three-dimensional model into which the three attributes of a color can be represented, plotted, or recorded.
Colorimeter	device that provides correlated color temperature, luminance readings, and CIE x, y coordinates.
Common Mode Rejection (CMR)	the rejection of external noise by a balanced circuit, due to the two main conductors being 180° out of phase
Communications protocol	set of procedures which controls how a data communications network operates.
Compact Flourescent Light (CFL)	flourescent light in a form factor similar to an incandescent lamp
Composite analog video signal	video signal which carries all image and sync information on a single coaxial cable
Compression tool	hand tool used to terminate coaxial cable with F, RCA or BNC connector
Conductor	anything that electricity flows through.
Conduit	metal or fiber pipe or flexible tube used to enclose electrical cables; may be used for either high voltage or low voltage.
Conflict resolution	act of arbitrating differences of belief or opinion about a given set of conditions or circumstances.
Construction lines	on an architectural drawing, the lines used to aid the drafter setting up the drawing. Also called guide lines.
Contact closure	control or sensing based on the simple closing of a circuit
Continuity tester	an instrument that tests the ability of a circuit to conduct electrical current; does not indicate the level of performance
Contract	legally binding exchange of promises. Also, the document that spells out this promise in writing. Contracts are used as a management tool and as a safeguard for your financial interests.
Contrast ratio	ratio between peak white level and the darkest black level.
Convection	transfer of heat from one substance to another through circulatory molecular motion, the natural tendency of heat to rise
Cost benefit analysis	an analysis of the relationship between the costs of undertaking a task or project and the benefits likely to arise from the project.

Crossover network	combination of audio filters which directs different frequency ranges to different destinations
Crosstalk	A magnetic or electrostatic coupling, which causes energy transfer from one wire to another. Shielding or special twisting can solve this issue.
Current	a flow of electricity through a conductor. Measured in Amperes.
Customer relations	task of keeping customers informed, handling complaints efficiently, and truly listening to what customers have to say; being a customer-focused organization
D/A Converter	A device that converts a digital signal to an analog signal
D65	see Illuminant D.
D6500	see Illuminant D.
Daisy chain	cabling method that connects end points directly in line without returning to a common point.
DB-9	a D-subminiature connector with 9 pins, often used for serial communication
dBA-weighting	sound-pressure levels derived using A-weighting; denoted by "dBA," or A-weighted dB levels; standard weighting curve applied to audio measurements, designed to emulate the response of the human ear at lower levels
dBC-weighting	follows the frequency sensitivity of the human ear at higher levels; noise measurements made with the C weighting scale are designated dBC.
DBS	see Direct Broadcast Satellite.
DC	see Direct Current.
Decibel (dB)	unit of measurement which is used to measure the intensity of sound, either acoustically or in an audio signal. One tenth of a Bel.
Decoding	returning a coded signal to its original format
Dielectric	continuous layer of polymeric material in a coaxial cable that separates the center conductor (found inside the dielectric) from the ground shields located outside the dielectric material. It is an insulator, and modifying its thickness in any way affects performance.
Diffuser	device which reflects sound in many directions
Digital Direct Drive Image Light Amplifier (D-ILA)	video projection technology that uses a reflective LCD to create images by using a light source that is reflected off the LCD and sent through the lens to the screen.
Digital Light Processing (DLP)	video display technology which uses a proprietary chip with Digital MicroMirrors to create pixels on a screen.

Digital MicroMirrors (DMM)	tiny mirrors that are used in a chip design to reflect light towards a projection screen to turn a pixel on, or away from the screen to turn a pixel off. Used in DLP projection technology (Digital Light Processing)
Digital multimeter (DMM)	a multimeter or volt/ohm meter that displays results digitally. See multimeter.
Digital Rights Management (DRM)	a broad term describing various methods of preventing the unauthorized use or reproduction of intellectual property such as books, movies, and audio
Digital Signal Processing (DSP)	processing of an audio signal in the digital domain
Digital Subscriber Line (DSL)	service from a telephone company that uses a DSL modem connected a telephone line to provide Internet service, without the installation of a second phone line.
Digital Theater Systems (DTS)	a multi-channel digital surround sound format owned by Digital Theater Systems (DTS, Inc.) used for sound on film, DVD and Blu-ray formats
Digital to Analog convertor	device that converts a digital signal to an analog signal
Digital Visual Interface (DVI)	first digital connection standard for consumer devices launched in 1999 to connect LCD monitors to personal computers. Developed by the Digital Display Working Group.
D-ILA	see Digital Direct Drive Image Light Amplifier.
Dimmer	device that regulates the amount of electrical voltage sent to a luminaire. Controlled by a slider, knob or control signal
Direct costs	expenses directly attributable to certain elements of a project, ie. electronic equipment.
Direct Current (DC)	a steady value of current that does not change direction of flow or amplitude.
Direct sound wave	a sound wave that arrives at the listener's ear (or microphone) directly from the source, without reflections
Direct view TV	image viewing device in which the image is being emitted directly from the surface of the display; includes CRT "tube" sets, flat panel plasma and LCD displays
Disc	optical storage medium such as CD, DVD and Blu-ray which can be removed from the player and taken elsewhere. Some are read-only, some are re-writeable.
Disk	magnetic storage medium, such as floppy disk or hard drive. May be portable or built into a compuer or other component.
Display calibration	process that involves adjustment of display control parameters to ensure that the viewer sees images within the proper contrast, color, and gray scale tone range, as the content creator intended
Distortion	difference between the in the input and the output of an audio device
Distribution amplifier	device that accepts and distributes one audio or video input out to 2 or more destinations.

DLP	see Digital Light Processing
DMM	see Digital MicroMirrors.
Dolby Digital	a series of lossy audio compression technologies developed by Dolby Laboratories.
Dolby Digital Surround EX	a higher quality (than Dolby Digital) digital audio compression scheme used which enables the use of a surround center channel (6.1)
Dolby Pro Logic II	a surround sound processing technology used to decode Dolby Surround soundtracks.
Dolby TrueHD	an advanced lossless multi-channel audio codec mainly used for HD DVD and Blu-ray.
Domain Name System (DNS)	used to resolve a URL (Uniform Resource Locator) (such as www.cedia. org) to a specific IP address.
DRM	See Digital Rights Management
Dry contact closure	condition created when two contacts close to make electrical continuity in a circuit.
Dreyfuss Human Factors Charts	studies on the size and shape of people and related ergonomic factors
DSL	see Digital Subscriber Line.
DSP	see Digital Signal Processing
DTS	see Digital Theater Systems
DTS HD Master Audio	a surround sound format that supports a virtually unlimited number of channels (limited by storage medium)
DTS-ES (DTS Extended Surround)	a matrix centered surround sound audio format that includes two variants, DTS-ES Matrix and DTS-ES Discrete 6.1
DVD	See Digital Versatile Disc
DVI	see Digital Visual Interface
DVI-A	interconnect used to carry a DVI signal to an analog display, such as a CRT monitor or low-cost LCD.
DVI-D	interconnect used for direct digital connections between source video and digital monitors
DVI-I	interconnect capable of carrying either a digital-to-digital signal or an analog-to-analog signal.

Dynamic Host Configuration Protocol (DHCP)	process that allows for allocating IP addresses dynamically by a DHCP server
Dynamic load	a physical load which is moveable, therefore requiring a much more robust mounting structure. Example; a wall mounted video display which swings out and can be tilted
Dynamic Routing	network traffic management process that uses an internal routing table to identify the most efficient path based on overall load, line delay and bandwidth and automatically reconfigures the network to take advantage of these more efficient pathways; generally preferred over static routing for handling peak traffic conditions.
Early reflections	set of reflections which have only been reflected once, arrived at the listener just after the direct sound
EIA	see Electronics Industries Association.
Electric radiant floor	floor system that consists of electric cables built into the floor to transmit heat.
Electrical plan	scaled drawing of a floorplan that indicates the locations of receptacles, switches and lighting fixtures.
Electricity	the flow of electrons that is produced when an electron jumps from the outer shell of its atom to the outer shell of a new atom and thusly knocks another electron out of orbit from that second atom.
Electromagnetic interference (EMI)	external inerference in a signal caused by electromagnetic induction or radiation from such sources as electrical wiring, motors, or lighting
Electromagnetic noise	imperfections in AC current.
Electron	a negatively charged particle that orbits the nucleus in concentric layers called shells.
Electron Current Flow Theory	in a circuit, current moves from the negative side to the positive side because electrons are what current measures and that is the direction they move.
Electronic documents	documents created and often communicated by computers and computer software.
Electronic Systems Technician (EST)	an individual who installs, upgrades and services electronic systems in the field
Electronic Systems Designer (ESD)	an individual who designs electronic systems, including components, interconnection, and functionality
Electronics Industries Association (EIA)	individual organizations that together have agreed on certain data transmission standards such as EIA/TIA-232 (formerly known as RS-232).
Elevation view	architectural drawing representing a horizontal view of a wall or space in a plan.

EMI	see Electromagnetic interference
Energy monitoring	the process of gathering and analyzing data related to where and when energy is being consumed
Energy management	the process of actively and automatically reducing the use of energy by controlling appliances, lighting, shades, HVAC, and other subsystems
Envelopment	the ability of a surround system to fully immerse the listener in a three-dimensional space
Equal loudness contours	see Fletcher-Munson curves.
Equalization	the process of adjusting the relative energy in various audio frequency ranges
Equalizer	device used to cut or boost individual frequencies of an audio signal using a number of filters. May be analog or digital.
EST	see Electronic Systems Technician
Ethernet	a collection of technologies used for establishing connections in a network; standardized as IEEE 802.3.
Exterior elevation	drawing which shows the exterior surfaces of the home, as if the viewer was standing outside looking at the house
F-connector	coaxial connector type which uses the solid center conductor of the cable and the braided sheild as its two conductors.
Fiber optic cable	type of cable with a glass center core surrounded by a jacket made of layers of various forms of plastic that transmits data using light rather electricity
File Transfer Protocol (FTP)	protocol used to move files either individually or in groups to another computer on the network.
Firewall	security measures (hardware and/or software) that blocks unauthorized users from gaining access to a computer or network.
Fixed bid	a proposal wich defines the work to be done at a pre-agreed total price
Fletcher-Munson curves	represent an average of the frequencies that the human ear perceives as loudness at different levels; also called equal loudness curves or equal loudness contours.
Floor plan	widely used drawing type which shows rooms, doors, walls, etc. as viewed from above. Allows locations to be shown for most wall mounted devices such as outlets, controls, etc.
Flowchart	a graphical representation, primarily through the use of symbols, of the sequence of activities in a system (process, operation, function, comparison or activity).

Fluorescent lamp	device that is a gaseous discharge light source with a tubular electric lamp containing low-pressure mercury vapor and other gases.
Foil tape sensor	a device for sensing broken glass.
Foot fall	sounds created when someone is walking on the floor above; energy from the steps is transferred to the floor and from the floor into the space below
Frame	complete image on a TV screen.
Frame rate	the number of video images captured or displayed each second.
Frequency	number of cycles completed each second by a certain AC voltage.
Frequency mete	an instrument used to measure the frequency of signals being produced by various electronic equipment.
FTP	see File Transfer Protocol.
Fundamental frequency	the initial and lowest frequency of a resonance or complex wave
Gamut	see color gamut.
Gantt Chart	chart that illustrates the duration of tasks as compared to the project time-frame; highlights dependencies, milestones and resources associated with particular tasks.
Gateway	hardware or software that makes it possible to maintain communication between computers or computer networks that may use different communication protocols; also called a router.
Glass break detector	sensor that responds to the shattering of a pane of glass. They may be either audio sensitive or mechanical.
Graphical User Interface (GUI)	a way of communicating with a CPU by manipulating buttons(icons) and windows with a mouse or pointing device (your finger or stylus)
Graphic equalizer	an audio equalizer which adjusts individual frequency ranges by way of vertical sliders. Typically divides the spectrum into octaves or 1/3 octaves.
Ground	A conductive connection between the earth or other large conducting body and an electrical circuit to form a complete electrical circuit.
Ground loop	electrical condition where two cirtuits are connected which have different potentials to ground. Can cause unwanted noise and a safety hazard.
GUI	see Graphical User Interface.
Hardwired system	a system using wires to communicate rather than through the air (or wireless).

Harmonic distortion	unwanted harmonics, or overtones, introduced into an audio signal by an audio component. Measured in %. A lower distortion specification means a more accurate reproduction of sound
Harmonics	integral multiples of the fundamental frequency of a complex wave. Harmonics contribute to the unique sound of an instrument or voice.
HDCP	see High-bandwidth Digital Content Protection
HDTV	see High-definition digital television.
Heating, Ventilation, and Air Conditioning	the system which heats and cools the home and introduces fresh air into the system
Hexadecimal	see Hexadecimal numbering system.
Hexadecimal numbering system	a mathematical representation of counting sequentially in which all numbers are comprised of sixteen unique values: 0 thru 9, and A thru F.
High Definition Television (HDTV)	digital television format that provides a high-quality widescreen picture (16x9) with digital surround sound.
High voltage dimmer	traditional dimmer that works with a standard 120v house current.
High-bandwidth Digital Content Protection (HDCP)	form of Digital Rights Management developed by Intel Corporation that uses an advanced key exchange system which is highly secure
High-Definition Multimedia Interface (HDMI)	a compact audio/video interface used to transmit uncompressed digital streams, digital surround sound, control and network information and two channels of audio return signal.
Hub	communication device that contains multiple ports. Each computer in a network is directly connected with each port of the hub. Also, used to describe the point on a network where many circuits are connected.
Hue	specific color classification given to an object based on the seven colors found in the spectrum; red, orange yellow, green, blue, indigo, or violet.
Hum	an undesirable noise often caused by grounding problems or leakage of AC into a signal
HVAC	see Heating, Ventilation, and Air Conditioning
Hybrid lighting control system	combination of both distributed and central lighting control features.
Hydronic radiant heat	floor system that pumps heated water through tubing laid in a pattern underneath the floor.
Hyperlink	a text word or graphical icon or picture in an HTML document that links, or allows navigation to, another HTML document or file.

HyperText Transfer Protocol (HTTP)	protocol used to carry requests from a browser to a Web server in order to transport pages from the server back to the requesting browser.
IEEE 1394	standard developed by Apple for high speed data transmission on local cable networks (LANs) as well as home networks. Commonly referred to as FireWire.
Illuminance	method of light measurement where the light source is measured by pointing the light metering device at the source to determine the intensity of the light falling upon an object
Illuminant D3	illuminant upon which the video industry is standardized that represents that color of white most observers perceive as the most neutral shade of white; also corresponds to a color temperature of 6500 Kelvin; also called D6500 or D65.
Impact Isolation Class (IIC)	test used to measure foot falls
Impedance	the opposition that a cable, component or circuit offers to alternating current. Measured in Ohms (Ω)
Impedance meter	device used to measure opposition to AC current flow in a system
Incandescent lamp	a light device that uses a coiled tungsten-wire filament secured inside a glass bulb that glows when electricity is passed through it.
Indirect costs	expenses of a project not directly relatable to a tangible item; ie. rental fees or advertising costs.
Inductance	the capability of a coil to store energy in a magnetic field surrounding it.
Inductor	a device that stores electrical energy in a magnetic field.
Infrared	light with a wavelength longer than visible light. Commonly used to transmit data from remote controls. An LED is generally used to emit the IR pulse. May also be sent over wire and converted back to light at the destination using an emitter.
Infrasonic	sound that is below the human audible frequency range of sound.
Insulator	materials that do not carry electricity.
Integrated system	several sub-systems connected so that they work together and can be controlled easily
Integration	act of combining parts into a whole.
Interior elevation	a drawing which shows vertical surface of a home's interior. Especially useful for display or screen placement and acoustical treatment location
Interlaced scanning (i)	television format in which electron beams sweep the picture tube twice to create half the image on the first pass (on each alternating line) and the other half on the second pass.

International Telecommunication Union (ITU)	an agency of the UN (United Nations) which specializes in issues related to information and communication technologies
Internet	global network used to transmit the data/information using the TCP/IP protocols.
Internet Protocol Address (IP Address)	unique number that each device on a computer network uses for identification when using the TCP/IP communication protocol.
Internet Service Provider (ISP)	company that provides connectivity to the Internet.
Intranet	internal network of an organization that uses Internet protocols.
IPV6	Internet Protocol Version 6, the alpha-numeric system which dramatically expands the number of possible internet addresses in the world
IP Address	see Internet Protocol Address.
IR	see Infrared
ISP	see Internet Service Provider.
ITU	see International Telecommunication Union
Key System Unit	central processor on a multi-line telephone system which allows users to directly dial other users on the system as well as outside lines
Keypad	a set of buttons arranged in a block with numbers and or letters used for the control over telephony, security or audio/video systems and subsystems
KSU	See Key System Unit
Lamp	commonly called a light bulb; a device used in a light fixture that influences the amount and type of light produced by the fixture.
Legend	table or list on a drawing identifying what each drawing element or symbol represents
Lens shift	optical feature on a projector which allows the image to be shifted off of its native center without sacrificing its native geometry
Letterboxing	process of adding black horizontal bars at the top and bottom of the screen to place a wide image on a 4:3 display surface
LFE	see Low frequency effects channel
Light Emitting Diode (LED)	diode (semi-conductor) which emits light when current is applied. Used for common lighting as well as to back-light an LCD video display
Lighting control system	light fixtures and associated communications hardware and software which control the lighting system

Local Area Network (LAN)	computer network covering a local area, such as a home, office or small group of buildings.
Loudness	the overall sound level as perceived by a listener
Low frequency effects channel (LFE)	in a multichannel surround system, the name of the audio channel designated for special sonic effects (below 125 Hz) usually reproduced by a subwoofer
Low voltage transformer	transformer that comes in two forms: magnetic and electronic.
Lumen	international (SI) unit of luminous quantity of light that equals the amount of light spread over one square foot of surface by one candle's power when all parts of the surface are exactly one foot from the light source.
Luminaire	light fixture that holds the lamp or bulb and also the parts around it that help position, protect, and connect lamps.
Luminance	method that measures the amount of light being emitted or reflected off a particular surface. Units of measure is the candela per square meter
Lux	unit of light measurement that is equivalent to one lumen per square meter.
MAC address	Media Access Control address, globally unique 40-bit address consisting of a manufacturer's code and a serial number; a hardware address that uniquely identifies each node of a network.
Macro	a file containing a sequence of commands that can be executed as one command; often built to perform frequently used, as well as complex operations.
Magnetic field	results when atoms in a material are aligned in a particular way.
Masking	absorbtive panels which cover unused portions of a projection screen, depending on the aspect ratio of the content
Mesh network	network configuration where each node sends and receives its own data, plus serves as a relay for other nodes
Metadata	data that provides information about other data. Example; information about a recorded song such as artist, title, genre, etc.
Metal Oxide Varistor (MOV)	an electronic device which has low resistance at low voltage and high resistance at high voltage. The name is derived from "variable resistor". Commonly used to supress line voltage surges
Mission statement	a statement laying out the purpose of the company with the intent to keep all stakeholders aware of the mission
Mobile app	see App
Monitoring service	company that offers contracted services to receive information from security systems and pass on information to the appropriate people or authorities. AKA Central Station.

Motion detectors	a variety of technologies to sense movement.
MOV	See Metal Oxide Varistor
MPEG-2	software standard for the compression and transmission of digital data.
MPEG-4	collection of methods defining compression of audio and visual (AV) digital data.
Multimeter	an electronic measuring instrument that combines several functions in one unit
Multi-mode Optical Fiber	optical fiber whose core diameter is large relative to the wavelength (see wavelength) of the light, thus allowing a large number of modes (see modes) to propagate.
Multiplexer	device that allows multiple signals to be transmitted across a single physical channel.
Multi-source audio	audio system which allows the user to choose between different audio sources
Multi-zone audio	audio system which allows different sources to be sent to different parts of the home independently
Network Interface Device (NID)	the point of connection between networks. In a residence, the location of demarcation.
Networking	process of connecting computers so they can share information, resources and equipment.
Nibble	half of an 8-bit Byte. Equal to 4 bits.
NID	see Network Interface Device.
Noise Reduction Coefficient (NRC)	the mathematical average of absorption coefficient values (typically 125, 500, 1K, 2K, 4K & 8K Hz) for a given material. Intended for use in the speech range.
NRC	see Noise Reduction Coefficient
Octave	a doubling or halving of a frequency; a 2:1 ratio of two frequencies. In music two notes of the same name, 12 semi-tones apart
Ohm	the amount of electrical resistance, or opposition, between two points.
Ohm's Law	states that the voltage (V) across a circuit is equal to the product of the current (I) flowing through it and the resistance (R) of the circuit (E = I x R).
Ohmmeter	an instrument used to measure the DC electrical resistance in a circuit or component
Packet	unit of data that is routed between an origin and a destination on the Internet or any other packet–switched network.
Parallel circuit	a circuit with the same voltage across their ends resulting in identical polarities.

Parametric equalizer	an electronic multi-band audio processor that allows for control of amplitude, center frequency, and bandwidth
PDF	Portable document format; the native file format for Adobe Acrobat. PDF is the file format for representing documents in a manner that is independent of the original application software, hardware, and operating system used to create those documents. A PDF file can describe documents containing any combination of text, graphics, and images in a device-independent and resolution independent format.
PDP	see Plasma Display Panel.
Peak	high pressure point of a sound wave; also called a crest.
Peak-to-peak	the total height of a wave, from maximum positive to maximum negative
Phosphors	chemical compounds which radiate light when struck by a beam of electrons.
Photodiode	device capable of converting light to voltage or current.
Picture Line-Up Generation Equipment(PLUGE)	equipment used to generate a monochrome test pattern primarily intended for proper calibration of contrast (system gain) and brightness (black level).
Pink noise	a variation of white noise that has equal energy in all octaves. Commonly used to measure the performance of loudspeakers and audio systems
Pixel density	number of pixels per square inch; higher pixel density correlates to a higher overall resolution.
Pixels	single points of color and light on a video display
Plan view	drawing which shows a building or room as viewed from above
Plasma Display Panel (PDP)	type of television viewing device that uses a highly ionized xenon and neon gas in thousands or millions of micro-chambers
Plot plan	drawing wich shows buildings, roads, and other features of a project site, and other above-ground features, as viewed from above, on a scale allowing a large area of land to be covered.
PLUGE	see Picture Line-up Generation Equipment.
PO	see Purchase Order.
Point-to-point diagram	detailed wiring diagram for the assembly and service of equipment racks, head ends and area electronics. Often refered to as a schematic diagram.
Polarity	a description of either positive or negative applied to an electric charge, voltage, or speakers.
Polarity test	a test designed to make sure that each speaker in the system has the proper polarity

Polarity tester	a device to determine positive or negative response to signal by speakers in an audio system
Power	the energy used to operate an electrical device or devices.
Power over Ethernet (PoE)	the ability to pass power (up to 25.5W), as well as data, over an Ethernet network. Defined by IEEE 802.3at-2009
Powerline networking	home network that uses special modules plugged into electrical outlets to connect peripherals together to form a network.
Primary listening position	the location in a home theater which the system is calibrated to for optimal performance
Programming	the overall concept of designing, writing, testing, debugging, and documentation of computer programs.
Progressive scanning (p)	television format in which information to each pixel in a frame of video sequentially, from left to right and top to bottom to create an image.
Project Manager	person responsible for oversight on the project from the beginning planning stages to completion.
Project scope	extent or range of view, outlook, application, operation, effectiveness, etc.
Protocol	standard for transmitting data between two devices.
Pull tension	the amount of tension put on a cable as it is being installed or while suspended between two points. The maximum recommended pull tension is the amount (in lbs. or kg) beyond which performance may be compromised
Purchase Order (PO)	contract between a buyer and seller showing items to be obtained; each with an associated monetary value.
Quality assurance	systematic process of verifying that a product meets the specified requirements.
Radio frequency (RF)	radio or electrical waves in the frequency range of about 3kHz to 300GhZ.
Radio frequency interference	external inerference in a signal caused by high frequency radio or television signals
Rattle	a sound made by loose objects shaking or vibrating against one another
RCP	see Reflected ceiling plan
Reactance	a function of the inductance and/or capacitance within a circuit and their effect on the phase relationship of voltage and current within the circuit.
Real Time Analyzer (RTA)	a device which displays the frequency spectrum of an audio signal in real time.

Red line drawing	marked-up field drawing showing changes to the original documents by annotating the changes with red line markings. As-buit drawings are created from the red-line drawings.
Reflected ceiling plan	drawing which shows a building or room as viewed from above but detailing the ceiling plane rather than the floor. Includes details such as lighing, HVAC, in-ceiling speakers, ceiling tile configuration, etc.
Reflection	sound wave which reaches the ear after bouncing off a surface.
Refresh rate	rate at which the frames are displayed.
Relay	an electrical switch that is controlled (opens and closes) by another electrical switch.
Resistance	the opposition offered by an electrical conductor to the flow of a current through it, resulting in a conversion of electrical energy into heat.
Resistor	an electronic component that opposes an electric current by generating a voltage drop in proportion to the current between its two terminals (based on Ohm's law) and converting electrical energy into heat.
Resolution	measurement of the quality of a video image that is directly related to the number of pixels that make up the image.
Resolution test pattern	monochrome pattern traditionally used to evaluate television system bandwidth and frequency response.
Retrofit	to install something into an existing structure
Revenue	income that a company receives from normal business activities by the sale of goods or services to customers
Reverb	see reverberation
Reverberation	the continuing sound after the sound source has ceased to vibrate as a result of repeatedly reflecting off multiple surfaces; also called reverb
Revision	change to a document usually identified by a number or letter code such as Rev.1, Rev.2, etc. Each revision should include a timestamp and the name or initials of the person making the change.
Revision block	grid-like section on a drawing which lists and keeps track of the history of changes (revisions) made to the drawing.
RFI	see Radio frequency interference
RFI	see Request for Information.
RFP	see Request for Proposal.
RFQ	see Request for Quotation.

RH	see Relative humidity
RJ-31X	interface which allows security system's dialer to "sieze" the phone line and call out, even if a home phone is in use or off the hook.
RJ-45	standard 8 position/8 contact (8P8C) type of connector used for data and telephone. May be male (plug) or female (jack).
RJ-45 crimp tool	hand tool used to terminate twisted pair cable with RJ-45 plug
RMS (Root Mean Squared)	the average power an amplifier will deliver into a specific load. The most accurate way to compare power amplifier specs
Room mode	a frequency which resonates within an bounded space forming a standing wave; the rooms dimensions correlate to the range of many modal frequencies possible.
Router	a network device that connects two separate networks together. In the home, this will often serve as the 'Gateway' from the Internet to the home LAN.
RT60	means of expressing the measurement of the reverberation time in a physical space that indicates the elapsed time required for the reverberation to decay by 60 decibels
Satellite Television (DBS)	system that sends TV broadcasts digitally,directly from a communications satellite to home antennas, or dishes.
Saturation	relative brilliance with which a film (or print) reproduces a subject's colors.
Sawtooth wave	a non-sinusiodal waveform that resembles the blades of a saw.
Scale	ratio of measuring units expressing a proportional relationship between the drawing and the full size item or space it represents.
Schematic diagram	detailed diagram for the purpose of understanding the interconnection of devices and to provide graphical information for the assembly or service of a system.
Scope	objective(s) and extent of work to be accomplished by the project.
Scope creep	on-going addition of requirements to a project without corresponding adjustment of approved cost and schedule allowances.
Scope of Work document	agreement document among the project team and key stakeholders. It represents a common understanding of the project and sets authorities and limits for the project manager and team. It includes business objectives and the boundaries of the project including approach, deliverables, milestones, and budget.
Scope statement	short verbiage that addresses what a project will encompass.
SDTV	see Standard Definition Digital Television.
Seasonal Energy Efficiency Ratio (SEER)	quantitative measurement rating system that evaluates the amount of energy a particular HVAC system delivers to every dollar spent

Section view	part of a drawing that shows the vertical relationships of structural materials or a cut-view of an object.
Secure HTTP (HTTPS)	HTTP protocol which adds a layer of security encryption.
Security panel	the central processor of the security system.
SEER	see Seasonal Energy Efficiency Ratio.
Sensor	device that detects a specific condition, such as temperature, moisture, contact etc. so that another component can react to that condition
Series circuit	wiring confitguration where loads are connected end to end and their values, in ohms, are added together to get the total resistance of the circuit.
Server	computer or device on a network that manages a network's resources.
Shield	a part of many cable types which reduces signal loss of high frequencies and external interference such as RFI and EMI. Sometimes also provides an electrical return path.
Shielded Twisted Pair (STP)	twisted pair cable, usually 8 conductor, which includes a foil shield to protect from EMI and RFI
Short circuit	unintentional connection between + and - of a power source, without going through a load
Simple Mail Transfer Protocol (SMTP)	protocol used in the delivery of electronic mail (e-mail)
Sine wave	type of waveform signal in which the voltage or equivalent rises and falls smoothly and symmetrically; also called sinusoidal wave.
Single Mode Optical Fiber (SMF)	type of optical fiber designed to carry only a single ray of light (mode).
Single-source audio	distributed audio system which transmits audio from one single source to multiple rooms at the same time.
Sinusoidal wave	see sine wave.
Site plan	drawing that shows the location of items such as buildings, drives, parking, and water features of the property. Plot plan.
Skin effect	the tendency of electrons to flow near the outside (skin) of the conductor at 100 KHz+ frequencies
Slap echo	the back-and-forth reflection of primarily mid-range and high frequency sound between two parallel hard surfaces.
SMPTE Color Bars	an industry standard test pattern which includes seven vertical bars of 75% intensity. Developed by the Society of Motion Picture and Television Engineers

SMTP	see Simple Mail Transfer Protocol.
Sound isolation	process of reducing the amount of sound transmitted from one space to another.
Sound Pressure Level (SPL)	the strength or intensity of acoustic sound waves, measured in dB; a logarithmic ratio referenced to the threshold of audibility
Sound wave	The cyclical variation of air pressure at a frequency which stimulates hearing.
SoundTransmission Class (STC)	method used to quantify how well a boundary such as a wall, ceiling or floor attenuates air-borne sound traveling from one space to another, measured at 16 frequencies from 125Hz to 4kHz
Speaker	an electromechanical transducer that converts an electrical signal into audible sound (or an enclosure combining transducers into a system)
Speaker polarity test	a test to ensure that speakers are in phase with each other.
Speed of sound	distance traveled per unit of time by a sound wave. In dry air (68° F / 20° C) the speed of sound is 1,126 ft/sec (343.2 m/sec). It travels about 4 times as fast in water.
SPL	see sound pressure level.
Square wave	a non-sinusoidal waveform that is squared off at the peaks of the wave resulting from regular and instantaneous alteration of the current.
Stakeholders	interested parties in a project.
Standard Definition Digital Television (SDTV)	digital television format that includes 480-line resolution in both interlaced and progressively scanned formats in a 4:3 aspect ratio
Standing wave (room mode)	see Room mode
Star topology	network architecture configuration in which each node on a network is connected directly to a central hub or concentrator.
STP	see Shielded Twisted Pair.
Subtractive color process	photographic process in which all but the desired colors are removed by passing the light through subtractive filters.
Subwoofer	a loudspeaker designed to reproduce the lowest audible frequencies, typically the lower two octaves of human hearing
Switch	an Ethernet switch is a networking device that connects networked devices together.
Symbol	graphical representation of a real device or object used in a diagram or architectural drawing.

Systems Designer	industry professional who specifies the components and configuration of integrated electronic systems
Tag	abbreviated way to show additional properties of an item on a drawing. The tag can be a square, circle or hexagonal shape including a leader line connecting the tag to the item.
Task lighting	light that is directed to a specific area in order to accomplish a task
TCP/IP	collection of data communication protocols developed to standardize data transfer between systems and networks.
Telecommunications Industries Association (TIA)	national (US) trade organization that publishes TIA/EIA standards.
Telepresence	technology allowing people in different locations to see and hear each other for enhanced communication, both business and personal
Temperature	the measure of the heat
Termination	process of installing a connector or connection on a cable or wire to make it useable for the intended use. Examples; F Connector on RG6, RJ-45 or punch block on Cat5e.
Terrestrial signal	older source of TV signals for most home viewing; NTSC standard used for over-the-air (OTA) broadcasts.
Test pattern	special test pattern for adjusting TV receivers or color encoders
Threshold of hearing	lowest level of sound that is audible to the human ear. Typically assumed to be 0dB (much higher for very low frequencies)
Threshold of pain	point at which the loudness of sound causes pain to the human ear.
Throw distance	Length of the projection beam needed for a projector to produce image of a particular size.
THX	the trade name for a high-fidelity sound reproduction standard used in movie theaters, home theaters, gaming consoles and car audio systems.
TIA	see Telecommunications Industries Association.
Timeline	a linear representation of the progress of a project with important points marked according to when they occur.
Title Block	section of a drawing showing important project information such as company logo, clients name, project name, date, etc.
Toggle switch	electrical switch that uses a mechanical lever or handle to open or close the electrical contacts

Tone sweep	a test tone which sweeps smoothly from 20Hz to 20kHz; useful for finding rattles and resonances in a room
Touchscreen	deviced used to allow control via simply touching a graphical display through a resistive or capacitive screen
Touchscreen layout	plan and organization of the items shown on the touchscreen.
Transformer	a device that moves electrical energy from one circuit to another through inductively coupled electrical conductors.
Transmission Control Protocol/ Internet Protocol (TCP/IP)	set of protocols that allow communication between computers; used as the standard for transmitting data over networks and as the basis for standard Internet protocols.
Trigger	event that invokes a stored procedure on the basis of certain pre-determined data-related events.
Trigger event	see Trigger.
Trough	low pressure point of a sound wave
Tungsten-halogen lamp	gas filled tungsten incandescent lamp containing a certain percentage of halogens.
Twisted pair	two insulated copper wires twisted around each other to reduce induction (therefore interference); common type of transmission media.
UHF	see Ultra High Frequency.
Ultra high frequency (UHF)	radio frequency range from 300 MHz to 3 GHz.
Ultrasound	sound that is above the human audible range
Uninterruptible Power Supply (UPS)	a battery backup or mechanical device used to provide immediate power if electrical service is interrupted.
Unshielded Twisted Pair (UTP)	commonly used cable in the electronic systems industry used for telephone and networking. Usually 8 conductor, no sheild. Includes unsheilded CAT5e and CAT6
User interface (UI)	tools or software that allow people to interact with a machine, device, computer program, or other complex tool
UTP	see Unshielded Twisted Pair
Vertical resolution	number of distinct horizontal lines, alternately black and white, that can be seen in a TV image.
Very High Frequency (VHF)	radio frequency range from 30 MHz to 300 MHz.
VHF	see Very High Frequency.

Video scanning	general term for the process of drawing a picture on a TV screen.
Volt	the unit of measure for the electromotive force or EMF; the force that pushes the electrons through a wire and is referred to as electrical pressure.
Volt/ohm meter	see multimeter
Voltage	the difference of electrical potential between two points of an electrical or electronic circuit.
Voltmeter	an instrument used to detect and to measure electrical voltage
VOM	See volt/ohm meter
WAN	see Wide area network.
WAP	see Wireless access point.
Watt	unit in which power is measured.
Watt meter	an instrument used to measure total power consumption in watts.
Wavelength	distance between consecutive crests of a wave. Also, the distance a sound wave travels (at the speed of sound) during the time it takes to make one cycle. Also applies to visible and non-visible light.
White level	brightest part of a video signal
White noise	sound that includes every frequency in the human range of hearing with equal energy at each frequency
White paper	authoratative report or guide that details solutions to specific problems
Wide Area Network (WAN)	A network connecting computers over a very large geographical area such as states, countries, and the world. The Internet can be considered a WAN
Wi-Fi	IEEE 802.11 standard for wireless communications and products which conform to this standard
Wi–Fi Protected Access (WPA)	a Wi-Fi encryption standard. Now superseded by WPA2
Wire	a single conductor, solid or stranded, with or without insulation
Wire stripper	hand tool used to remove just the insulator from a single wire, either solid or stranded
Wireless access point (WAP)	a device that allows wireless devices to connect to a wired network. May be free-standing or part of a router which also has wired ports.
Wireless Local Area Network	network connecting two or more devices wirelessly, usually also providing access to the Internet

Wireless network	network that does not use hardwired cables to transmit information; instead uses wireless technology
Wireless system	a means of transferring information using radio, microwave or infrared technology.
Wiring schedule	see Cable schedule
WLAN	See Wireless Local Area Network
Work order	internal document used to define the exact scope of work to be done. Often specifies who is to do the work, and the time allotted.
World Wide Web Consortium (W3C)	international governing body that addresses the development of platform-independent web standards and specifications.
WPA	see Wi–Fi Protected Access.
X-10	open industry communication standard that is used by electronic devices to communicate with each other using the power line for communication. Developed in 1975, no longer supported.
Zigbee	wireless communication protocol capable of utilizing mesh networks, best suited for intermittent data
Zone	group of lights, speakers, sensors or other devices that work together and are controlled as one group.
Z-Wave	RF-based, two-way mesh networking protocol that allows for the control of many home appliances.

QUESTIONS AND ANSWERS

Chapter 1

1. The first home theaters were made possible because of the introduction of:
   a. The DVD player
   b. The VCR
   c. The Laserdisc
   d. The DLP projector

2. The first CEDIA EXPO took place in _____.
   a. 1980
   b. 1990
   c. 2000
   d. 2010

3. A distributor:
   a. Stocks product
   b. Provides manufacturer training
   c. Sell wholesale to dealers and retalers
   d. All of the above

4. The person who installs, upgrades, and services electronic systems in the field is a (an)___:
   a. Electronic Systems Technician (EST)
   b. Electronic Systems Designer
   c. Systems engineer
   d. Control system programmer

5. The person who ensures success by tracking cost, time, and resources is the:
   a. Designer
   b. Programmer
   c. Administrator
   d. Project Manager

Chapter 2

1. Professionalism results in:
   a. More efficiency
   b. Referrals
   c. Customer satisfaction
   d. All of the above

2. Documents which utilize the elements of time, cost, and scope to ensure success are:
   a. Architectural drawings
   b. Project management documents
   c. System design documents
   d. Floor plans

3. An amendment to a work order, which conveys modifications to cost, scope, and some times design, is known as a:
   a. Change order
   b. Work order
   c. As-built drawing
   d. Scope statement

4. A Cable Schedule should include the following information:
a. Location ID
b. Cable ID
c. Check or initials to show each run has been tested
d. All of the above

5. The type of drawing that shows the screen wall and acoustical panels as if the viewer is looking directly at the walls is a:
a. Floor plan
b. Exterior elevation
c. Interior elevation
d. Reflected Ceiling Plan (RCP)

Chapter 3

1. A SWOT analysis outlines Strengths, Weaknesses, Opportunities, and ______.
a. Technologies
b. Taxes
c. Threats
d. Tendencies

2. Current assets are:
a. Cash
b. Accounts receivable (AR)
c. Inventory
d. All of the above

3. Gross profit divided by income =
a. Gross margin
b. Markup
c. Net profit
d. Real profit

4. Once your business is established and licensed as a retail merchant, you will be able to:
a. Advertise
b. Buy wholesale
c. Get liability insurance
d. Hire employees

5. The _________ represents the financial activities of the company related to sales and cost of sales over a specific period of time.
a. Financial statement
b. Mission statement
c. P&L
d. Retained earnings

Chapter 4

1. What is the square footage of the room in the following diagram?
   a. 81
   b. 72
   c. 90
   d. 100

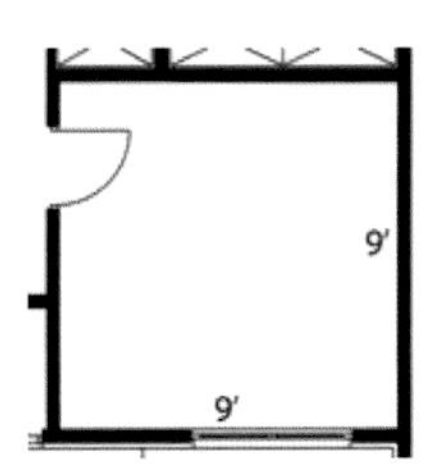

2. Convert 6 ½ inches to decimal equivalents in feet:
   a. 0.541 ft.
   b. 0.5 ft.
   c. 0.641 ft.
   d. 0.6 ft.

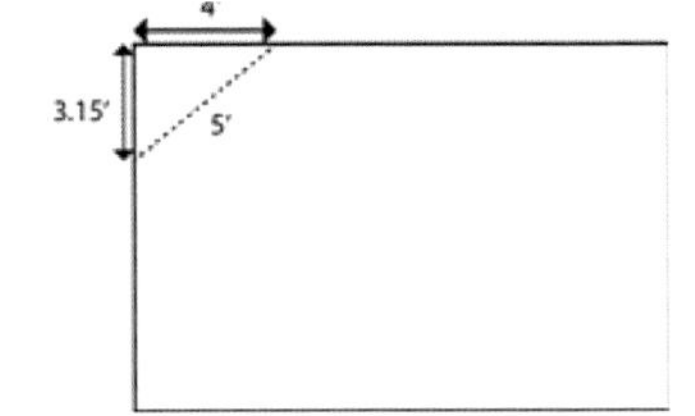

3. In the following diagram, are the walls square?
   a. Yes
   b. No

4. Assuming the speed of sound to be 1,130 ft/sec, what is the frequency of a sound wave with a wavelength of 2 ft.?
   a. F=565 Hz
   b. 56.5 Hz
   c. 2260 Hz
   d. 226 Hz

5. An amplifier channel is rated at 100 watts into an 8 ohm load. You decide to use two 8 ohm speakers, in parallel. What is the total load the amplifier channel will receive in this configuration?
   a. 8 ohms
   b. 4 ohms
   c. 16 ohms
   d. 2 ohms

Chapter 5

1. A formal set of industry or company-accepted guidelines intended to promote better performance and efficiency is called a___________.
   a. Standard
   b. Code
   c. Recommended practice
   d. Regulation

2. Which of these could be the Authority Having Jurisdiction on a project?
   a. State Board of Licensure
   b. City Building Inspector
   c. Building Superintendent
   d. Any of the above

3. ___________ should be used any time a power tool or hammer is in use.
   a. Steel toe shoes
   b. Eye protection
   c. Respirator
   d. Gloves

4. Which of these should be done when using a step ladder?
   a. Face the ladder
   b. Make sure the sides are fully extended
   c. Avoid standing on the top platform or highest step
   d. All of the above

5. Violating a ____________ can result in fines and/or stoppage of work.
   a. Standard
   b. Recommended Practice
   c. White paper
   d. Code

Chapter 6

1. The one document that contains all critical information about the prewire (cables, loca tions, rough-in devices, etc.) is the:
   a. Floorplan
   b. Cable schedule
   c. Interior elevation
   d. Electrical plan

2. Outlet height should always be:
   a. At 12”
   b. The same in every room of the home
   c. Consistent with electrical outlets already installed on the same wall
   d. “Hammer-height”

3. The wooden framing members which is horizontal and supports a floor or ceiling are called:
   a. Studs
   b. Joists
   c. Rafters
   d. Headers

4. Unshielded Twisted Pair cable can be used for:
   a. Telephone
   b. Data
   c. Control signals
   d. All of the above

5. Observing minimum bend ratio is especially critical with coaxial cable because:
   a. Compressing the dielectric changes the cable’s impedance and affects performance
   b. Bending too tight can damage the center conductor
   c. Stretching the jacket can make it susceptible to moisture
   d. The braid can be easily torn

Chapter 7

1. The connector which is universally used to terminate coaxial cable for television signal distribution (antenna, cable, satellite) in the home is the _____ connector.
   a. BNC
   b. F
   c. RCA
   d. RG59

2. The RJ-45 jack configuration which is compatible with both data application and telephone lines 1 and 2 is:
   a. TIA T568B
   b. TIA 570
   c. TIA T568A
   d. BICSI

3. The highest bandwidth and throughput is found in _______.
   a. Fiber optic cable
   b. Coaxial cable
   c. Unshielded Twisted Pair cable
   d. Shielded Twisted Pair cable

4. The simplest type of communication cable testing (continuity and/or mapping) is called __________.
   a. Certification
   b. Verification
   c. Qualification
   d. Clarification

5. The crimp-style F-connector is not recommended because:
   a. It is not compatible with all F-type female jacks
   b. It doesn't have adequate bandwidth for digital signals\
   c. It allows air and moisture to oxidize the conductors inside
   d. It has a left-handed thread

Chapter 8

1. A professional equipment rack has what advantage over a typical cabinet?
   a. Available accessories
   b. Ventilation management
   c. Accessibility
   d. All of the above

2. The recommended method of securing interconnects in an equipment rack is:
   a. Velcro™
   b. Nylon wire ties
   c. Electrical or gaffers tape
   d. None, leave unsecured for ease of access

3. According to UL 1678, when a piece of equipment is mounted so that it is moveable (dynamic load), the entire mounting system must be able to support;
   a. The weight of the device
   b. 2x the weight of the device

c. 4x the weight of the device
d. 8x the weight of the device

4. Most failures of electronic equipment are due to:
   a. Faulty manufacturing
   b. Excessive heat over time
   c. Design flaws
   d. User error

5. On a fully adjustable projector mount, the "yaw" adjustment refers to:
   a. Vertical tilt up and down
   b. Left and right tilt with the same center point
   c. Left and right movement on a horizontal plane
   d. Movement toward, and away from, the screen

Chapter 9

1. Which of these converts acoustical energy to an electrical signal?
   a. A/D convertor
   b. D/A convertor
   c. Microphone
   d. Loudspeaker

2. A high-efficiency loudspeaker is one with a sensitivity of:
   a. 90dB and above
   b. 85dB and below
   c. 75dB and above
   d. 75dB and below

3. Pink noise has;
   a. Equal energy in each octave
   b. Equal energy at all frequencies
   c. Equal energy in all parts of the room
   d. Frequencies adjacent to infra-red signals

4. In surround sound nomenclature, the ".1" in "5.1" stands for:
   a. One center channel output
   b. One subwoofer
   c. One Low Frequency Effects channel output
   d. One source input

5. High frequency sound has:
   a. Higher amplitude
   b. Lower amplitude
   c. Shorter wavelengths
   d. The ability to pass through walls easily

Chapter 10

1. What is the desired maximum level variation within one room of a distributed audio system?
   a. 3 dB
   b. 6 dB
   c. 10 dB
   d. 1 dB

2. Outdoor speakers require more power because:
   a. They are usually less efficient than indoor speakers
   b. They often require long speaker cable runs
   c. They don't have the advantage of room reflections
   d. All of the above

3. A system that allows the choice of both radio and internet music, with all rooms receiving the same programming, is:
   a. Multi-zone/Multi-source
   b. Single-zone/Multi-source
   c. Single-zone/Single-source
   d. Constant voltage

4. When the load (impedance) on an amplifier channel output is increased:
   a. Its output (in Watts) goes down
   b. Its output becomes louder
   c. Its output section is likely to overheat
   d. Its output exhibits a significant loss of high frequencies

5. If the "Zone Two" output of a surround receiver is used to provide audio to two additional rooms, what precaution should be taken to ensure stable performance?
   a. Make sure the output on the internal amplifiers is at maximum, to provide needed headroom
   b. Use smaller gauge speaker wire to both rooms, keeping impedance steady
   c. Use the line outputs and a separate power amplifier to drive the extra rooms
   d. Compensate for line loss by using the equalizer to boost high frequencies

Chapter 11

1. HDMI V 2.0, introduced in 2013, added what feature?
   a. 4K support
   b. Up to 32 channels of audio
   c. 4 simultaneous audio streams
   d. All of the above

2. Which HDMI communication allows the source device to understand the capabilities of the display device?
   a. Hot Plug
   b. DDC
   c. EDID
   d. ARC

3. The standard aspect ratio for High Definition content is:
   a. 4:3
   b. 16:9
   c. 16:10
   d. 2.35:1

4. Which display technology utilizes millions of tiny mirrors to create an image?
   a. DLP
   b. OLED
   c. CRT
   d. LCD

5. When a rear projector system is installed is it important to remember:
   a. The area behind the screen must be kept dark for the picture to look good
   b. The throw distance will be exactly half that needed for front projection
   c. A matte white screen must be used for best contrast
   d. All of the above

Chapter 12

1. The ideal primary viewing distance from the image is:
   a. 2 X the image width
   b. 3 X the image diagonal
   c. 3 X the image height
   d. 1.75 X the image width

2. When designing a dedicated home theater, the best way to ensure good sightlines for viewers in two rows is to:
   a. Make the screen larger
   b. Raise the position of the screen
   c. Offset the seating
   d. Elevate the second row on a riser

3. When a wavelength is longer than a room dimension it can cause:
   a. Flutter echo
   b. Long decay times
   c. A Standing wave
   d. Comb filtering

4. The international standard which is the basis for surround sound speaker configuration was originally established by ______.
   a. The ITU
   b. TIA 570-A
   c. Dolby Laboratories
   d. VESA

5. The ideal way to compensate for image keystoning is with:
   a. Digital keystone correction
   b. Optical vertical lens shift
   c. Video processing
   d. A curved screen

Chapter 13

1. How many bits are in a byte?
   a. 2
   b. 4
   c. 8
   d. 12

2. Which service cannot run via IP
   a. Video Distribution
   b. Telephony
   c. Control
   d. All can use IP

3. TCP/IP is:
   a. the basic communication language or protocol of the Internet
   b. the part of a system that converts analog to digital
   c. the only protocol that enables computers to communicate with each other
   d. None of the above

4. Which of these is needed for two or more devices to access the internet at the same time?
   a. Switch
   b. Hub
   c. Modem
   d. Router

5. A packet contains two things, a header and :
   a. Contents
   b. A footer
   c. A payload
   d. ASCII text

Chapter 14

1. Lighting control provides:
   a. Convenience
   b. Security
   c. Energy savings
   d. All of the above

2. LED lamps can have a lifespan up to:
   a. 1,000 hours
   b. 5,000 hours
   c. 10,000 hours
   d. 50,000 hours

3. The development which marked the beginning of widespread use of lighting control in the home was the:
   a. Solid-state dimmer
   b. MOV (Metal Oxide Varistor)
   c. Incandescent lamp
   d. LED lamp

4. CFL lamps should not be carelessly disposed of, because they contain:
   a. Lead
   b. Mercury
   c. High voltage
   d. Phosphorus

5. A free-standing lamp used specifically for reading would be considered:
   a. Ambient lighting
   b. Accent lighting
   c. Task lighting
   d. None of the above

Chapter 15

1. Motorized blinds can:
   a. Provide protection from UV light
   b. Transform a room for a different use
   c. Save energy
   d. All of the above

2. Cables which will be flexed repeatedly should;
   a. Be plenum rated
   b. Have stranded conductors
   c. Be attached tightly to the mechanism in many places
   d. Bundled together

3. Which type of motorized blinds would be the best choice in a retrofit situation?
   a. Wireless
   b. IP
   c. Venetian
   d. Hard-wired

4. Projector screen masking:
   a. Covers unused portions of the screen, depending on the aspect ratio
   b. Covers the entire screen when not in use
   c. Is used to protect the screen when the surrounding walls are being painted
   d. Is one of the few subsystems which cannot easily be motorized

5. Power cables and signal cables should:
   a. Be different colors
   b. Cross only at a right angle
   c. Be tied together for convenience
   d. Have the same jacket material

Chapter 16

1. An electricity billing rate which increases based on the amount the customer uses is known as a _____.
   a. Time Of Use (TOU) rate
   b. Tiered rate
   c. Flat rate
   d. Demand rate

2. A "phantom load" describes a situation where:
   a. The user turns on a device remotely
   b. Electricity is being used but cannot be measured
   c. Electricity is being used even though the device is turned off
   d. A device operates without using any electricity at all

3. The "Prius Effect" is when:
   a. A person uses less energy simply by being aware of their usage
   b. Fossil fuels and electricity work together
   c. A user is unaware of energy use because they cannot hear it
   d. A client wants energy savings solely because it is the "right thing to do"

4. The first step toward controlling energy use is _______.
   a. Dimming lights
   b. Adjusting the thermostat
   c. Knowing how it is being used
   d. Being willing to make some sacrifices

5. A Kilowatt Hour is a measure of what?
   a. Current
   b. Power
   c. Energy
   d. Time

Chapter 17

1. Which is NOT an input device in a security system?
   a. Annunciator
   b. Motion detector
   c. Photoelectric beam
   d. Keypad

2. Which input device requires no additional power to operate?
   a. Photoelectric beam
   b. Pass Infrared (PIR) motion detector
   c. Magnetic contact
   d. Glass break sensor

3. Which communication method is quickly becoming the primary method of central station communication?
   a. Dialer
   b. Long range radio
   c. Cellular
   d. Internet

4. An electric lockset which includes a keypad, to control who can open the door, is called:
   a. A cipher lock
   b. A combo lock
   c. A biometric lock
   d. An IP lock

5. An IP camera may have the capability to:
   a. Serve as a motion detector, triggering a message or recorder
   b. Pan, Tilt, and Zoom to better view the area
   c. Send images to mobile devices, TVs and touchpanels
   d. All of the above

Chapter 18

1. An increase in voltage of 10-35% which lasts up to a few minutes is called a:
   a. Spike
   b. Surge
   c. Over-voltage
   d. Sag

2. A voltage regulator keeps voltage constant by
   a. Filtering out changes
   b. Adding voltage from another electrical source
   c. Saving overages to use when voltage drops
   d. Changing the current draw to compensate for voltage

3. Drawing different amounts of current to compensate for change in output voltage A good surge protector should have a response time of:
   a. 1 second
   b. 1 millisecond
   c. 1 nanosecond
   d. 1 minute or less

4. Uninterruptable Power Supplies have a battery which _______.
   a. Need to have water added annually
   b. Have a 5 year service life
   c. Is usually not replaceable
   d. Can be picked up at any convenience store

5. Virtually all domestic electrical services in the world are:
   a. 230 volts
   b. Balanced systems
   c. AC
   d. Grounded

Chapter 19

1. An IR signal carried over a cable requires a minimum of ___ conductors?
   a. 2
   b. 3
   c. 4
   d. 8

2. The most important advantage of an RF remote system is:
   a. Batteries last longer
   b. It doesn't affect nearby homes
   c. It does not require direct line of sight
   d. It can store and issue macro commands

3. In order to address devices requiring IR or RF commands from an IP based control system, a ________ is needed.
   a. Router
   b. Hub
   c. Bridge
   d. Connecting block

4. A "macro" ___________.
   a. Allows a remote to communicate with several different devices at once
   b. Is a set of instructions, in sequence, initiated by one user interaction
   c. Sends the IR signal out in all directions equally
   d. Is a program that can be used moved from one device to another, regardless of manufacturer

5. Serial control:
   a. Includes RS-232 and RS-485
   b. Is carried only over cable
   c. Is highly dependable
   d. All of the above

Chapter 20

1. Equipment racks should be:
   a. Loaded and tested at the shop, to minimize the impact on the homeowner
   b. Installed and loaded onsite, to ensure proper interaction with installed components
   c. Completely sealed and covered, to prevent client access
   d. Made of wood, to prevent ground loops

2. Left, right, and center speakers should be aimed at the primary listening position so that:
   a. The listener gets the highest possible SPL
   b. They sound "in phase"
   c. The listener is hearing "on-axis" response from each speaker
   d. The listener gets optimum bass response

3. The contrast control on a video display adjusts the_______.
   a. Black level
   b. White level
   c. Sharpness
   d. Color intensity

4. ___________ should be stored onsite in a container with your company information on it.
   a. TV pedestals
   b. Factory remotes
   c. User manuals
   d. All of the above

5. Distribut ed audio within a room should vary in SPL by no more than _____.
   a. 6 dB
   b. 3 dB
   c. 12 dB
   d. 10 watts

Index

Symbols

A

B

C

N

O

P

Q

R

S

T

U